MW01011574

Water and the Cell

Water and the Cell

Edited by

Gerald H. Pollack

Department of Bioengineering, University of Washington, Seattle WA, USA

Ivan L. Cameron

Department of Cellular and Structural Biology, UTHSCSA, San Antonio, TX, USA

Denys N. Wheatley

BioMedES, Aberdeen, UK

A C.I.P. Catalogue record for this book is available from the Library of Congress.

ISBN-10 1-4020-4926-9 (HB)
ISBN-13 978-1-4020-4926-2 (HB)
ISBN-10 1-4020-4927-7 (e-book)
ISBN-13 978-1-4020-4927-9 (e-book)

Published by Springer,
P.O. Box 17, 3300 AA Dordrecht, The Netherlands.

www.springer.com

Printed on acid-free paper

Cover design: The structural relationship of water molecules is shown in a cross-sectional view of a collagen fibril in a fully hydrated tendon; the tight hexagonal packing in a total cluster of 18 water molecules around each tropocollagen molecule forms a firm (tightly associated) monolayer. The design was taken with the kind permission of Professor Gary Fullerton from his paper (Fullerton GD and Amuroa MR. *Cell Biology International* **30**, 56–65, 2006 – Figure 2), in agreement with the publishers, Elsevier Press. It was selected and redrawn by Denys Wheatley, and finally placed against the same background image in white against the pale blue of the cover with the assistance of Springer-Verlag (Graphics).

CONTENTS

PREFACE

This edited volume deals with the state of water in the vicinity of biological interfaces, both intracellular and extracellular. This issue is of critical importance, for the cell is extremely crowded with interfaces, and as a result practically all cell water is interfacial. The character, or state, of this water may therefore be central to cell function.

What is meant by the 'state of water?' Few would question that water coming out of a household tap is a liquid, but water in an ice cube is something altogether different: it is a solid that floats on tap water (also known as bulk water). It is water in the solid state.

The fact that ice floats is an indication that it is less dense than water. Clearly, the physical properties are different. Water molecules below 0 °C form a crystal. In this crystal, the two positively charged hydrogen atoms of water bind to the double negative charges of oxygen atoms of two adjacent water molecules. The resulting crystal lattice is arranged in such a way as to be less dense than tap water, and constituent water molecules are also less mobile.

But what of water adjacent to surfaces, be they biological or inanimate? Does this represent yet another state of water, distinct from solid or liquid? Water's positively charged hydrogen atoms might be expected to bond to a negatively charged surface, whereas its negatively charged oxygen atom ought to bond to a positively charged surface. One might then ask if the dipolar orientation of the first bound water layer might provide an arrangement of charges that causes a second layer of dipolar water molecules to form. If so, how many layers might build? How might the physical properties of this water differ from those of bulk water?

These issues are dealt with in depth in the chapters of this volume. Several chapters imply that the ordered interfacial zone may extend considerably farther than generally envisioned.

There is appreciable background for such thinking, both theoretical and experimental. For example, consider the surface of a polished silver chloride crystal. Positive Ag^+ and negative Cl^- charges are spaced at a distance of about 3 Angstroms, or about the diameter of a water molecule. Thus, the surface charges form a checkerboard, whose spacing is equal to the size of a water molecule. In this case the positive charge of one water-hydrogen atom could bind to a negatively charged surface-Cl^-, while the negatively charge of oxygen of another water molecule could bind to a positively charged surface-Ag^+. Bound water molecules in this first layer could then hydrogen bond to one another to form a highly polarized monolayer that could serve as the nidus for formation of a second layer. In this situation, it is predicted that numerous water layers could form (Ling 2003). In fact Hori (1956) has demonstrated that water between the surfaces of two quartz or polished glass plates spaced less than 100 μm apart does not freeze at temperatures as low as −90 °C. The inability to freeze is an indication of structuring. The span in question is 30,000 water-molecule diameters. Clearly, at these inanimate surfaces, multiple layers of water molecules are perturbed relative to water molecules in bulk.

What about the water in biological systems? Is the state of water in cells similar to bulk water, or is it organized? When a ciliated protozoan, such as *Tetrahymena*, growing in a dilute proteose peptone media is smashed between a microscope slide and a cover slip, the cortex of the cell ruptures, and a water-immiscible substance flows out and separates as a drop of cytoplasm (Cameron, unpublished). Further smashing of such extrusions can fractionate the drop into smaller immiscible droplets. This is a common observation, seen in various forms by many others. It indicates that protoplasm in the cell is immiscible in a dilute solution, and that retarding the flow of fluid/water from the cell protoplasm does not necessarily require an intact cell membrane; it is inherent in the physical features of the protoplasm itself.

These observations alone would appear to cast doubt on the tenet of the cell membrane as the critical water-diffusion barrier between the extracellular environment and the cytoplasm, where the intracellular water is assumed to be relatively free to exit the cell upon membrane disruption. Even in the absence of a membrane, the protoplasm does not dissipate.

The chapters in this monograph deal with water at the interfaces of both inanimate and biological systems. The biological systems include both filamentous and globular proteins, hydrophilic and hydrophobic surfaces, extracellular materials, and cells. What evolves is that water within cells is to a major extent ordered differently than bulk water, and functions not as an inert solvent, but as an active player. Most of intracellular water is adsorbed onto surfaces, which themselves are dynamic.

Understanding water order in biological systems is key to an understanding of life processes and to an understanding of diseases.

The material in the book should be of value to any person interested in the role of water inside the cell. This includes professionals in the area of cell biology, chemistry, and biochemistry. It also includes students interested in understanding the underlying basics of life.

The reader will be richly rewarded with insights difficult or impossible to obtain in current textbooks, which generally treat water merely as a background carrier with limited significance.

It was Albert Szent-Gyorgyi, Nobel Laureate, who stated that 'Life is water dancing to the tune of macromolecules.' Szent-Gyorgyi's famous pronouncement is borne out in the contents of this volume. Water is definitely a major player in the biology of the cell.

REFERENCES

Hori, T. (1956) Low Temperature Science A15:34 (English Translation) No. 62 U.S. Army Snow, Ice and Permafrost Res. Establihment, Corps of Engineers, Wilmetti & Il. Aee Ling 2003
Ling G.N. (2003) A new theoretical foundation for the polarized multiplayer theory of cell water and for inanimate systems demonstrating long-range dynmamic structuring of water molecules. Physical. Chem. Phy. and Med. NMR 35:91–130

CHAPTER 1

A CONVERGENCE OF EXPERIMENTAL AND THEORETICAL BREAKTHROUGHS AFFIRMS THE PM THEORY OF DYNAMICALLY STRUCTURED CELL WATER ON THE THEORY'S 40TH BIRTHDAY

GILBERT N. LING
Damadian Foundation for Basic and Cancer Research, Tim and Kim Ling Foundation for Basic and Cancer Research, c/o Fonar Corporation, 110 Marcus Drive, Melville, NY 11747,
E-mail: gilbertling@dobar.org

Abstract: This review begins with a summary of the critical evidence disproving the traditional membrane theory and its modification, the membrane-pump theory – as well as their underlying postulations of (1) free cell water, (2) free cell K^+, and (3) 'native'-proteins being truly native.

Next, the essence of the unifying association-induction hypothesis is described, starting with the re-introduction of the concept of protoplasm (and of colloid) under a new definition. Protoplasms represent diverse cooperative assemblies of protein-water-ion – maintained with ATP and helpers – at a high-(negative)-energy-low-entropy state called *the resting living state*. Removal of ATP could trigger its auto-cooperative transition into the low-(negative)-energy-high-entropy *active living state* or *death state*.

As the largest component of protoplasm, cell water in the resting living state exists as polarized-oriented multilayers on arrays of some fully extended protein chains. Each of these fully extended protein chains carries at proper distance apart alternatingly negatively charged backbone carbonyl groups (as N sites) and positively charged backbone imino group (as P sites) in what is called a NP-NP-NP system of living protoplasm. In contrast, a checkerboard of alternating N and P sites on the surface of salt crystals is called a NP surface.

The review describes how eight physiological attributes of living protoplasm were duplicated by positive model (extroverts) systems but not duplicated or weakly duplicated by negative model (introverts) systems. The review then goes into more focused discussion on (1) water vapor sorption at near saturation vapor pressure and on (2) solute exclusion. Both offer model-independent quantitative data on polarized-oriented water.

Water-vapor sorption at physiological vapor pressure ($p/p_o = 0.996$) of living frog muscle cells was shown to match quantitatively vapor sorption of model systems containing exclusively or nearly exclusively fully extended polypeptide (e.g., polyglycine, polyglycine-D,L-alanine) or equivalent (e.g., PEO, PEG, PVP). The new Null-Point Method of Ling and Hu made studies at this extremely high vapor pressure easily feasible.

G. Pollack et al. (eds.), Water and the Cell, 1–52.

Solute exclusion in living cells and model systems is the next subject reviewed in some detail, centering around Ling's 1993 quantitative theory of solute distribution in polarized-oriented water. It is shown that the theory correctly predicts *size dependency* of the q-values of molecules as small as water to molecules as large as raffinose. But this is true only in cases where the excess water-to-water interaction energy is high enough as in living frog muscle (e.g., 126 cal/mole) and in water dominated by the more powerful extrovert models (e.g., gelatin, NaOH-denatured hemoglobin, PEO.) However, when the probe solute molecule is very large in size (e.g., PEG 4000), even water 'dominated' by the weaker introvert model (e.g., native hemoglobin) shows exclusion.

Zheng and Pollack recently demonstrated the exclusion of coated latex microspheres 0.1 μm in diameter from water 100 μm (and thus some 300,000 water molcules) away from the polarizing surface of a poly(vinylalcohol) (PVA) gel. This finding again affirms the PM theory in a spectacular fashion. Yet at the time of its publication, it had no clear-cut theoretical foundation based on known laws of physics that could explain such a remote action.

It was therefore with great joy to announce at the June 2004 Gordon Conference on Interfacial Water, the most recent introduction of a new theoretical foundation for the long range water polarization-orientation. To wit, under ideal conditions an 'idealized NP surface' can polarize and orient water *ad infinitum.* Thus, a theory based on laws of physics can indeed explain long range water polarization and orientation like those shown by Zheng and Pollack.

Under near-ideal conditions, the new theory also predicts that water film between polished surfaces carrying a checkerboard of N and P sites at the correct distance apart would not freeze at any attainable temperature. In fact, Giguère and Harvey confirmed this too retroactively half a century ago

Keywords: water, cell water, polarized multilayers, association-induction hypothesis, AI Hypothesis, polarized multilayer theory, polarized oriented multilayer theory, PM theory, long-range water structure, water, vapor pressure, super-cooling, non-freezing water, silver chloride crystals, glass surface, BET theory

Symbols and Abbreviations: a, amount of water (or other gas) adsorbed per unit weight of adsorbent; α, polarizability; BET Theory, the theory of multilayer gas adsorption of Brunauer, Emmett and Teller (1938); d, distance between nearest neighboring sites on an NP surface; E^n, (negative) adsorption or interaction energy of water molecules polarized by, but far removed from an idealized NP surface (see Figure 27); μ, permanent dipole moment; NO surface or system, a checkerboard of alternatingly negatively charged and vacant sites; NO-NO-NO system, a matrix of arrays of properly-spaced negatively charged N sites and vacant O sites; NP surface, a checkerboard of alternatingly negatively charged N sites and positively charged P sites; NP-NP system, two juxtaposed NP surfaces; NP-NP-NP system, a matrix of more or less parallel arrays of linear chains of properly spaced N and P sites; p/p_o relative vapor presssure equal to existing vapor pressure; p, divided by the pressure at full saturation under the same condition; PEG, poly(ethylene glycol); PEO, poly(ethylene oxide); PVA, polyvinyl alcohol; PVP, polyvinylpyrrolidone; PM Theory, the Polarized-Oriented Multilayer Theory of Cell Water; PO surface, a checkerboard of alternatingly positive P sites and vacant O sites; PP surface, a checkerboard of uniformly positively charged sites; q-, or q-value, the (true) equilibrium distribution coefficient of an ith solute between water-containing phase of interest (e.g., cell water) and a contiguous water-containing phase such as the bathing medium; r, the distance between nearest neighboring water molecules; ρ-, or ρ-value, the apparent equilibrium distribution coefficient may include bound solute in addition to what a q-value represents

For not telling the whole truth, Martha Stewart went to jail. Many know that. In contrast, few are aware that many more than one scientist, teacher, textbook writer etc. have been engaged knowingly or unknowingly in telling half-truth and untruth. But they don't go to jail. Instead, they are blissfully honored and rewarded for passing half-truths and untruths as the whole truth and teaching them to generation after generation of young people now living and yet to come. Why does a civilized society built on the laws of equal justice, openly condone the opposite?

A moment of reflection would reveal an obvious cause: a rarely discussed 'Achilles heel' in even the finest forms of governments in existence. That is, the vast number of our species whose wellbeing and even survival hang on what we decide to do or not to do today have no say in making those decisions – since they are not born yet.

Martha Stewart went to jail because not telling the whole truth caused some monetary and related losses to people now living. And these living people *had* votes and voices. As a result, government officials took action. Yet those same government officials or their equivalents would probably only shrug – if that, – when told that many scientists and science teachers were doing what Martha Stewart did – only on a much grander scale.

For, as a rule, what a sound basic science can offer lies in the *future* – e.g., in practical applications built upon new knowledge that basic research brings to light. Those future applications would be the modern equivalents of the steam engines, the electric motor, the electric generator, and the wireless telegraphy. None of these was invented out of thin air. They grew out of the progress made in earlier basic science.

Only by now, our need for further progress in basic science, especially basic physiological science, has far surpassed that of the past. For Mankind will soon face problems it has not faced before: overpopulation, exhaustion of natural resources, increasingly more deadly diseases beyond what our make-believe understanding of living phenomena could cope with – to mention only three.

But seen from the viewpoints of the research-funding agencies, members of school boards and even the Nobel Prize committees, research and teaching based on the most up-to-date valid new knowledge or based on some popular, but erroneous idea might not seem to matter that much.

To begin with, they usually do not have the up-to-date expertise or adequate time to know and understand the difference. And the few who did find out are alone. The majority, who may see little gains but more headaches for themselves in rocking the boat, easily outvotes them.

Nonetheless, the condoned blurring of what is right and what is wrong cannot continue indefinitely. Look at Enron, the seventh largest US corporation before its downfall and A.B. Anderson, the once gold standard of accounting worldwide. For in a global capitalist economy, individual nations and even the world as a whole have become bigger versions of corporations like Enron and A.B. Anderson. They too cannot long endure if the line between what is truth and what is falsehood is being blatantly ignored.

At this juncture, I like to quote Andy Grove, one time CEO of Intel, who wrote the book: *Only the Paranoid Survive* (Grove 1996.) For what separates a paranoid from a normal counterpart is the preoccupation of the paranoid with the *future* (and the preoccupation of the normal with *now*.) As a self-diagnosed paranoid, Andrew Grove saved Intel by making drastic changes in the makeup of the company and in transforming the world's largest semiconductor maker to the premier manufacturer of microprocessors.

That is why in Andrew Grove and those who think and act like him lies the real hopes of the future. They live in the present but they keep their eyes open to what lies ahead. They are the alert bus drivers on a treacherous mountain road. In some ways, they are Plato's philosopher kings.

It is on this note of hope, that I write the following review on the basic science of life, or cell physiology, which had seen a profound (but artificially hidden) change that Andrew Grove would have called a *strategic inflection point*. Only this one occurred half a century ago.

1. THE FIRST UNIFYING THEORY OF CELL PHYSIOLOGY AND THE SUBSEQUENT VERIFICATION OF ITS ESSENCE

Fundamentally speaking, cell physiological research is like solving a gigantic crossword puzzle. Like the crossword puzzle, cell physiology also has just *one unique solution*. But to reach out to that unique solution, cell physiologists of the past faced an insurmountable obstacle.

That is, when the study of cell physiology began, the physico-chemical concepts needed to construct the correct unifying theory were not yet available. An *incorrect guiding theory* was doomed to be introduced and it was (see below.) And as time went on, this incorrect theory would either kill that branch of science, or worse: it would be taught as unqualified truth to younger generations living and yet to come.

Meanwhile, the study of cell physiology broke up into smaller and smaller fragments or specialties. In time each specialty spawned its own lingo, its own methodology and its own subspecialties; the contact of each specialty with other specialties become less and less frequent and more and more perfunctory. The cumulative result is as Durant described: 'We suffocated with uncoordinated facts, our minds are overwhelmed with science breeding and multiplying into speculative chaos for want of synthesis and a unifying philosophy.' (*The Story of Philosophy*, Durant 1926, reprinted repeatedly till at least 1961, p 91).

Now, Durant's complaint addressed the lack of a correct unifying philosophy or theory, which alone can bind together and make sense out of the senseless fragments. Then, often quietly and little by little, the obstacles to produce a correct unifying theory of cell physiology gradually melted away – when the most relevant aspects of physics and chemistry reached maturity in the late 19th and early 20th century.

Therefore, in broad terms it was not entirely surprising (although it has never ceased to be surprising to me) that some forty years ago a unifying theory of cell physiology built upon mature physics and chemistry made its debut. It bears the name, the *association-induction* (AI) *hypothesis* (Ling 1962). Worldwide experimental testing and confirmation of its essence followed rapidly – as chronicled in three additional monographs published respectively in 1984, 1992 and 2001 (Ling 1984, Ling 1992, Ling 2001).

It would seem that the day would soon arrive when swift progress would light up another new age in science (of the living) like the one (of the dead) in the 17th – early 20th centuries. Unfortunately, forty years afterward, it has not happened yet.

As it stands today, few biomedical researchers, teachers or students here and abroad have ever heard of these books and what they tell, let alone understanding or teaching them. Instead, obsolete ancient ideas called the membrane theory, or its later version called the membrane-pump theory, are still universally taught as proven truth at all level of education – long after both have been thoroughly and resoundingly proven to be wrong.

For the details of the widely taught, but incorrect misguiding theory and alternatives, you must consult my most recent book, *Life at the Cell and Below-Cell Level*. It is the only book that takes you through the complete history of cell physiological research, beginning with the invention of microscopes and the first perception of the living cell as the basic unit of all life forms.

Here I offer a short cut to a part of the hidden scientific history as well as a list of references to the original sources of publication. But, above all, this article ends with an account of some important new discoveries that occurred *after* the publication of *Life* in the year 2001.

2. THE COMPLETE DISPROOF OF THE MEMBRANE (PUMP) THEORY AND ITS ANCILLARY POSTULATIONS

According to the membrane theory, each living cell is a small puddle of ordinary liquid water. In this ordinary liquid water is freely dissolved small salt ions of various kind, mostly potassium ions (K^+), and large molecules, mostly proteins (and some RNA and DNA).

Substances like K^+ and sodium ion (Na^+) are, as a rule, found in the cell at concentrations different from their counterparts in the surrounding medium (Figure 1). This type of asymmetrical solute distribution was seen as the consequences of a sieve-like cell membrane. With rigid pores of exactly the same and correct size, the cell membrane permits the intra-, extra-cellular traffic of ions and molecules smaller than the membrane pores but keeps out ions and molecules larger than the pores – absolutely and permanently.

When the sieve membrane idea failed to explain the asymmetrical distribution of K^+, Na^+ and other solutes, an *ad hoc* membrane pump theory was installed in its place. Then, it is a battery of submicroscopic pumps in the cell membrane that are installed to maintain the *status quo*. The sodium (potassium) pump, was a prominent example. Located in the cell membrane, this pump is postulated to push sodium ion (Na^+) out of the cell and to pull potassium ions (K^+) into the cell,

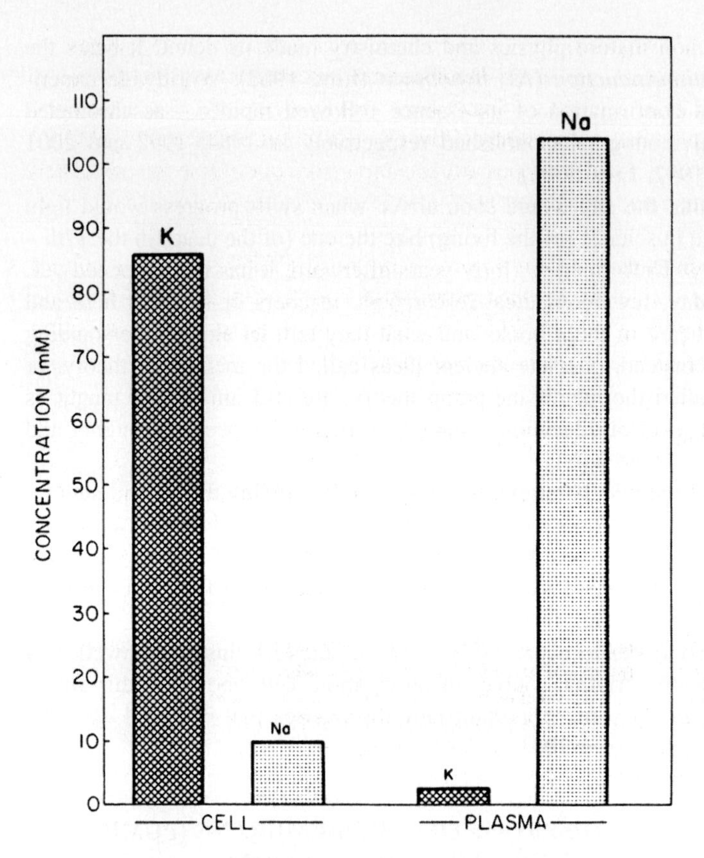

Figure 1. Potassium ion (K) and sodium (Na) concentration in frog muscle cells and in frog blood plasma. Concentrations in frog muscle cells and in frog blood plasma are given respectively in millimoles per liter of cell water or plasma water (from Ling 1984 by permission of Plenum Press)

24 hours a day, 7 days a week without stop. As mentioned above already and to be elaborated some more below, this membrane pump model did not fare better than the original sieve membrane theory.

The following is an itemized list of the decisive experimental findings. These findings have passed the final verdicts on the fate of both the original sieve membrane theory and the membrane-pump theory – as well as on the fate of the ancillary assumptions on which both the sieve membrane and the membrane pump theory were built.

2.1 Disproof of the Sieve-like Cell Membrane Concept

The sieve concept separates ions and molecules into two categories. Those that are able to pass through the membrane barrier and those that are (permanently and absolutely) unable to do so. This concept of all-or-none segregation reached the peak of its development with the publication of the famous paper by Boyle and Conway on page 1(to page 63) of the 100th volume of the prestigious (English) Journal of Physiology (Boyle and Conway 1941.) However, even before the paper appeared in print, contradictory experimental evidence were rapidly collecting. Included were those from Conway's own laboratory (Conway and Creuss-Callaghan 1937) – showing that ions and molecules supposedly to be too large to traverse the postulated membrane pores, in fact, can enter and leave the cells with ease (Ling 1952, pp 761–763).

Table 1 taken from a more recent paper of Ling et al. (1993) shows that solutes from the small, like water, all the way to raffinose (molar volume, 499 cc) can all traverse the cell membrane without difficulty. Clearly, the asymmetrical distribution of solutes is not due to a sieve-like mechanism.

2.2 The Disproof of the Membrane Pump Theory

In 1952 I first presented results of my earlier study on the (would-be) energy requirement of the hypothetical sodium pump in metabolically inhibited frog muscle. To halt respiration, I used pure nitrogen (in addition to sodium cyanide). To halt glycolysis, the alternative route of energy metabolism, I used sodium iodoacetate.

Table 1. The time required for each of the 22 solutes investigated to reach diffusion equilibrium in isolated frog muscle cells. The data as a whole show that with the exception of three pentoses an incubation period of 24 hours at 0 °C is adequate for all the other solutes studied. The three pentoses took about twice as long or 45 hours to attain equilibrium (from Ling et al. 1993, by permission of the Pacific Press, Melville, NY)

Solute	Equilibration time (hours)
water	$\ll 1$
methanol	<20
ethanol	<20
acetamide	<10
urea	<24
ethylene glycol	<10
1,2-propanediol	24
DMSO	<1
1,2-butanediol	24
glycerol	<20
3-chloro-1,2-propanediol	24
erythritol	<20
D-arabinose	<45
L-arabinose	<45
L-xylose	<45
D-ribose	<24
xylitol	24
D-glucose	<15
D-sorbitol	<10
D-mannitol	<24
sucrose	<8
raffinose	10

The results showed that the minimum energy need of the sodium pump would be at least 400% of the maximum available energy.

In years following, the technique was steadily improved so that by 1956, I was able to achieve the highest accuracy in the last three sets of experiments, the results of which are shown graphically in Figure 2. Now, the minimum energy need of the sodium pump was shown to be no longer 400% as from early studies, but at least 1500% to 3000% times the maximum energy available (Ling 1962, 1997). Clearly, the asymmetrical distribution of solutes is not due to membrane pumps either.

2.3 The Disproof of the Free Cell Water Postulation

The free cell water postulation was disproved when Ling and Walton showed that centrifugation at 1000 g for 4 minutes quantitatively removes all free water found in the extracellular space of the isolated frog sartorius muscle. Yet the same centrifugation treatment failed to extract any water from within the cells (Ling and Walton 1976) – after (part of) the cell membrane has been surgically removed and electron microscopy revealed no membrane regeneration following surgery (Cameron 1988).

2.4 The Disproof of the Free Cell Potassium Postulation

The free cell potassium postulation was also fully disproved on at least four accounts.

First, in healthy cells, the diffusion coefficient of $K^+(D_K)$ was found to be only 1/8 of that in an isotonic solution, while in the same preparation the diffusion coefficient of labeled water was reduced only by a factor of 2. Killing the muscle by prior metabolic poisoning increased the K^+ diffusion coefficient to close to that in an isotonic KCl solution; injury produced a D_K in-between that of the healthy living cell and that of the dead cell (Ling and Ochsenfeld 1973).

Second, if the bulk of cell K^+ is free, an impaling intracellular K^+-sensitive microelectrode should register a uniform activity coefficient of cell K^+in all types of cells probed. And that uniform activity coefficient should match the activity coefficient of free K^+ in a KCl solution of similar ionic strength. In truth, the activity coefficients actually measured among different cell types varied from as low as 0.3 to as high as 1.2 (Table 8.2 in Ling 1984).

Third, if cell K^+ is free, its location in frog muscle should be higher in the I bands where the water content is higher than in the adjacent A bands. Instead, the great majority of K^+ is located at the edges of the A bands and at the Z line. (For in depth, definitive work, see Edelmann 1977, 1984, 1986; for earlier and less-than exhaustive work, see Macallum 1905; Menten 1908; Ling 1977; Tigyi et al. 1980–81; von Zglinicke 1988.) (Figures 3, 4).

Fourth, this regionally-accumulated, radioactively-labeled K^+ could be 'chased away' by adding competing alkali metal ions like Rb^+ or Cs^+ to the external

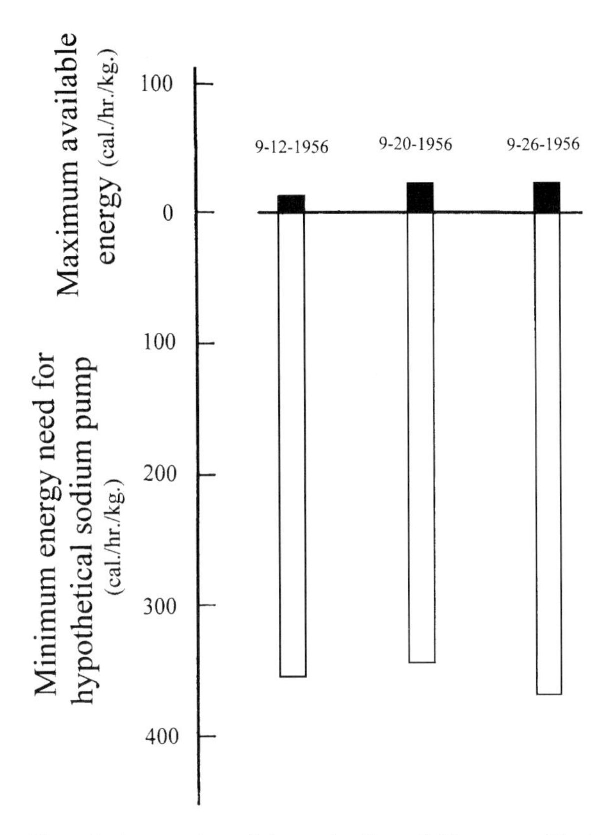

Figure 2. A comparison of the maximally available energy of (poisoned) frog sartorius muscle cells at 0 °C (upward black bars) and the minimum energy need to pump Na^+ against both (measured) electric potential gradient and a concentration gradient. Duration of the experimental observation for experiment (9-12-1956) lasted 10 hrs; Experiment 9-20-1956, 4 hrs; Experiment 9-26-1956, 4.5 hrs. Active oxidative metabolism was suppressed by exposure to pure nitrogen (99.99%, in addition to 0.001 M NaCN); glycolytic metabolism, by sodium iodoacetate and doubly insured by actual lactate analysis before and after the experiment. Other detailed studies reported in 1952 (Ling 1952, Table 5 on page 765) and in 1962 (Ling 1962, Table 8.4) showed respectively that under similar conditions of 0 °C temperature and virtually complete inhibition of active energy metabolism the K^+ and Na^+ concentrations in frog muscle, nerves and other tissues remain essentially unchanged for as long as the experiments lasted (5 hrs. for the 1952 reported experiment, and 7 hrs 45 min in the 1962 reported findings). (For additional details, see Ling 1962, Chapter 8 and Ling 1997. Since the book referred to here as Ling 1962 has been out of print, its entire Chapter 8 has been reproduced as an Appendix in the article, Ling 1997 bearing the title: *Debunking the Alleged Resurrection of the Sodium Pump Hypothesis.*) In the computations, it was assumed that the frog muscle cell does not use its metabolic energy for any other purpose(s) than pumping sodium ion and that all energy transformation and utilization are 100% efficient (from Ling 2004, by permission of the Pacific Press, Melville, New York)

medium. The extent of displacement varied with the *short-range attributes* (e.g., size) of the displacing ions – indicating that the K^+ ions are engaged in close-contact adsorption and not free in the cell water (Ling and Ochsenfeld 1966).

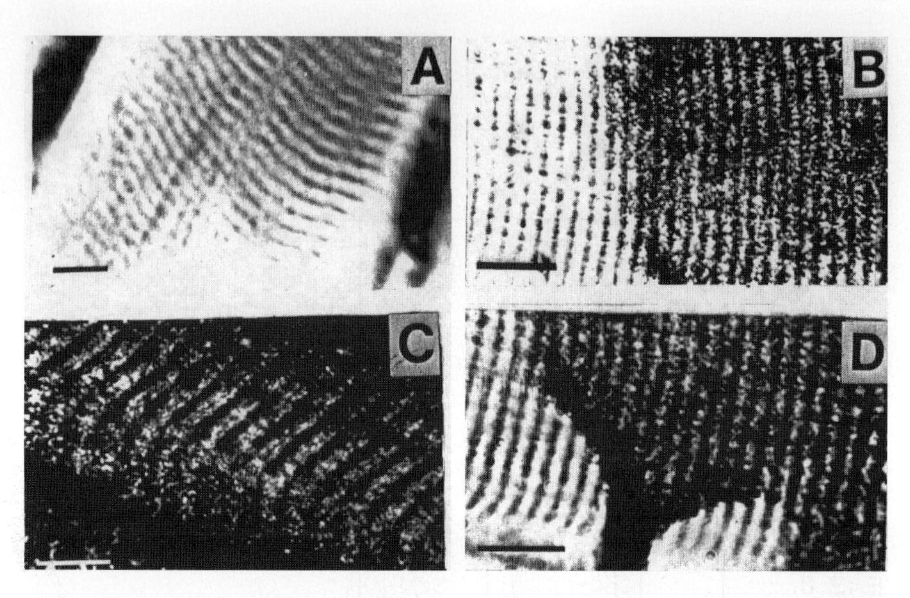

Figure 3. Auto-radiographs of dried single muscle fibers. (A) Portion of a single muscle fiber processed as in all the other auto-radiographs shown here but not loaded with radioisotope. (B), (C) and (D) were auto-radiographs of dried muscle fibers loaded with radioactive ^{134}Cs while living and before drying. (B) and (D) were partially covered with photo-emulsion. Muscle in (C) was stretched before drying. Bars represent 10 micrometers. Incomplete coverage with photo-emulsion in B and D permits ready recognition of the location of the silver grains produced by the underlying radioactive ions to be in the A bands. Careful examination suggests that the silver grains over the A bands are sometimes double. A faint line of silver grains also can be seen sometimes in the middle of the I bands, corresponding to the position of the Z line (from Ling 1977, by permission of the Pacific Press, Melville, NY)

(For additional confirmatory work from X-ray adsorption fine structure of cell K^+, see Huang et al. 1979; for first order quadrupole broadening of Na^{23} in K^+-depleted frog muscle of Cope and Ling, see Ling 2001, p 187–190.)

2.5 The Disproof of the Postulation of the Existence of All Intracellular Proteins in the Conformation Conventionally called 'Native'

From section 2.4 above, we know now that K^+ in living cells is not free. That is just another way of saying that virtually all cell K^+ is in some way bound. In mature human red blood cells, which have no nucleus, nor significant amount of DNA or RNA, the only macromolecular component large enough in size and amount to provide enough binding sites for cell K^+ is proteins. And of the proteins in mature mammalian red blood cells, fully 97% is hemoglobin. This leaves hemoglobin as the only bearer of binding sites in mature mammalian red blood cells for cell K^+ as well as the bulk of cell water – also shown to be not free in section 2.3 above.

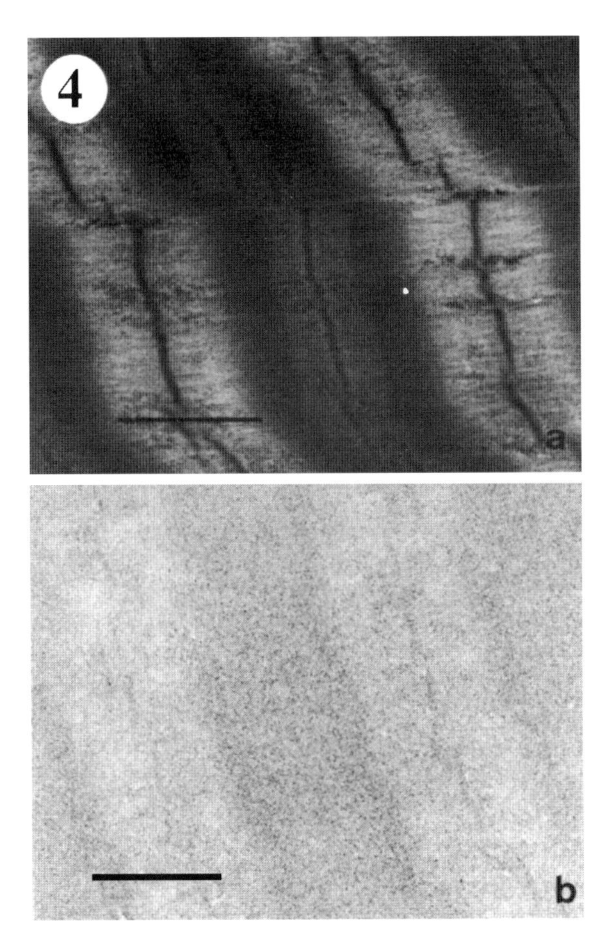

Figure 4. Cut sections of frog sartorius muscle 'stained' with a solution containing 100 mM LiCl and 10 mM CsCl by procedure described in Edelmann 1984. In a, the section was obtained by freeze-drying and embedded. In b, the muscle was fixed with glutaraldehyde and then embedded. Note that selective uptake was only observed in the freeze-dried preparation. Taken together, this type of studies has demonstrated the successful capturing of the *resting living state* of the muscle cells by the adsorption staining procedure introduced by Edelmann (from Edelmann 1986, by permission of Scanning Electron Microscopy International)

The conclusion that both cell K^+ and cell water must be bound to hemoglobin in mature human blood cells, offers an unusual opportunity. That is, an opportunity to put to test the widely-accepted idea that intracellular proteins exist in what is conventionally called native state and as such can be obtained from any biochemical supply house in a bottle – often in crystalline forms. However, there is so far no evidence that what we call native hemoglobin really means what it is supposed to mean, i.e., as it exists in living red blood cells.

For if this popular but unproved idea is correct, a water solution of mammalian hemoglobin at the concentration that it occurs in red blood cells (35%) should selectively bind K^+. In addition, the bulk of surrounding 65% water should have low solvency for Na^+ sulfate.

To put this prediction to a test, a 35% hemoglobin solution was enclosed in a dialysis sac, and allowed to reach diffusion equilibrium with K^+ and Na^+ in the solution bathing the sac. Analysis of the ionic concentration in the bathing solution revealed no or virtually no accumulation of K^+ by the hemoglobin in the sac (Beatley and Klotz 1951; Table 1 in Ling and Zhang 1984). Nor does that 65% water in the sac show reduced solvency for Na^+ (as sulfate) (Table 2A, also Table IX in Ling and Ochsenfeld 1989). In other words, store-bought native protein is not native in the true sense of the word.

Now, if we expose human red cells to a hypotonic lysing solution containing ATP, the red cells hemolyze, losing varying amounts of its hemoglobin as well as most of its K^+ and gained Na^+. If we now 'reseal' the *hemolyzed red cells* or '*ghosts*' by adding sucrose to make the lysing solution isotonic, they would regain more or less their original volume and their lost K^+. In addition, they would extrude the extra Na^+ gained. Most significant was that the amount of K^+ gained as well as the Na^+ extruded are directly proportional to the hemoglobin retained and/or recaptured in the ghosts. In ghosts with no hemoglobin, neither was K^+ regained nor Na^+ extruded (Figure 5).

In summary, cell K^+ and cell water are not free but are 'bound'. In mature mammalian red blood cells, the only major cell component that could bind these small molecules and ions is hemoglobin. Yet store-bought hemoglobin called native does not work. In contrast, hemoglobin in healthy living red blood cells as well as in 'resealed' ghosts – in the presence of ATP – does work. So there is a profound difference between what is conventionally called 'native' and what is truly native – that is, as it occurs in normal living cells. The following simple experimental finding does offer a clue as to the cause of this difference.

In this simple experiment, we titrated the native hemoglobin with NaOH (Ling and Zhang 1984). As the added OH^- neutralizes the positive charges of the ϵ-amino groups of the lysine side chains and the guanidyl groups of the arginine side chains, a profound change takes place in the hemoglobin.

As a result of this change, the up-to-now impotent hemoglobin not only can now adsorb selectively large amount of K^+ (or other alkali metal ions), but also profoundly alters the solvency of the bulk phase water. In the end, the mix of NaOH-titrated hemoglobin and its adsorbed K^+ and water begin to look like what it might be like inside normal red blood cells. But that is not all that make them look similar.

In addition, the NaOH-treated hemoglobin solution is no longer the free-flowing liquid the simple hemoglobin solution once was. The viscosity of the solution has gone up so much that it now takes on the form of a solid gel.

With proper micro-dissecting tools, one can cut up a red blood cell into small fragments without losing its hemoglobin. This retention indicates that hemoglobin is not free but attached to the red stroma proteins (Best and Taylor 1945, p 7.) Certainly there is no question that fresh meat (muscle cells) is in the form of a fairly rigid gel and so is axoplasm of a squid axon (Hodgkin 1971, p 21).

For those used to preparing protein solutions, pure crystalline store-bought native hemoglobin is remarkable in that even at a concentration of 40% (w/v), a hemoglobin solution still flows freely like water. Of course, this is in keeping with the well-known fact that the 'native' hemoglobin molecules are tightly folded and more or less spherical structures (Perutz et al., 1968).

Table 2. The apparent equilibrium distribution coefficient or ρ-value of Na^+ (as sulfate) in water containing native proteins (A), gelatin (B) and PVP (C, E.) and PEO (D.). The ρ-value differs from the (true) equilibrium distribution coefficient or q-value in that the solute in the cell or model water may not all exit in cell water as it is the case with the q-value. However, the ρ-values shown here are all at, or below unity. This means that if some of the solute is adsorbed on the protein or polymer, its quantity was minimal. a, $NaSO_4$ medium; b, Na citrate medium (from Ling et al. 1980 by permission of the Pacific Press, Melville, NY)

Group	Polymer		Concentration of medium (M)		Number of assays	Water content (%) (mean ± SE)	ρ-Value (mean ± SE)
(A)	Albumin (bovine serum)		1.5	a	4	81.9 ± 0.063	0.973 ± 0.005
	Albumin (egg)		1.5	a	4	82.1 ± 0.058	1.000 ± 0.016
	Chondroitin sulfate		1.5	a	4	84.2 ± 0.061	1.009 ± 0.003
	α-Chymotrypsinogen		1.5	a	4	82.7 ± 0.089	1.004 ± 0.009
	Fibrinogen		1.5	a	4	82.8 ± 0.12	1.004 ± 0.002
	γ-Globulin (bovine)		1.5	a	4	82.0 ± 0.16	1.004 ± 0.004
	γ-Globulin (human)		1.5	a	4	83.5 ± 0.16	1.016 ± 0.005
	Hemoglobin		1.5	a	4	73.7 ± 0.073	0.923 ± 0.006
	β-Lactoglobulin		1.5	a	4	82.6 ± 0.029	0.991 ± 0.005
	Lysozyme		1.5	a	4	82.0 ± 0.085	1.009 ± 0.005
	Pepsin		1.5	a	4	83.4 ± 0.11	1.031 ± 0.006
	Protamine		1.5	a	4	83.9 ± 0.10	0.990 ± 0.020
	Ribonuclease		1.5	a	4	79.9 ± 0.19	0.984 ± 0.006
(B)	Gelatin		1.5	a	37	57.0 ± 1.1	0.537 ± 0.013
(C)	PVP		1.5	a	8	61.0 ± 0.30	0.239 ± 0.005
(D)	Poly(ethylene oxide)		0.75	a	5	81.1 ± 0.34	0.475 ± 0.009
			0.5	a	5	89.2 ± 0.06	0.623 ± 0.011
			0.1	a	5	91.1 ± 0.162	0.754 ± 0.015
(E)	PVP	Q	0.2	b	4	89.9 ± 0.06	0.955 ± 0.004
		S*	0.2	b	4	87.2 ± 0.05	0.865 ± 0.004
		Q	0.5	b	3	83.3 ± 0.09	0.768 ± 0.012
		S	0.5	b	3	81.8 ± 0.07	0.685 ± 0.007
		Q	1.0	b	3	67.0 ± 0.26	0.448 ± 0.012
		S	1.0	b	3	66.6 ± 0.006	0.294 ± 0.008
		Q	1.5	b	3	56.3 ± 0.87	0.313 ± 0.025
		S	1.5	b	3	55.0 ± 1.00	0.220 = 0.021

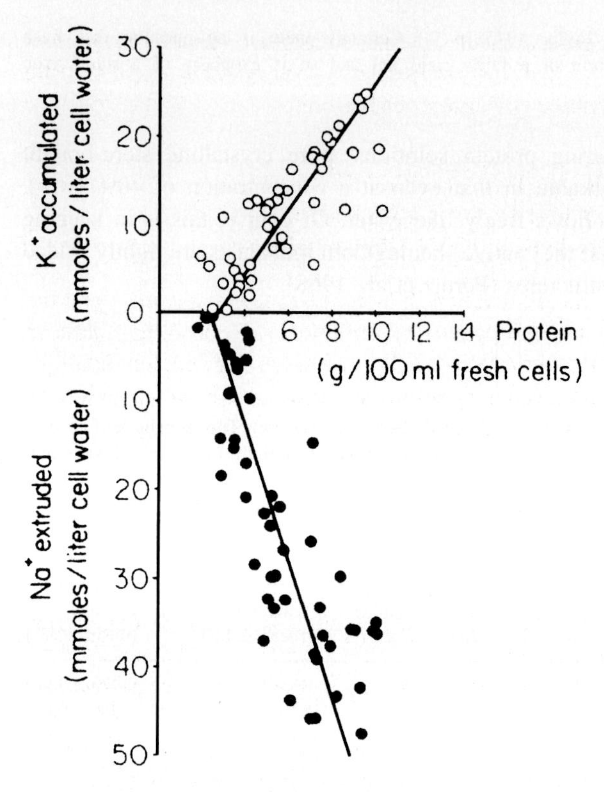

Figure 5. Re-uptake of K^+ and extrusion of Na^+ from red-blood-cell ghosts prepared from washed human red blood cells. The study followed rigorously a procedure described in (Freedman 1976). Freshly drawn blood was obtained (mostly) from different donors. When blood from the same donors was used, it was drawn at least 6 weeks apart. Each data point represents the difference of K^+ or Na^+ concentration in samples of the ghosts at the beginning of incubations and after 18 hours of incubation in the presence of ATP (37 °C). Straight lines shown in the graph were obtained by the method of least squares. Total protein content was obtained by subtracting the sum of the weights of lipids, phospholipids, salt ions, and sucrose from the dry weights of the ghosts (from Ling, Zodda and Sellers 1984, by permission of the Pacific Press, Melville, NY)

From this starting point, the observation that titration with NaOH should bring about a drastic increase in viscosity, there is only one reasonable explanation: *the tightly-folded 'native' hemoglobin molecules have dramatically unfolded in consequence*. Such an unfolded protein assumes the conformation know as *fully extended conformation*. And it is in this conformation, often also called denatured conformation, that it can adsorb K^+ and reduce the solvency of bulk-phase water for Na^+.

Ironically, this finding shows that we have the thing largely turned up side down. Not only is truly native protein not what we have been calling 'native'; what we commonly referred as 'denatured' is, at least in the case of hemoglobin, in fact closer to being truly native.

Indeed, a vast amount of experimental data has collected in the last forty years in support of this conclusion. I shall discuss them in a section below under the title of the Polarized-Oriented Multilayer Theory of Cell Water.

3. A BRIEF OUTLINE OF THE UNIFYING THEORY OF CELL AND SUB-CELLULAR PHYSIOLOGY: THE ASSOCIATION-INDUCTION HYPOTHESIS

As its title clearly indicates, the association-induction hypothesis is built upon the fundamental concepts of close-contact *association* among its constituent parts in the form of ions and molecules so that electrical polarization and depolarization (or *induction*) could link them into a coherent whole. To see how association-induction works, we begin by invoking an old concept, the concept of protoplasm.

3.1 The Restoration of the Concept of Protoplasm

In 1835 Felix Dujardin (1801–1860) described what he saw under the microscope: a gelatinous substance oozing out of the broken end of a protozoon (then called Infusoria.) Dujardin described this 'living jelly' as a 'pulpy, homogeneous, gelatinous substance without visible organs and yet organized...' (Dujardin 1835). Though he gave this gelatinous substance the name, *sarcode*, the name *protoplasm* was broadly adopted in the end.

Thirty-three years later, in his famous Sunday evening lecture in Edinburgh on November 8, 1868, Thomas Huxley called protoplasm 'the physical basis of life.'

The discovery of protoplasm inspired the introduction of the idea of colloid and colloid chemistry. Unfortunately, the understanding of both protoplasm and of colloid were handicapped by the lack of depth in our understanding of (relevant) physics and chemistry at that time – as I have already alluded to in broader terms at the beginning of this communication. This is one reason how the simpler membrane theory gained dominance. Indeed, by the beginning of the 21st century, even the word, protoplasm has become all but forgotten.

Nonetheless in my opinion, protoplasm has been there since life began. So it is a great honor for me to re-introduce this most basic knowledge of biology to the world again.

Given the substantial progress made in the revelant parts of physics and chemistry in the late 19th and early 20th centuries, the AI Hypothesis came into existence and with it, a new definition of protoplasm was born.

3.2 A New Definition of Protoplasm

According to the association-induction hypothesis, protoplasm remains the physical basis of life as Thomas Huxley first and rightly pointed out.

Only protoplasm is no longer defined by its appearance. True, protoplasm may exist in the form Dujardin described as 'pulpy, homogeneous, structureless and yet structured...', but it may also assume a wide variety of other forms as well. What it looks like is only a superficial facet of its existence. What underlies protoplasm to make life possible defines protoplasm.

Except 'ergastic' matter such as the watery solution in the central vacuole of mature plant cells and the inclusions inside food vacuoles of protozoa, etc., all the living part of cells and their living appendages are made of protoplasm. An example of the makeup of automobiles may make the definition easier to understand.

The precise composition, properties and functions of different steel vary. They vary because each kind of steel must serve its specific function in an automobile. For the same reason, the precise composition, properties and functions of different protoplasm vary – in order to serve the specific function of that part of the protoplasm.

Nonetheless, all kinds of steel are steel. That is, they all contain as its major constituents, iron, carbon, other metals and nonmetals. Protoplasm is primarily a system of proteins, water, ions and other small and big molecules functioning as controlling *cardinal adsorbents*. As the *principle cardinal adsorbent*, ATP plays a critically important role in making the living alive.

A correct though variable chemical composition is only one common feature shared by all living protoplasm. Just as vital is how all these constituents are linked together electronically in what physicists called ferromagnetic cooperativity or more precisely what I call auto-cooperativity (Ling 1980). Thus the protein-water-ion-cardinal adsorbent system exists together at a low energy-low entropy state, or what I prefer to describe as high (negative) energy-low entropy state called the *living state*.

3.3 The Living State

Consider a chain of soft-iron nails joined end-to-end with bits of string (Figure 6A). Bring a strong horseshoe magnet close to the end of one of the terminal nails, a chain reaction follows. As a result, the loosely tethered chain of soft-iron nails assumes a more rigid configuration. And with that change, they also pick up the randomly scattered iron filings in the vicinity. Take away the magnet, the system more or less returns to its earlier more random configuration. (Similarly, an electronic rather than magnetic model can be constructed as shown in Figure 6B.)

What the magnet does in this model is to transform the system from a low (negative) energy-high entropy state to a high (negative) energy-low entropy state. According to the AI Hypothesis, protoplasm may also exist in these two alternative states.

However, instead of the tethered chain of soft-iron nails, we have the proteins with their partially resonating and highly polarizable polypeptide chains. And instead of iron filings, we have K^+ and water molecules. And instead of the horseshoe magnet,

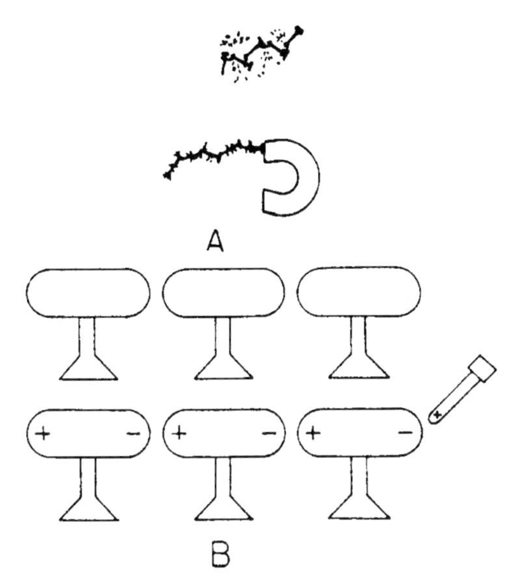

Figure 6. Two models demonstrating information and energy transfer over distances due to propagated short-range interactions. (A) A chain of loosely tethered soft-iron nails is randomly arrayed and does not interact with the surrounding iron filings. The approach of a magnet causes propagated alignment of the nails and interaction with the iron filings. (B) Electrons in a series of insulators are uniformly distributed before the approach of the electrified rod, R. Approach of the rod relocates the electrons by induction such that the insulator becomes polarized with regions of low electron density and regions of high electron density (from Ling 1969, by permission of Intern Rev Cytol)

we have the *principle cardinal adsorbent*, ATP (and its helper the protein Z). (See Figure 7 below.) Only here, the high (negative) energy, low entropy-state with ATP adsorption on the appropriate *cardinal site* constitutes what is known as the *resting living state*. The alternative low-(negative)-energy, high-entropy state is either the *active living state* (as in all reversible transitions) or *dead state* (in an irreversible transition). (Figure 7).

In the next section, I shall point out that according to the AI Hypothesis, protoplasm is basically an electronic machine. Surprising as it may seem, that recognition made in 1962 was also one of history's firsts.

The theory is that diverse variety of protoplasm all existing in the resting living state makes up the entire living cells. This in turn implies that *all cell water must also exist in a physical state different from that of normal liquid water*.

Of course, I have already shown in section 2.5 decisive evidence that no free water exists in typical cells like frog muscle. In the next section I shall go into a little more detail in reviewing the polarized-oriented multilayer (PM) theory of cell water and of inanimate systems demonstrating long-range dynamic water structuring.

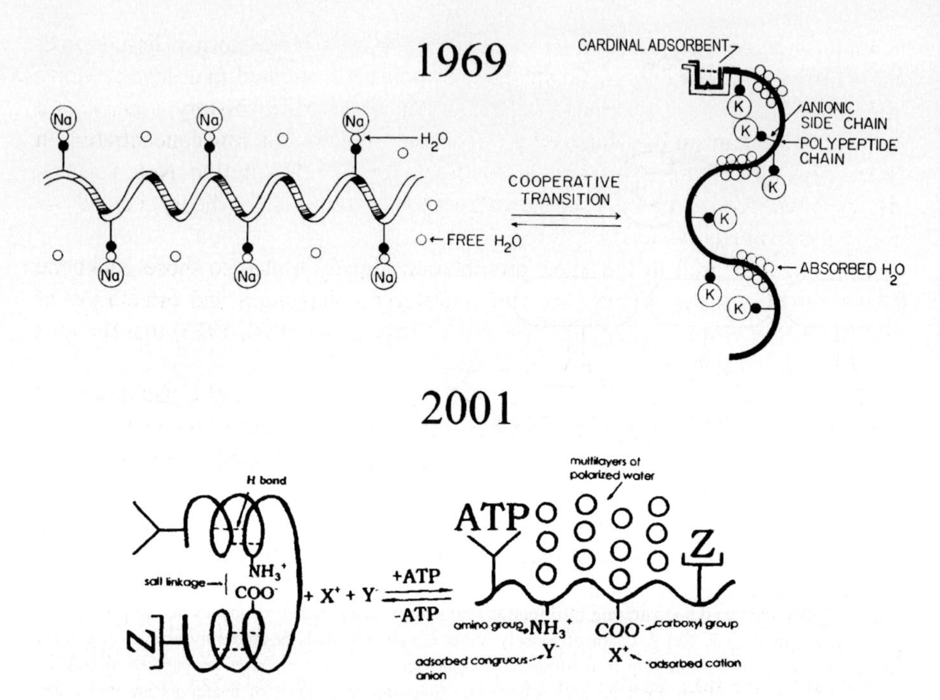

Figure 7. Two diagrammatic illustrations published respectively in 1969 and in 2001. The original legend of the 1969 presentation reads: 'Diagram of a portion of a protein molecule undergoing an auto-cooperative transformation. For simplicity, adsorbed water molecules in multilayers are shown as a single layer. W-shaped symbol represent a (principle) cardinal adsorbent like ATP.' (The 1969 figure shown above is a slightly modified version of the original to correct an illustration error). (from Ling 1969 by permission of the Intern Rev Cytol). The original legend to the 2001 version reads: 'Diagrammatic illustration of how adsorption of the cardinal adsorbent ATP on the ATP-binding *cardinal site* and of 'helpers' including the *congruous anions* (shown here as 'adsorbed congruous anion' and Protein-X (shown as Z) unravels the *introverted* (folded) secondary structure shown on the left-hand side of the figure. As a result, selective K^+ adsorption can now take place on the liberated β-, and γ-carboxyl groups and multilayer water polarization and orientation can now occur on the exposed backbone NHCO groups. The resting living state is thus achieved and maintained' (from Ling 2001 by permission of the Pacific Press, Melville, NY)

4. THE POLARIZED-ORIENTED MULTILAYER THEORY OF CELL WATER AND MODEL SYSTEMS

4.1 A Brief Sketch of the Theory

In 1965, three years after the publication of the association-induction hypothesis proper, the polarized multilayer theory – recently modified to read *polarized-oriented multilayer (PM) theory* of cell water and model systems was introduced. Figure 8A reproduces the key figure in my first public presentation at the New York Academy of Sciences symposium on the 'Forms of Water in Biological Systems' (Ling 1965).

What Figure 8A represents is twofold.

First, it suggests that all the water in all living cells is not normal liquid water but water assuming the dynamic structure of polarized-oriented multilayers.

Second, this picture diagram – again for the first time in history – presents a molecular mechanism by which solutes like Na^+ are kept at a low concentration in living cells on account of an unfavorable free energy of distribution. Note that this theory would not have been possible without the first part of the theory, i.e., *all the cell water* is altered water.

The language used in the 1965 presentation already hinted to those backbone NHCO groups as the primary sites of multilayer polarization and orientation of cell water. But it was not until 1970 and still later (Ling 1970, 1972) that the idea became firmly established in my thinking.

However, long before 1965, J. H. de Boer and C. J. Dippel (1933) had described their idea that multiple layers of water molecules could be adsorbed on the backbone NHCO groups of gelatin. Their original illustration is reproduced here as Figure 9. I did not know about the existence of this paper until last year.

Figure 10 is a reproduction of a figure published in 1972 in an article bearing the title 'Hydration of Macromolecules' in the monograph, *Water and Aqueous Solutions* (Ling 1972). As indicated by the small arrows, Figures 10a, 10b and 10c emphasize that nearest neighboring sites bearing electric charges of the same polarity, orient water dipoles in the same direction. Since water molecules oriented in the same direction repulse one another, multilayer water polarization would not occur on this type of surfaces.

It is only when nearest-neighboring sites bear alternatingly positive (P) and negative (N) electric charges that multiple layers of water molecules can be polarized and oriented in consequence of the attractions among all nearest neighboring water molecules. And to the best of my knowledge, it was the same deBoer mentioned above – with co-author, C. Zwikker – who first pointed out in print this idea (see below).

A checkerboard of alternating N and P sites are what I later designated as an NP system while two juxtaposed NP surfaces are called an NP-NP system (Figure 10d). When either the N or P sites is replaced by a vacant O site, we have what are called an NO-NO system (Figure 10f) or PO-PO system. Not shown in this illustration is what I call a NP-NP-NP system or NO-NO-NO system, which are *parallel arrays of linear chains carrying alternating N and P sites or alternating N and O sites respectively*. They are of central importance in water polarization in living cells because within living protoplasm, there are no *bona fide* flat surfaces like those on salt crystals.

4.2 Four Pre-existing Theories on Multilayer Gas Adsorption and their Respective Shortcomings

At the time when the PM theory was first introduced in 1965, there were four quantitative theories known to me for the multilayer adsorption of gaseous molecules.

C.P. de Boer and C. Zwikker offered the first quantitative theory of multiple layer adsorption of gases on the surface of salt crystals (de Boer and Zwikker 1929).

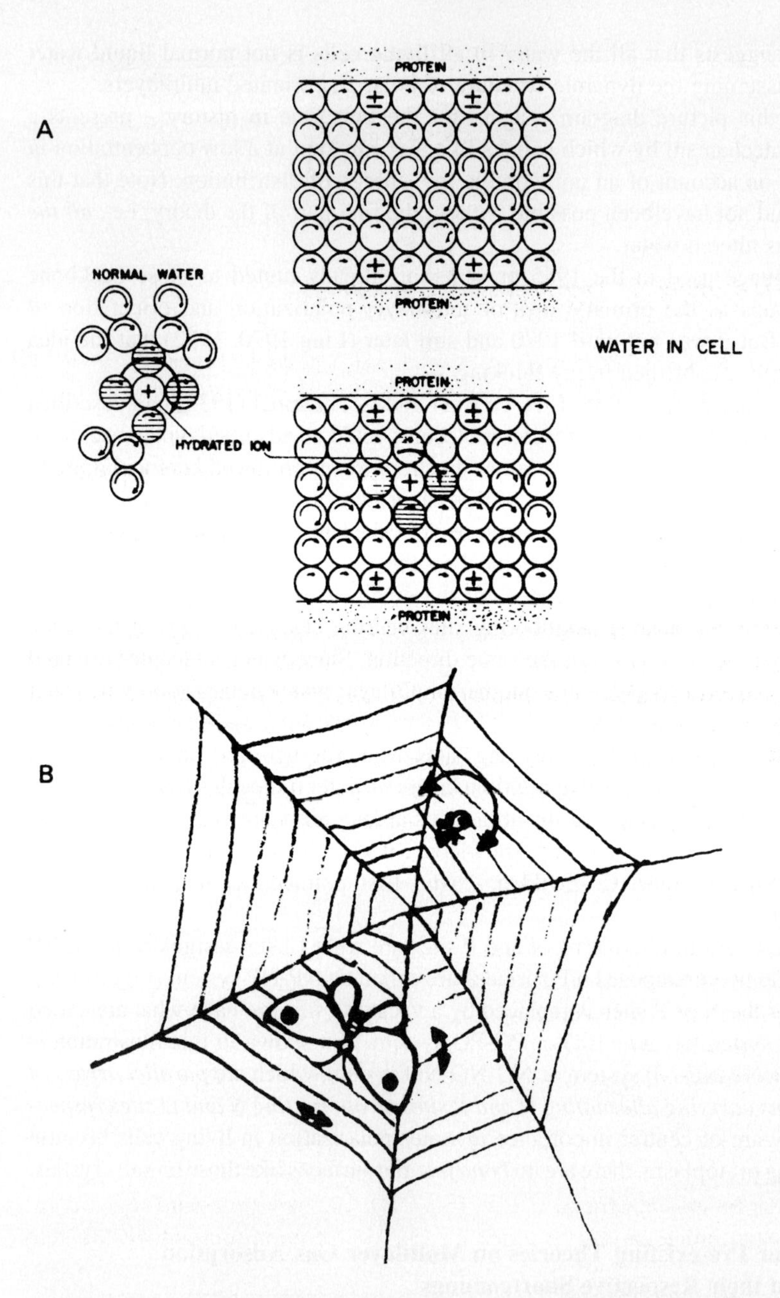

Figure 8. Motional reduction in polarized-oriented water. (A) Diagrammatic illustration of the reduction of rotational (and translational) motional freedom of a hydrated Na^+ ion in water assuming the dynamic structure of polarized multilayers. Size of the curved arrows indicates degree of rotational freedom of both the water melodies (empty circle) and hydrated cations. Reduced motional freedom is indicated by the smaller sizes of the arrows. (This part of the figure was taken from an early paper, Ling 1965). Now we know that one aspect of this diagram is less applicable to living cells. Thus, the degrees of

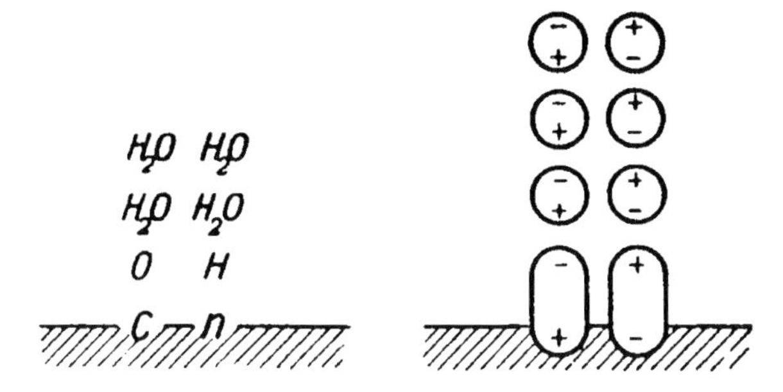

Figure 9. Theoretical model of de de Boer and Dippel showing how dipolar NH and CO groups of gelatin can polarize and orient multiple layers of water molecules (from de Boer and Dippel 1933)

They suggested, as mentioned above, that the presence of alternatingly positive-, and negative electrically charged sites allow the formation of deep layers of gas molecules on the surface of salt crystals as illustrated in their figures reproduced here as Figure 11.

de Boer and Zwikker's *polarization theory* was intended to describe the multilayer adsorption of all types of gas molecules, some without a *permanent dipole moment* like non-polar nitrogen, others with a permanent dipole moment like water vapor.

Within a decade or so after the publication of the de Boer-Zwicker theory, Bradley added two more theories of his own, one specifically for the multilayer adsorption of non-polar gas molecules without a permanent dipole moment (Bradley 1936a) and the other for polar molecules with permanent dipole moments (Bradley 1936b). Each of these three theories can be expressed by an equation of the same form:

$$(1) \qquad \mathrm{Log}_{10}(p_o/p) = K_1 K_3^{\,a} + K_4$$

where a is the amount of gas adsorbed by a unit weight of the adsorbent. (p_o/p) is the reciprocal of the relative vapor pressure. K_1, K_3 and K_4 are all constants at the same temperature. The meanings of each of the three constants vary from one theory to the other but are all too complicated to provide quantitative insights into the adsorption process. Equation (1) can be written in the double log form:

$$(2) \qquad \log_{10}[\log_{10}(p_o/p) - K_4] = a\ \log_{10} K_3 + \log_{10} K_1$$

If the data on the gas adsorbed (a) at different relative vapor pressures (p/p_o) are such that rational numbers can be found for each of the three constants so that the

Figure 8. polarization of water molecules far and near tend to be more uniform than that indicated in the diagram. (from Ling 1965, by permission of the Annals of New York Academy of Sciences). (B) Illustration of the greater degree of motional restriction of larger butterflies snared in a spider web. (Larger) size of arrow represents (greater) degrees of motional freedom (from Ling 1992, by permission of Krieger Publ. Co.)

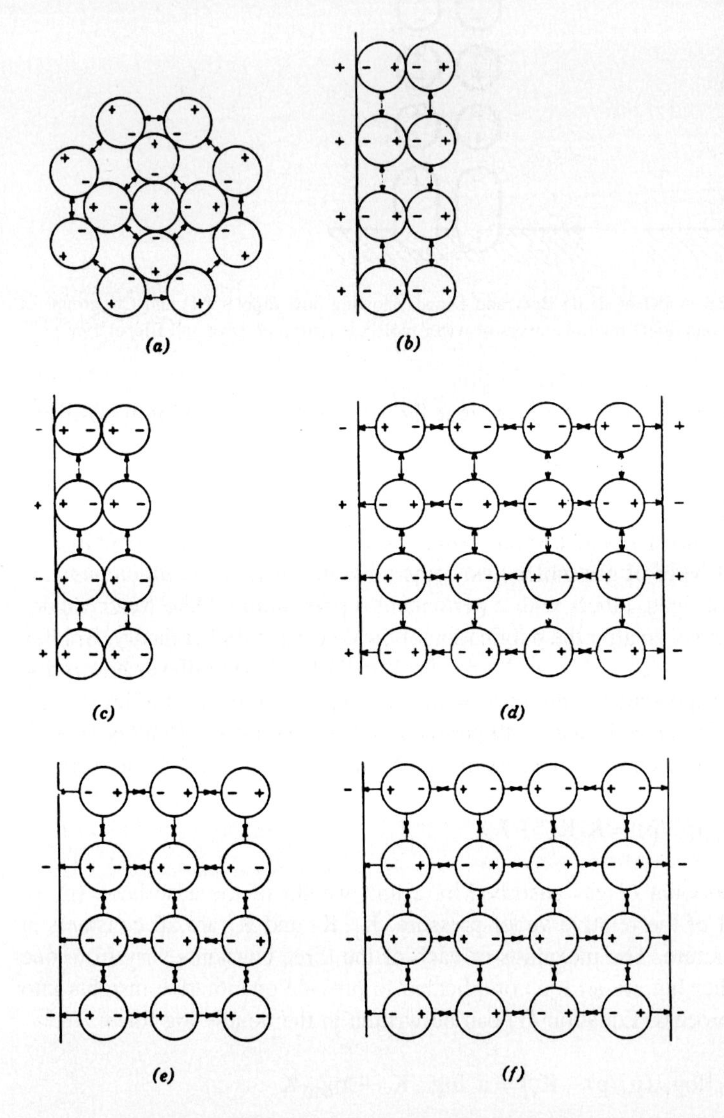

Figure 10. Diagrammatic illustration of the way that individual ions (a) and checkerboards of evenly distributed positively charged P sites alone (b) or negatively charged N sites alone (c) polarize and orient water molecules in immediate contact and farther away. Emphasis was, however, on uniformly distanced bipolar surfaces containing alternatingly positive (P) and negative (N) sites called an NP surface (d). When two juxtaposed NP surfaces are facing one another, the system is called an NP-NP system. If one type of charged sites is replaced with vacant sites, the system would be referred to as PO or NO surface (e). Juxtaposed NO or PO surfaces constitute respectively a PO-PO system or NO-NO system. Not shown here is the NP-NP-NP system comprising parallel arrays of linear chains carrying properly distanced alternating N and P sites (modified after Ling 1972, reproduced with permission of John Wiley and Sons., Inc.)

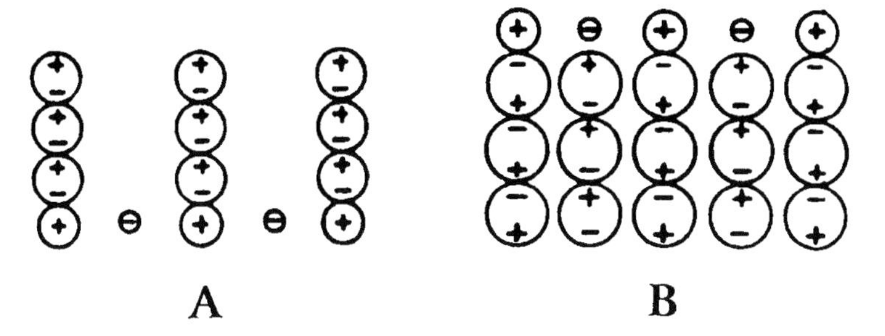

Figure 11. Reproduction of figures presented in the paper by de Boer and Zwikker in 1929 showing their vision of a checkerboard of alternating N and P sites and two different ways they saw how water molecules might be reacting to the charged sites. Although de Boer and Zwikker's Polarization Theory had encountered serious problems, their earlier explicit presentation of what I call the NP surface concept has been of great importance in the subsequenct development of the PM theory. Their reproduction reminds us of their contribution (A and B are respectively what de Boer and Zwikker labeled as Figure 3 and 4 respectively in their original paper)

values of a's can be plotted against the entire left hand side of Equation (2) to yield a straight line, one then regards the data as fitting the equation.

A decade later, Brunauer, Emmett and Teller (the Edward Teller of the Hydrogen bomb fame) demolished the de Boer and Zwikker polarization theory (Brunauer et al. 1938). They did this on the ground that electrical polarization alone – as it is the case in the polarization theory – cannot bring about adsorption of more than one layer of gas molecules even for a gas like argon with one of the largest *polarizability* among the noble gases. Brunauer, Emmett and Teller then offered their own theory, which was nicknamed BET theory after the first letters of their names.

Unfortunately, the BET theory is of no help to me either since virtually all the water that forms multiple layers in their model is simply normal liquid water. As such, it could not fulfill the need of polarized-oriented water demanded by the PM theory with altered physico-chemical properties. This leaves only Bradley's theory for polar gases with a permanent dipole moment like water. Again, despite the endorsement by Brunauer, Emmett and Teller (ibid, p 311) Bradley's theory for gases with permanent dipole moments also has serious problems. But until something better turns up (see below) that was all we had to work with.

4.3 Colloid, Its Birth, Its Unjustified Abandonment and Its Eventual Restoration – with a New Definition Based on the PM Theory

Thomas Graham (1805–1869) was an English chemist. He spent a good part of his life studying diffusion. He was not a cell physiologist and did not even use the word, protoplasm at all in his famous defining paper of 1861 (Graham 1861). Nonetheless, there is little question that he had initiated an important exploration into the nature of protoplasm by introducing the name and concept of colloid, the namesake of gelatin or glue (κολλή).

Under the heading of colloid chemistry and the leadership of a number of capable scientists like Dutch scientist, H.G. Bungenberg de Jong, a great deal of highly valuable information has been obtained. Nevertheless, colloid chemistry like the concept of protoplasm has lost its bearing and has become all but extinct from the cell physiology – until the emergence of the AI Hypothesis but especially its subsidiary, the PM theory of cell water.

A near-fatal mistake was a wrong-headed definition of colloid – in terms of the size of the colloidal particles. So when macromolecular chemistry came into being, colloid chemistry lost its identity.

It was in 1985, when I offered (my first) new definition of colloid (Ling 1985). This and a later more up-to-date definition quoted below were the offspring of two new developments:

(i) The full elucidation of the amino acid composition of gelatin (Estoe 1955), revealing large percentages of proline (13%) and hydroxyproline (10%), both lacking the positively H of the usual peptide NH group, cannot form an intra- or inter-macromolecular H-bonds; an even larger percentage of glycine (33%), which has a very low *α-helical potential* (see Table 4 in Ling 2001, p 145) and thus low propensity to form such intra- or inter-macromolecular H bonds.

(ii) The introduction of the PM theory, which suggested that exposed NHCO or NCO of polypeptide chains polarize and orient in multilayers of cell or model water. Combined, these new ideas and facts led eventually to the latest new definition of colloid as follows:

A colloid is a cooperative assembly of fully extended macromolecules (or large aggregates of smaller units) carrying properly-spaced oxygen, nitrogen or other polar atoms and a polar solvent like water, which at once dissolves and is polarized and oriented in multilayers by the macromolecules or equivalents. (See Ling 2001, p 84.)

In this new definition to colloid chemistry, the difference between macromolecular chemistry and colloid chemistry requires no further explanation. This new definition also affirms the insight and genius of Thomas Graham in being able to see the importance of gelatin as a *bona fide* colloid. Thus, most isolated proteins exist in the folded conformations with their backbone NHCO groups locked in H-bonds and are thus unavailable to polarize and orient the bulk-phase water. Gelatin, in contrast, is at least (13% + 10% + 33% =) 56% in the *fully extended conformation* in consequence of its inherent amino-acid makeup. The fact (to be presented below) that gelatin hydration matches those of PEO, PEG, PVP, each with its entire collection of NO sites fully exposed to bulk-phase water suggests that more than 56% of the NCO and NHCO groups of gelatin are fully exposed to the bulk-phase water.

4.4 Experimental Verification of the PM Theory of Cell Water and Model Systems

For a full account of the experimental verification of the PM theory, the reader is strongly advised to go to my 1992 and/or 2001 books. The limited presentation below emphasizes vapor sorption and solute exclusion studies with all the remaining subjects mentioned by names only in section 4.4.1 immediately following.

4.4.1 *Overall triple experimental verification of the PM theory on eight sets of physico-chemical attributes of cell water and their models*

An inanimate model is called a *positive model* or a *negative model,* depending on whether or not it can duplicate effectively a cell physiological phenomenon due to the presence or absence of a critical feature or property. The affirmation of the respective comparisons between the living cell and a positive model, between the living cell and a negative model and between the positive and negative inanimate models is called a *triple confirmation* (Ling 2003). While the term, positive and negative models apply to all models, more specific names were given to the specific models for cell water. Thus a positive model for polarized-oriented water is called an *extrovert* model; a negative model is called an *introvert* model.

Since the introduction of the PM theory in 1965, worldwide testing resulted in the triple confirmation of all eight sets of basic attributes of cell water so far studied in depth: (1) osmotic activity; (2) swelling and shrinkage; (3) freezing point depression; (4) vapor sorption at near saturation; (5) NMR rotational correlation time $\{\tau_r\}$; (6) Debye dielectric reorientation time $\{\tau_D\}$; (7) rotational diffusion coefficient from quasi-elastic neutron scattering; (8) solute exclusion. (For references of all these experimental studies, see Ling 1992, p 108; 2001, p 78.) In what follows, I shall discuss in more detail only two of these subjects: vapor sorption at near saturation and solute exclusion.

Space limitation does not allow more details. Nonetheless, it gives me both pleasure and honor to cite investigators whose scientific insights, skill and dedicated efforts have made all these possible: Freeman Cope, Carlton Hazlewood, Raymond Damadian, Jim Clegg, Miklos Kellermayer, Bud Rorschach, E. Ernst, A.S. Troshin, Dimitri Nasonov and many others. Then there are still others whose work would be discussed in greater detail below.

4.4.2 *Vapor sorption of living cells and model systems at relative humidity ranging from near zero to 0.99 and higher*

Throughout history, many studies of the adsorption of water by proteins and other biomaterials have been reported. Almost all of these studies do not go beyond 95% saturation on the high end. This is unfortunate because the *physiological vapor pressure*, that is the vapor pressure of a typical Ringer's solution or plasma is way beyond 95%. For example, the relative vapor pressure of (frog) Ringer's solution is 0.996.

However, there is one notable exception. Namely, the comprehensive study by J.R. Katz (1919) of water vapor sorption on a wide variety of chemical substances and biomaterials at relative humidity as high as 100% saturation.

(1) *The earliest theory known to me of a common origin of water sorption in protoplasm and in gelatin by Heinrich Walter based on J.R. Katz's vapor sorption data*

J.R. Katz was not that certain about the accuracy of his data obtained at what he labeled as 100% saturation. So he gave these numerical figures behind ± sign. Nonetheless, the difference between the water vapor uptake of (denatured collagen or) gelatin at 100% relative humidity (±4.6 grams of water per gram of dry protein) and so-called native proteins like hemoglobin (0.796 grams of water per gram of dry protein) could not be more striking. This subject will be brought up again below.

However, at the next highest relative vapor pressure Katz studied (96.5%), the uptake of water vapor by gelatin was only 0.64 grams per gram. Thus fully 86% of the water uptake of gelatin occurs above 96.5%, which is above the usual upper limit at 95%.

Four years after Katz's paper was published, Heinrich Walter (1923) at the Botanical Institute at Heidelberg reported that the volume of (vacuole-free) protoplasm from various algae, when immersed in sucrose solutions of different strengths, exhibit a certain quantitative pattern of variation. Walter then suggested that this pattern of swelling or shrinkage bears resemblance to the pattern of water vapor uptake by gelatin, starch etc., which Katz had demonstrated earlier. Walter's illustration is relabeled and shown here as Figure 12.

Alas, this perfectly reasonable idea of Walter was also ignored and so rarely cited that once again I was totally unaware of its existence until very recently. Thus to

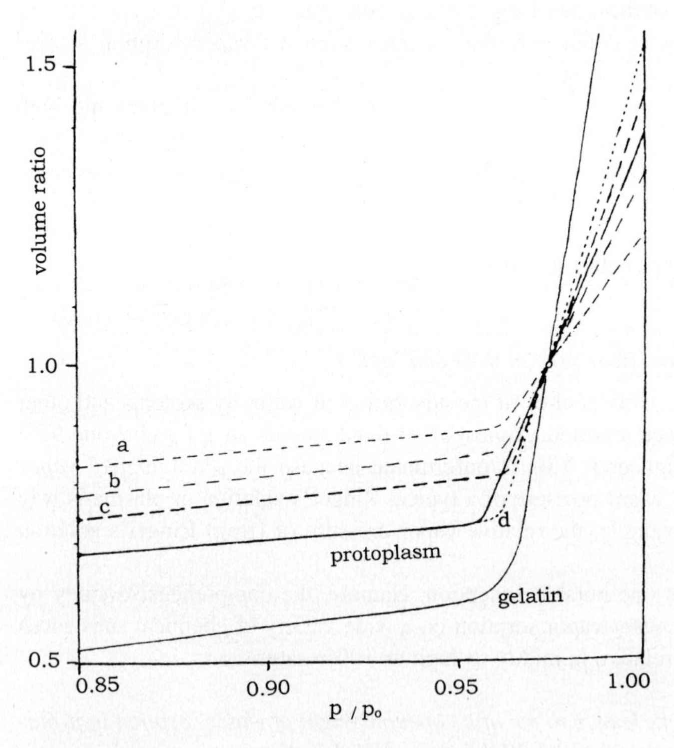

Figure 12. Heinrich Walter's demonstration of parallelism between the swelling and shrinkage of algae protoplasm in sucrose solutions of different strengths and water vapor sorption of gelatin, starch (a), nucleic acids (b) and casein (c) at different relative vapor pressure from J.R. Katz (1919) (from Walter 1923)

the best of my knowledge on this day (February 29, 2005), Walter was the first to suggest in 1923 that water in protoplasm and in gelatin share a common origin.

But Walter offered no idea on what that common origin is. When de Boer and Dippel did suggest ten years later that the polypeptide NHCO groups of gelatin could provide the seat of multilayer adsorption of water, they made no connection between their idea and Walter's. Fortunately, other facts known were able to lead me to make the several connections that spelled out as the PM theory.

(2) *Ling and Negendank's study of water sorption of surviving frog muscle cell strips at relative humidity from near zero to 0.996*

Having given due credit for Walter's idea, I must now point out that there is a gap of experimental knowledge. This gap lay between J.R. Katz's water vapor sorption data of gelatin (casein, starch and nucleic acids) and Walter's data, which he claimed to be water sorption on plant protoplasm obtained by soaking the plant cells containing the protoplasm in sucrose solutions of various strength.

There is no question that the cell membrane involved like all other cell membranes is quite permeable to sucrose (Table 1). So his data of the water uptake was not that of the protoplasm but that of protoplasm *plus* varying amount of sucrose. That was the best he could do at the time but it was in need of improvement. In fact such an improvement was made (in ignorance of Walter's earlier idea) and reported by Ling and Negendank in 1970 (Ling and Negendank 1970). And it consisted of making a similar vapor sorption study as J.R. Katz had done on gelatin and other materials in 1919, but now directly on living protoplasm under sterile conditions.

First, by dissecting frog muscle into narrow strips and exposing them to a vapor phase kept at different relative vapor pressure – all under sterile conditions – Ling and Negendank found that the time of water vapor sorption equilibrium at 25 °C was reached in about 5 days. Based on this knowledge, Ling and Negendank obtained the vapor sorption isotherm of living frog muscle cells at relative humidity ranging from 0.043 to 0.996 as shown in Figure 13.

The data can be sorted out into two parts. A small fraction making up about 5% of the total water uptake is taken up strongly at the lowest vapor pressure range. The interpretation we offered, that the 5% strongly bound water was taken up by polar side chains of cell proteins is in agreement with conclusions of the later work of Leeder, Watt and others (Leeder and Watt 1974).

The remaining 95% of the water adsorbed could fit the Bradley isotherm (shown for example in Figure 21 in Ling 2001) – much as Hoover and Mellon (1950) had demonstrated similar fitting of the Bradley isotherm of their data on water sorption of casein, cotton and especially polyglycine.

(3) *Hoover and Mellon's study of vapor sorption of polyglycine, proteins and other polymers*

I was not able to find out the molecular weight of the polyglycine Hoover and Mellon used. Nonetheless, it is not difficult to arrive at the conclusion that the

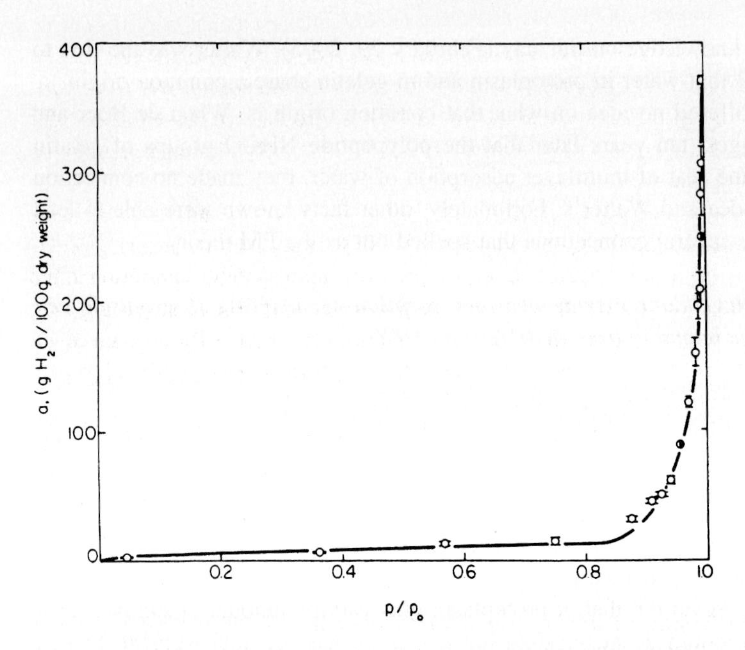

Figure 13. Water vapor sorption of surviving frog muscle cell strips at relative vapor pressure from 0.043 to 0.996. Small muscle cell strips were isolated under sterile conditions and incubated at 25° for 5 days to reach equilibrium. To prevent condensation of water on the wall of the wide glass tubes in which the muscle strips ware hanging, the entire tubes were immersed in a constant temperature bath maintained at 25 °C±0.05 °C. Different relative vapor pressures were provided by different concentrations of solutions of H_2SO_4 or NaCl (from Ling and Negendank 1970, by permission of the Pacific Press, Melville, NY)

polyglycine they used could not have been a small polypeptide. Or else these authors would not have referred to the polymer studied as high polymer in the title of the article.

On the assumption that they studied a high molecular weight polyglycine, one can reason that the terminal carboxyl and amino groups are trivial in number and that virtually all its water-sorbing sites were in the form of backbone NHCO groups. Whether this interpretation is completely right or only partly right, Hoover and Mellon themselves had inferred from their data that the backbone NHCO groups offer important sites of water vapor sorption. In addition, they also showed that the vapor uptake of polyglycine exhibits a sigmoid-shaped water sorption curve. This in turn has a two-fold significance.

First, it shows that the backbone NHCO groups are indeed important water sorption sites as pointed out earlier by Lloyd, Sponsler and others (Lloyd 1933; Lloyed and Phillips 1933; Sponsler et al., 1940).

Second, since there were no polar side chains in polyglycine, the abundant 'extra' uptake of water at very high humidity in this case at least was not built upon monolayers of water adsorbed on polar side chains as Leeder and Watt once suggested for proteins (1974, p 344). *Rather, the entire sigmoid-shaped uptake begins and ends as polarized oriented multilayers on the backbone NHCO groups.*

To gain more insight on the role of backbone NHCO groups, we move to the water sorption data of Katchman and McLaren (1951) on a closely similar polypeptide, polyglycine-DL-alanine.

Like polyglycine, polyglycine-DL-alanine also has virtually all its potentially water-sorbing sites in the form of backbone NHCO groups. And its strongly accelerating sorption of water at relative vapor pressure approaching saturation once more reminds us of a similar pattern seen in the water sorption of surviving frog muscle cells shown in Figure 13. This close quantitative similarity is made even clearer in Figure 14, taken from Ling 2003, which in turn was a modification of a still earlier illustration first published in 1972 (Ling 1972).

This quantitative matching of water sorption of intact living frog muscle and of polyglycine-DL-alanine (1:1) on the upper end of vapor sorption is highly significant. It confirms the PM theory of cell water as being polarized and oriented primarily by the exposed NHCO groups of the fully-extended proteins in living cells.

(4) *Ling and Hu's introduction of a new fast technique and their study of vapor sorption of PEO, PEG, PVP, gelatin and several native proteins at physiological, and still higher vapor pressure*

A skeptic critique may ask, 'How do you known that the NHCO groups of either polyglycine or polyglycine-DL-alanine are not engaged in α-helical or in other intra-, or inter-macromolecular H bonds and thus not free to adsorb water?' The answers are as follows.

First, there is no problem with polyglycine. It is well known to exist in the fully extended form in water (Bamford et al. 1956, p 310.) In contrast, a block polymer of poly-L-alanine (of 130 residues) is known to be entiely water insoluble. Indeed, so tight is its α-helical folding that it would not unfold even in 8 M urea (Doty and Gratzer 1962, pp 116–117). The fact that the co-polymer, polyglycine-DL-alanine containing one part glycine and one part D,L-alanine is *fully water soluble* as well as its strong water sorption make it a good bet that the polyglycine-DL-alanine investigated by Katchman & McLaren does not contain large block of alanine residues but contain more or less randomly distributed mixed polymer and as such, fully extended.

However, to leave no doubt about the fully extended nature of these poly-amino acids, Ling and Hu undertook the studies of some other *extrovert models* of the NO-NO-NO type, which being without P sites cannot engage in α-helical of other intra-, or inter-macromolecular H-bonds as polyamino acids can.

Furthermore, in these newer studies there was yet another improvement over the Katchman-McLaren data, which did not go beyond 95% relative vapor pressure at its high end. As already pointed out above, when we talk cell physiology, it refers to its normal physiological environment. As mentioned above, the physiological vapor pressure for frogs is 0.996.

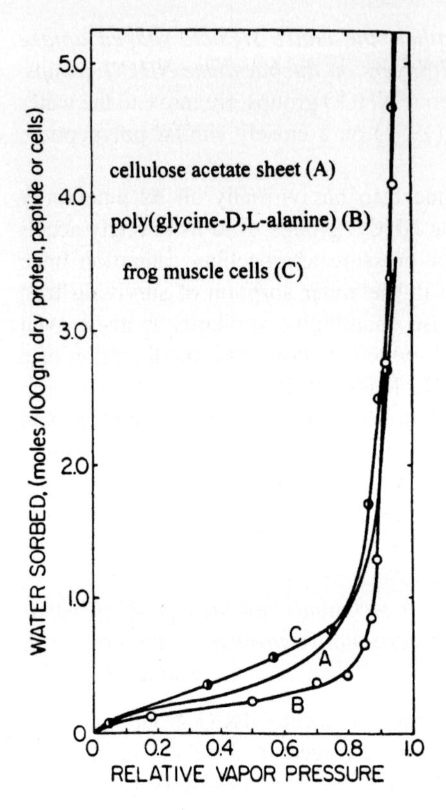

Figure 14. The strikingly similar steep uptake of water molecules at relative vapor pressure close to saturation of surviving frog muscle cells (marked C) and of the synthetic polypeptide, poly(glycine-D,L-alanine) (marked B), shows evidence that the backbone NH and CO groups (which are virtually all the functional groups that can interact with water molecules in the synthetic polypeptide) are the major seats of (multilayer) water adsorption in living cells as suggested in the polarized-oriented multilayer theory of cell water. The third curve from unpublished work of Palmer and Ling shows that water taken up by commercial cellulose acetate sheets is similarly adsorbed on dipolar sites – a matter of great significance because later work of Ling (1973) shows that the permeability of this membrane strikingly resembles the permeability of a live membrane (inverted frog skin) (figure reproduced from Ling 1972, by permission of John Wiley and Sons, Inc.)

To introduce our next subject, I ask the question, 'Why was J.R. Katz able to study water vapor sorption on biomaterials up to 100% relative vapor pressure and yet later work on vapor sorption, including that of Katchman and McLaren just quoted, shied away from relative vapor pressure higher than 95%?'

First, the continued dominance of the membrane (pump) theory with its free water and free K^+ assumptions has given a (false) reason to consider irrelevant the study of water sorption at physiological vapor pressure. But there is a second more legitimate reason.

At saturation or near saturation vapor pressure, the attainment of adsorption equilibrium is extremely slow and this slowness not only has made studies difficult but also uncertain in accuracy. Thus in Katz's work, it took several months to reach what he considered as equilibrium but behind ± signs. Ling and Hu showed that in the case of water sorption of the polymer, polyvinylpyrrolidone (PVP), at a near saturation vapor pressure, the equilibrium level was still not reached after three hundred and twenty (320) days of incubation.

It was to overcome this forbidding difficulty that Ling and Hu introduced their new method, called *nullpoint-method.* This new method shortened the equilibrium time to only five days and with this gain, there was also a gain in full reliability. To save space, Figure 15 with a detailed legend will give the reader some idea on how it is done.

Using this new null-point method, Ling and Hu studied the vapor sorption of gelatin, three oxygen-containing polymers that belong to what we call NO-NO-NO extrovert systems and several 'native' proteins including hemoglobin. Among the NONONO extrovert models studied are polyvinylpyrrolidone (PVP), poly(ethelene oxide) (PEO) and polyethylene glycol (PEG). The vapor pressures studied were immediately below, at, and above the physiological vapor pressure for frog tissues, 0.9969. Also included in Figure 16 is the upper end of the water sorption of surviving frog muscle data of Ling and Negendank presented in full in Figure 13.

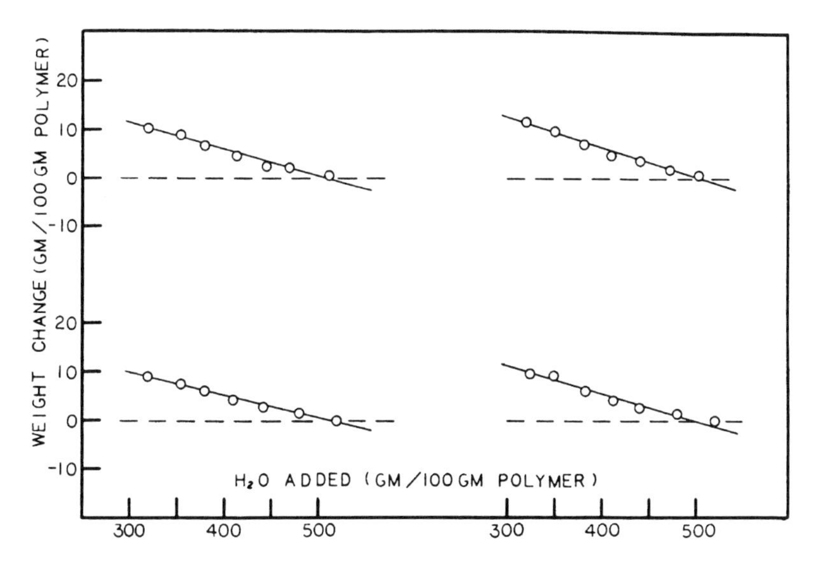

Figure 15. Quadruplicate determinations of the equilibrium water sorption in an atmosphere containing water vapor at a relative vapor pressure equal to 0.99775, of polyethylene glycol (PEG-8000) by the Null Point Method. The ordinates represent the percentage gains (or losses) of water in the different PEG samples after 5 days of incubation at 25 °C following the addition of different amounts of water to each sample (in grams of water added per 100 grams of dry sample weight) shown as abscissa. The zero gain (or loss) intercepts allows one to obtain from the quadruplicate set of data: 513.7, 505.0, 510.8 and 508.9 averaging 509.6 ± 3.65 (S.D.) (from Ling and Hu 1987, by permission of the Pacific Press, Melville, NY)

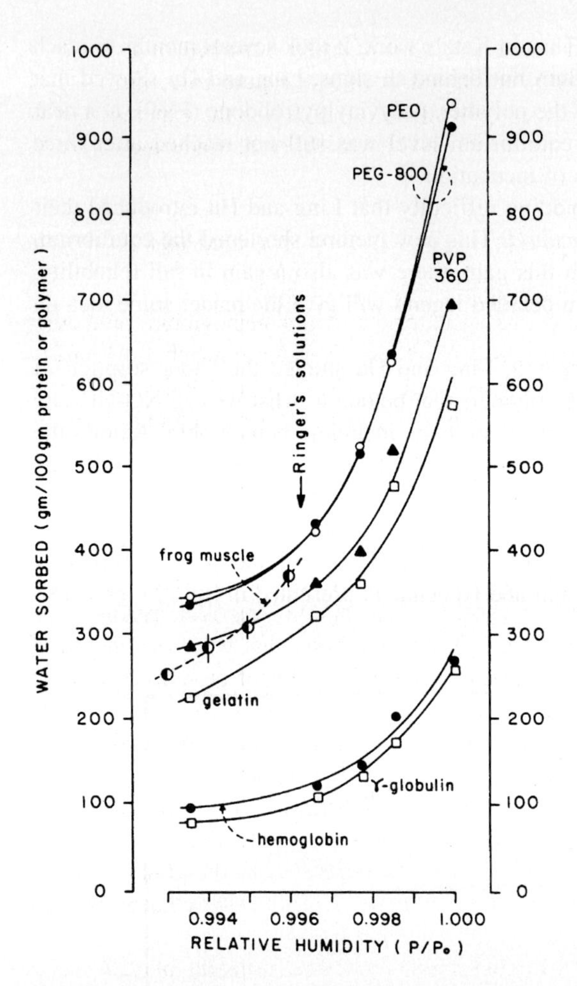

Figure 16. Equilibrium water vapor sorption obtained with the aid of the Null Point Method at relative vapor pressures from 0.99354 to 0.99997 of polyethylene oxide (PEO), polyethylene glycol (PEG-8000) polyvinylpyrrolidone (PVP-360), gelatin, γ-globulin, hemoglobin at 25 °C. (from Ling and Hu 1987, by permission of the Pacific Press, Melville, NY)

Note that at this extremely high physiological vapor pressure range, the uptake of water of these extrovert models has gone far above the water uptake at the upper limit of relative vapor pressure (95%) in the poly-glycine-DL-alanine study. As such, they match or even exceed the water uptake of living frog muscle cells. Since these extrovert NONONO models carry negatively charged oxygen atoms as N sites, but no positively charged P site in-between the N sites, they cannot form intra-, or inter-macromolecular H bonds. Accordingly, each of these oxygen atoms (together with its two negatively charged lone pair electrons and its adjacent vacant or O site) is free to polarize-orient multilayers of water as predicted in the PM theory.

The quantitative accord between the amount of water adsorbed of living frog muscle and these three NONONO polymers as shown in Figure 16 confirm one of the main postulates of the PM theory of cell water. That it is the exposed NHCO groups that polarize and orient by far the greater majority of water molecules in living cells.

(5) *Conflict with traditionally accepted low water binding data and a possible reconciliation*

Parenthetically, the demonstration that the so-called 'native hemoglobin' (and other 'native' proteins) do(es) not polarize and orient enough water to match that found in the living cells – even though these folded 'native' proteins do polarize and orient far more water than conventionally accepted by many first rate protein chemists (Table 3). The question may be raised: 'Which is right?'

The answer is probably that both are right. Thus, different investigators could be focussed on different portions of the affected water. The data obtained by equilibrium vapor sorption is direct and not model dependent. In contrast the methods used to obtain the data presented in Table 3 are almost all model-dependent and the values determined usually relied on certain simplifying assumptions.

In some of these cases at least, there is the possibility that what was measured was the more tightly held water bound to polar side chains of proteins mentioned earlier. Thus, you may recall, tightly bound water emerges as a separate fraction in the water vapor taken up by frog muscle cell strips already completed at very low vapor pressure. Its total amount was estimated to be about 5% of the total water content of the muscle cells. At a total water content of about 80%, that 5% amounts to 4 grams of sorbed water per 100 grams of fresh muscle. Taking a value of 20% total cell proteins, this fraction would be equivalent to 0.2 grams of water per gram of protein. This is a figure not far from the average water content listed in Table 3.

In the next section, I shall examine another physical property of cell water, its solvency for solutes of different sizes. This too is model independent. If at equilibrium a specimen of water accommodates only half of what that solute can dissolve in normal liquid water, at least 50% of the water in this specimen must be different from normal liquid water.

4.4.3 *Solute exclusion from polarized-oriented water in living cells and in model systems*

(1) *Theory of solute exclusion based on the PM theory*

Now, we are in a position to look at Figure 8A once more. You recall that that figure shows how proteins can polarize and orient multilayers of water. The figure to the left in Figure 8A also shows how restricted motional freedom might offer an (entropic) mechanism for the exclusion of solutes like the hydrated Na^+ – as illustrated in the analogy of a butterfly caught in a spider web shown in Figure 8B.

Table 3. Hydration of proteins

Protein	Reference listed at end of table	Technique identified at end of table	Hydration ($g\ H_2O/g$ protein)
Serum albumin	5	A	0.2
	6	B	0.30–0.42
	10	C	0.19–0.26
	11	D	0.31
	7	E	1.07[a]
	7	F	0.75[a]
	8	G	0.43
	12	F	0.40
	12	H	0.48
	13	A	0.18–0.64
	14	A	0.15
	9	A	0.23
Avg.			0.32
Ovalbumin	2	I	0.18
	5	A	0.1
	6	B	0–0.15
	8	G	0.31
	7	E	0.45[a]
	9	A	0.18
Avg.			0.17
Hemoglobin	6	B	0.20–0.28
	11	D	0.10
	8	G	0.45
	7	E	0.36[a]
	7	F	0.69[a]
	9	A	0.14
	10	A	0.2
Avg.			0.25
β-Lactoglobulin	6	B	0–0.20
	7	E	0.72[a]
	7	F	0.61[a]
	5	A	0.4
	9	A	0.24
Avg.			0.25

A Dielectric dispersion; B X-ray scattering; C Sedimentation velocity; D Sedimentation equilibrium; E Diffusion coefficient; F Intrinsic viscosity; G NMR; H Frictional coefficient; I ^{18}O diffusion.

1. Fisher 1965; 2. Wang 1954; 3. Adair and Adair 1963; 4. Adair and Adair 1947; 5. Oncley 1943; 6. Ritland et al., 1961; 7. Tanford 1961; 8. Kuntz et al., 1969; 9. Buchanan et al., 1952; 10. Cox and Schumaker 1961a; 11. Cox and Schumaker 1961b; 12. Anderegg et al., 1955; 13. Grant et al., 1968; 14. Haggis et al., 1951; 15. Miller and Price 1946. (For details of sources of the publications, see Reference at end of Ling 1972) (from Ling 1972, by permission of John Wiley and Sons, Inc.).

However, to facilitate a more quantitative approach, I must first introduce a parameter called the q-value.

The q-value stands for the (true) *equilibrium distribution coefficient* of a solute between cell (or model) water and the usually normal liquid water in the surrounding medium. The q-value refers exclusively to a solute (like sucrose) in the bulk-phase water of a living cell. As a rule, the q-value is at or below unity. Since according to the PM theory, solute distribution in living cell water and in model systems follows the Berthelot-Nernst distribution law (see Glasstone 1946, p 735), a plot of the equilibrium concentration of the solute against its concentration in the bathing medium should yield a straight line. And the slope of the straight line is equal to the q-value of that solute in the cell water or model water as the case may be.

In 1993, a full quantitative PM theory of solute exclusion involving both enthalpic and entropic mechanisms was published (Ling 1993; Ling et al. 1993). Working together, each mechanism provide its own size-dependent solute exclusion. Under the name, the 'size rule', the theory predicts that the larger the solute molecule, the lower is the q-value of that solute. Equation (A3), which summarizes the theory, is reproduced here as Appendix 1 given toward the end of this communication.

Take a large molecule like sucrose (or hydrated Na^+) for example. In order to transfer a molecule of sucrose from an external bathing solution made up of normal liquid water into a living cell, a hole must be excavated in the cell water to accommodate the sucrose. Since the average water-to-water interaction in the polarized-oriented cell water is stronger than that in the normal liquid water outside, more energy would be spent in excavating the hole in cell water than the energy recovered in filling the hole left behind in the normal water of the bathing solution.

This enthalpy or energy difference per molecule is the enthalpic (or energy) component of the size-dependent solute exclusion. Figure 17 shows the theoretical q-value for solutes of different size in cell water or model water with different levels of water-to-water interaction energy (alone). This is the size-dependent *bulk phase* enthalpy component of the free energy difference that determines the q-value of a solute.

Next to discuss is the entropic component. Here too, the larger the solute molecule the greater is its structural complexity The greater the structural complexity, the larger is its rotational entropy. The greater its rotational entropy, the greater is its propensity to lose a significant part of it due to the more 'sticky' polarized-oriented cell water with stronger water-to-water interaction than in normal liquid water of the surrounding medium. This is illustrated in the spider web analogy of Figure 8B. It is the larger butterfly that suffers proportionally the more motional restriction, while a smaller butterfly may not lose that much motional freedom to allow it to fly away.

Then, there is a third component in the makeup of the full expression of exclusion given in Equation (A3) mentioned above. This third component is a measure of how well the surface of the solute molecule fits the water structure surrounding the solute molecule. For solute that can fit into the water structure, that favorable component of (negative) energy gained would be added to the q-value, making it higher than that due to size alone. On the other hand, if the surface component is unfavorable, it would make the q-value still lower.

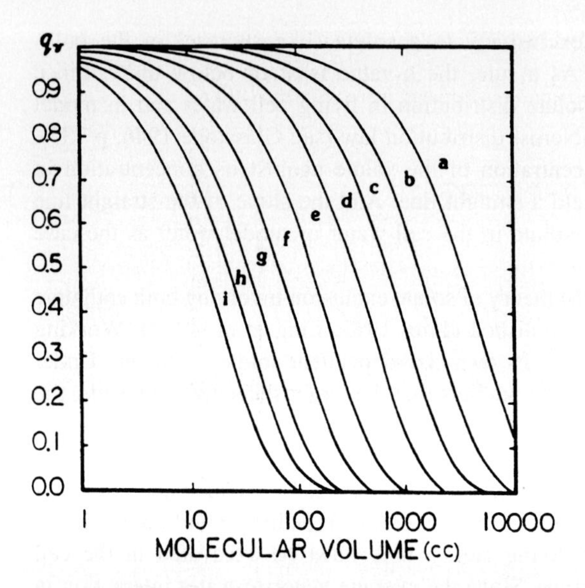

Figure 17. The theoretical volume (or solvent) component of the equilibrium distribution coefficient (q_v) for solutes of different molecular volume in water polarized at different intensity (See Equation A3 in Appendix 1). The intensity of water polarization due to the volume component of the polarization energy is given as the specific solvent polarization energy, ΔE_v. The specific value of ΔE_v in units of RT per cm^3 is indicated by the letter near each curve, where a represents 0.0002; b, 0.0005; c, 0.001; d, 0.002; e, 0.005; f, 0.01; g, 0.02; h, 0.03; i, 0.05. R is the gas constant and T the absolute temperature. At room temperature (25 °C), RT is equal to 592 cal./mole (from Ling 1993, by permission of the Pacific Press, Melville, NY)

(2) *Experimental testing of solute exclusion theory based on the study of small probe molecules with molar volume of 500 cc or less*

Based on what has been explained above, the PM theory could make certain predictions. Thus in normal resting living cells, the q-value of solutes should decrease with increasing size of the solute's molar volume, Figure 18 shows that this is true for frog muscle cells. Note that the slopes of the straight lines decrease with the increasing size of the solute. That slope, as mentioned earlier, is equal to the q-value of that solute.

According to the PM theory, water in extrovert models should demonstrate similar size-dependent q-values. Figure 19 and Figure 20 respectively show that this is true with two extrovert models studied in some detail, poly(ethylene oxide) or PEO and NaOH denatured hemoglobin.

On the other hand, water in a solution of so-called native hemoglobin – a well-established *introvert* model, should according to the PM theory, demonstrate a q-value close to unity for the same solutes of varying size that are excluded to varying degrees by normal living cells and by the extrovert models. Figure 21 shows that this expectation too is realized for solutes 500 cc or less in molar volume.

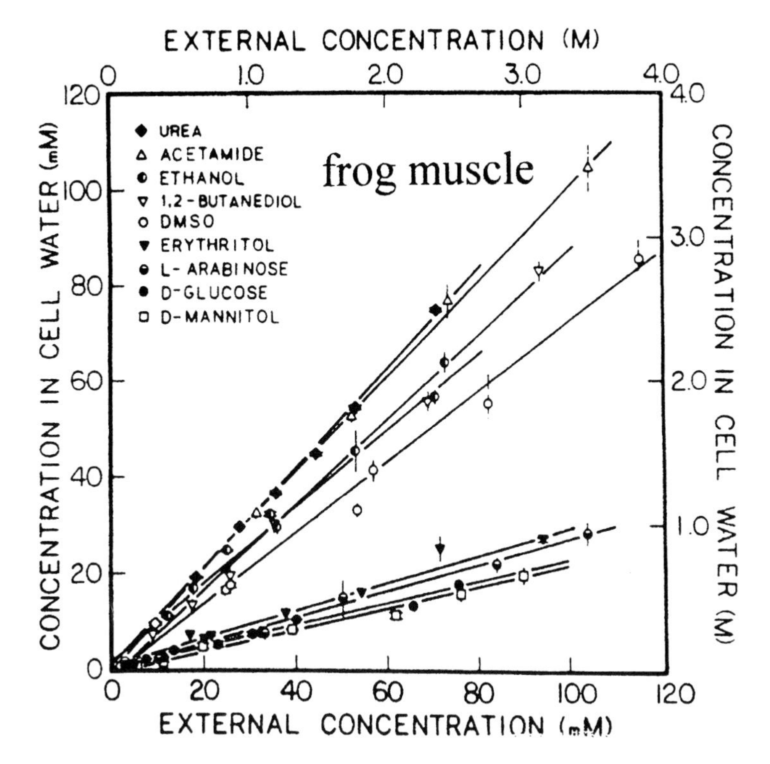

Figure 18. The equilibrium distribution of nine non-electrolytes in frog muscle cell water at 0 °C. Each point represents the mean ± S.E. of at least four samples. Incubation time (enough or more than enough to insure diffusion equilibrium in all cases) was 6 days for L-arabinose and L-xylose but 24 hours only for all others. Each of the straight lines going through or near the experimental points is obtained by the method of least squares, its slopes yielding the respective (true) *equilibrium distribution coefficient* or q-value of that solute (from Ling et al. 1993, by permission of the Pacific Press, Melville, NY)

The slopes of these straight line plots in Figures 18 to 21 yield respectively the q-values of the different solutes in muscle cell water, in the water containing NaOH-denatured hemoglobin, PEO and native bovine hemoglobin. These q-values are in turn plotted against the molecule weight of each solute in Figure 22 and its insets. Included in Inset B is also a set of data of gelatin, which were not our own but taken from Gary-Bobo and Lindenberg (1969). Unlike our data, their degree of exclusion was not obtained from linear plots from many experimental points at different concentrations. Rather, each point was from a single set of determinations at a specific concentration. Thus, this set of data is not rigorously obtained as are q-values. But judging from the general conformity, the deviations if any could not be too large.

The lines going through or near the experimental data points of the main figure and figures in the insets of Figure 22 were all obtained by visual inspection. All told, two kinds of curves were obtained. The curve in the form of a straight flat

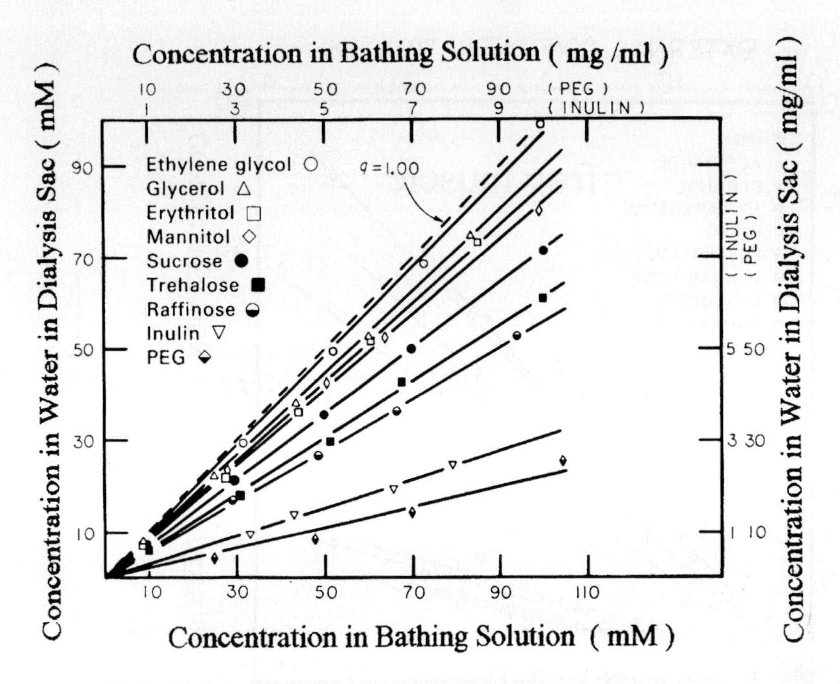

Figure 19. Equilibrium distribution of nine non-electrolytes in dialysis sacs containing bovine hemoglobin (18%) dissolved and incubated in, and denatured by 0.4 M NaOH. Incubation lasted 5 days at 25 °C. Symbol for each nonelectrolyte as indicated in figure (from Ling and Hu 1988, by permission of the Pacific Press, Melville, NY)

line was from the introvert model, native hemoglobin. The remaining four sets of data one from living frog muscle and the others from the three extrovert models all show a Z-shaped curve depicting decreasing q-values with increasing molecular weight of the solute. Note also that seven of the 21 experimental points from frog muscle do not fit in with the rest of the points, which provides the basis for the main Z-shaped line.

Next, we attempt to obtain quantitative data on the underlying mechanism of the solute exclusion in the living frog muscle cells. To achieve that, we need to put our theoretically derived equation (Equation A3) to work. That is, the frog muscle data points are plotted a second time in Figure 23. Here, instead of molecule weights of each solute studied, the q-values are plotted against their respective molecular volumes in cc as it is in Equation (A3).

More important, the lines going near and through the data points are not drawn by visual inspection. Instead, they are theoretical based on Equation (A3) cited in the Appendix. There are now two theoretical curves instead of one with seven apparently aberrant points.

One theoretical curve fits data points lower down and the other fitting the seemingly aberrant higher points in Figure 22. Only we now know that they are not aberrant at all. Indeed, the two theoretical curves were based on the same bulk-phase

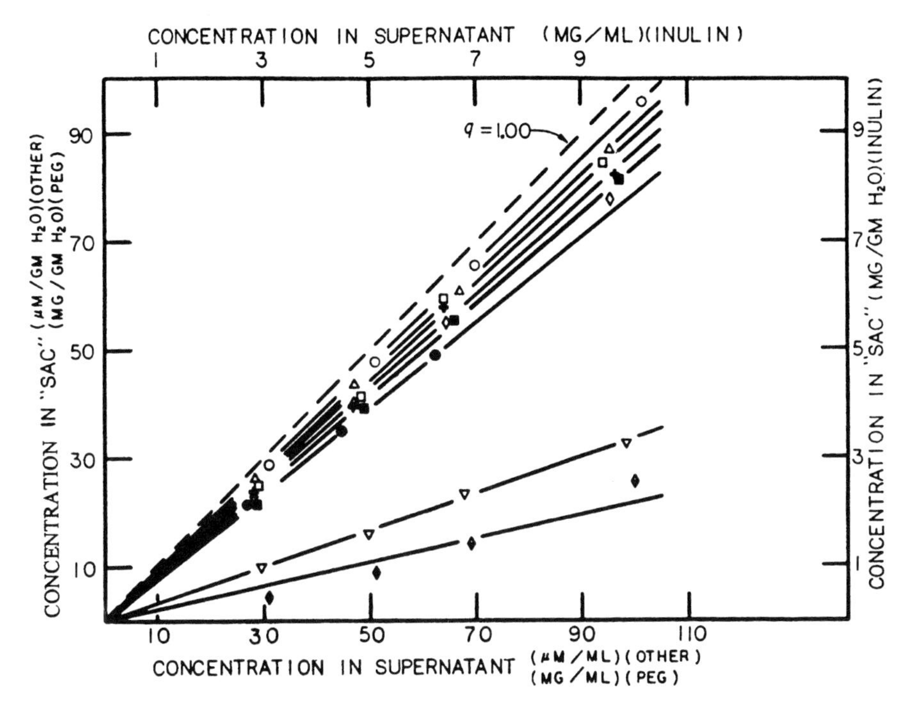

Figure 20. Equilibrium distribution of nine non-electrolytes in a solution of polyethylene oxide. Final concentration of the polymer was 15%. In addition, the solution also contained 0.4 M NaCl. The symbols used for all the non-electrolytes are the same as in Figure 19. D-xylose, which was absent in Figure 19 but present here, is represented by + (from Ling and Hu 1988, by permission of the Pacific Press, Melville, NY)

exclusion intensty, $\mathcal{U}_{vp}$, equal to 126 cal per mole. But the surface polarization energy $\mathcal{U}_s$ are not the same. It is 119 cal/mole for the lower curve but a good deal higher at 156 cal/mole for the upper curve.

This much higher surface polarization energy, $\mathcal{U}_s$, of the curve that better fit the seven 'aberrant' points is exciting because five of these seven solutes are chemicals known as *cryoprotectants.*

Now, cryoprotectants are chemicals, which when added to the culture medium protects the cells or embryos from freezing damages during storage in liquid nitrogen. One suspects that a sixth 'aberrant' chemical (urea) would have been yet another cryoprotectant if it were not for the fact that urea is also a protein denaturant, which might harm the cells.

(3) *Experimental demonstration of the exclusion of large macromolecules with a molar volume of 4000 cc by weakly polarized-oriented water*

At this juncture, it is timely to mention some new results of a study that we have just published (Ling and Hu 2004). It demonstrated that the water in a solution of 35% native bovine hemoglobin that shows no exclusion at all for solutes as small

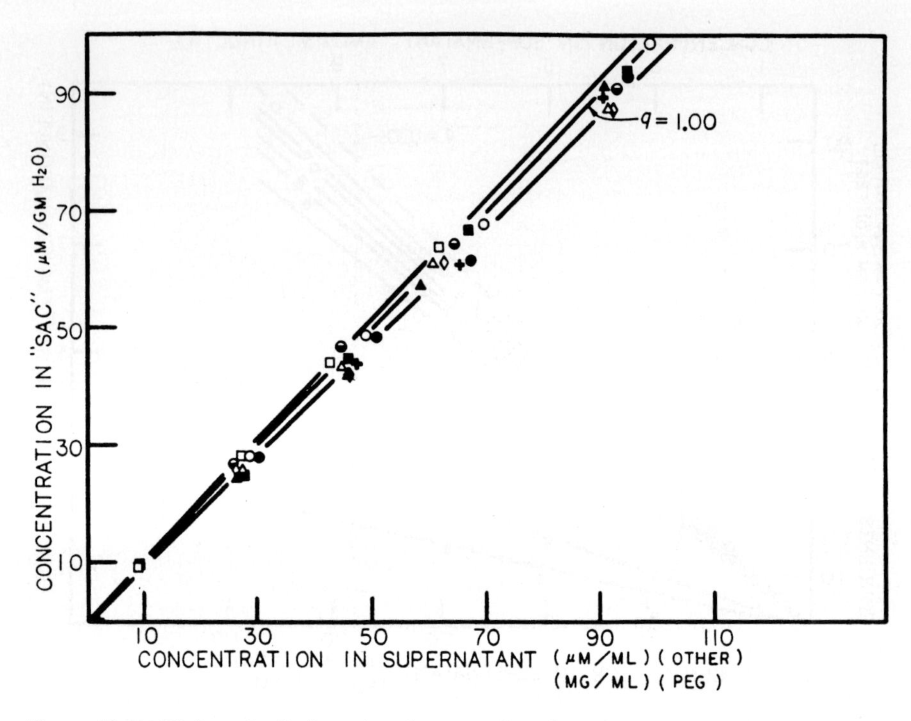

Figure 21. Equilibrium distribution of various non-electrolytes in a neutral solution of bovine hemoglobin (final concentration, 39% ± 1%) after 5 days of incubation at 25 °C. Solution also contains 0.4 M NaCl. Symbols used are the same as shown in Figure 19, except D-xylose, which is not in Figure 19 but present here. It is represented by + (from Ling and Hu 1988, by permission of the Pacific Press, Melville, NY)

as water to solutes as large as raffinose, is not normal liquid water either. Indeed, when the probe molecule used was poly (ethylene glycol) with a molecular volume of 4500 cc, it shows a q-value of only about 0.2.

The surprisingly low q-value for very large probes like PEG-4000 reminds us that the introvert models we studied so far are rarely, if ever the ideal introvert models (with the water-to-water interaction energy of perfectly normal liquid water). Rather, they are merely models with much lower "excess water-to-water interaction energy" than that of the extrovert models. Nonetheless, it is not normal liquid water. After all, some of the NHCO groups of *bona fide* globular native proteins are not all engaged in forming intra- or inter-macromolecular H bonds. Those NHCO groups existing as *free coils* or even *turns* as well as polar side chains (associated with the right kind of cations) may have some impact on water structuring as their counterparts in fully extended polypeptide chains.

Nonetheless, how large molecular size of the probe molecules can compensate for the weakness of the degree of polarization-orientation of the bulk phase water is clearly demonstrated here. This fact also puts us in a position better to assess the even more spectacular exclusion of really large probe molecules as reported by Gery Pollack's groups from Seattle, which will be discussed next.

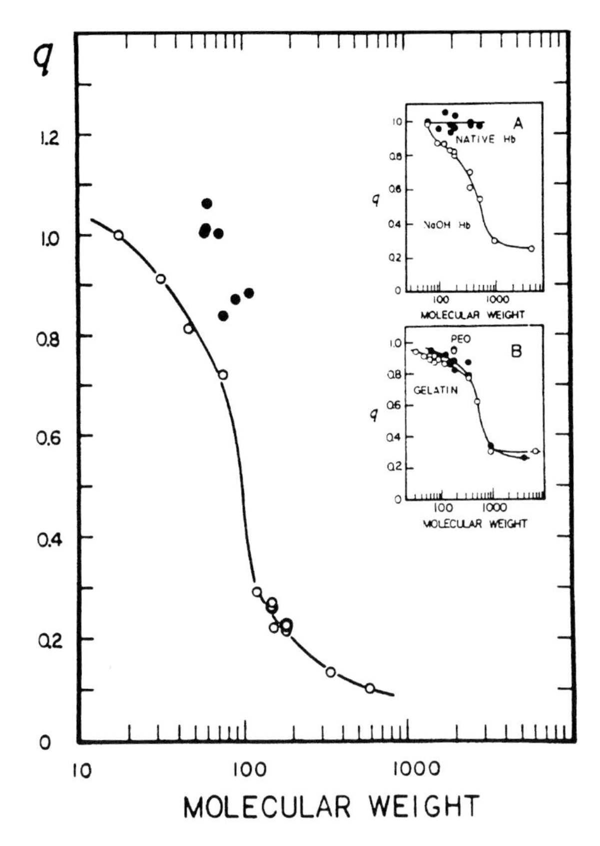

Figure 22. The (true) equilibrium distribution coefficients (or q-values) of 21 non-electrolytes in muscle cell water shown on the ordinate plotted against the molecular weights of the respective non-electrolytes. For comparison, q-value *vs* molecular weight plots of similar non-electrolytes in solutions of NaOH-denatured bovine hemoglobin (18%), gelatin gel (18%), poly (ethylene oxide) or PEO (15%) and native bovine hemoglobin (39%) are shown in the insets. Data on gelatin gel represent ρ- rather than q-values (see text) and were taken from Gary-Bobo and Lindenberg 1969 (from Ling et al. 1993, by permission of the Pacific Press, Melville, NY)

(4) *The exclusion of giant probe, latax-coated microspheres 1 μm in diameter*

The exclusion of latex-coated microspheres 1 μm in diameter by water dominated by the NONONO (or NPNPNP) surface of a polyvinyl alcohol (PVA) gel as revealed in the time sequence photography of Zheng and Pollack (2003) offered spectacular confirmation of the PM theory (Figure 24.) Because, like the saying goes, seeing is believing. There is no question here that the power of the NONONO surface could extend as far out as 100 μm or some 300,000 water molecules away from the PVA gel surface.

Polyvinyl alcohol, or PVA was a very good choice. As shown in Figure 25 taken from McLaren and Rowen's review. PVA has a strong affinity for water molecules.

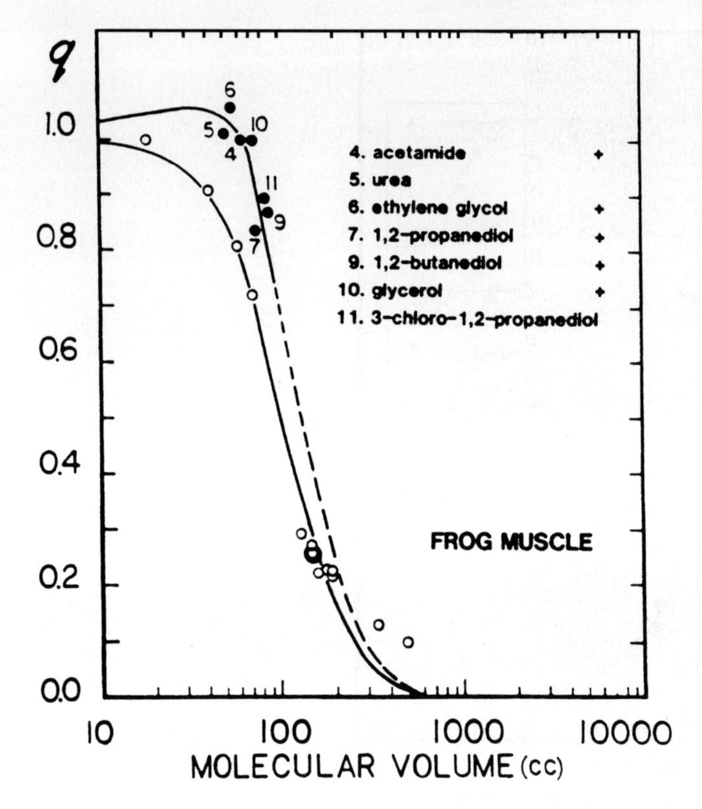

Figure 23. The equilibrium distribution coefficients (or q-values) of 21 non-electrolytes in muscle cell water plotted against the *molecular volumes* of the respective non-electrolytes. The points are experimental and are the same as shown in Figure 22. The lines going through or near the experimental points are theoretical according to Equation A3 in Appendix 1. For both the lower and upper theoretical curves, $\mathcal{U}_{vp}$, the 'exclusion intensity' for one mole of water, is the same at 126 cal/mole. $\mathcal{U}_s$, the surface polarization energy, on the other hand, is 119 cal/ mole of water for the lower curve and 156 cal/mole for the upper curve. Chemicals marked with the + signs are established cryoprotective agents (from Ling et al. 1993, by permission of the Pacific Press, Melville, NY)

From its structure, {-CH_2 CH(OH)-}$_n$ but in fact an NPNPNP system because the OH groups can function both as a proton acceptor and proton donator. There are two interesting features that this astonishing finding has demonstrated.

First, the Zheng and Pollack figure reproduced as Figure 24 shows a clear-cut boundary between a zone of exclusion and a zone of admittance.

The most important reason could be what the PM theory has long maintained that the water polarization-orientation is strongly *auto-cooperative* (Ling 1980, p 39). The polarization-orientation of one water molecule greatly enhances the neighboring water molecules to undergo the same change. In other words, there is a strongly positive *nearest neighbor interaction energy*. This confirmation of the theory is the most important.

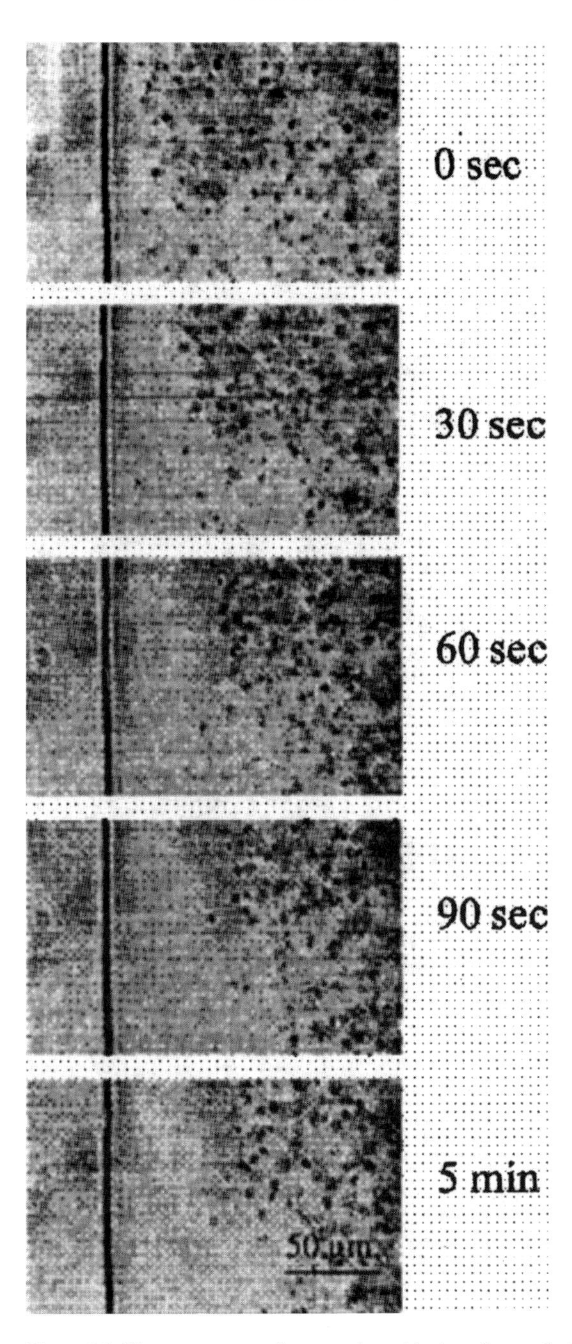

Figure 24. The emergence and progressive widening of an exclusion zone containing no microspheres as a function of time in a cylindrical channel of a polyvinyl alcohol (PVA) gel. Carboxylate groups-covered latex microspheres measured 2 μm in diameter (Zheng and Pollack 2003, by permission of Physical Review)

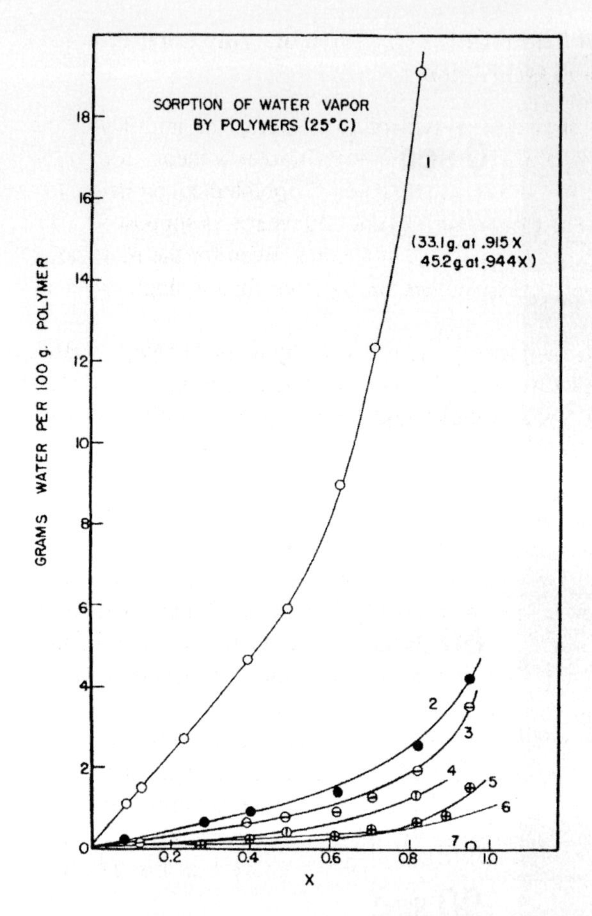

Figure 25. Sorption of water vapor by various polymers at 25 °C. Polyvinyl alcohol (1); polyvinyl butyral (2); ethylene vinyl alcohol (3); rubber hydrochloride (4); vinylidine chloride-acrylonitrile (5); chlorinated polyethylene (points not shown) (6); polyethylene (7) (from McLaren and Rowen 1951, by permission of J. Polymer Science)

But there might be another contributing factor, which is trivial in nature but nonetheless contributes to what we see. When the molecular volume of the probe is very large as it certainly is the case here, the visual image might not look significantly different when the q-value of the microspheres varied say between 0.01 to 0.0001. This effect tends to enhance the all-or-none appearance.

Having said that, I must return to the most important feature of the observation: its extreme reach of the influence of the PVA surface. To the best of my knowledge, this spectacular demonstration of a distance of 100 μm-wide clear zone was not predicted by any of the theory of gas adsorption known at the time when Zheng and Pollack published this work in 2003.

4.5 To the Rescue, a New Theoretical Foundation of Truly Long Range Water Polarization and Orientation

In an earlier section, I have shown how two of the theories of multilayer gas adsorption, de Boer-Zwikker's polarization theory and Bradley's theory for multilayer adsorption of non-polar gases fell on the roadside. As pointed out by Brunauer, Emmett and Teller in 1938, polarization alone cannot do what it is supposed to do: polarize and immobilize multiple layers of gas molecules. Even for the most polar gases, the mechanism proposed could produce much more than a single layer of adsorbed gas.

This trio then proposed their own theory often known by its nickname, the BET theory. Unfortunately, the BET theory has its own weaknesses. First, it is hard to see why the allegedly normal liquid water – free from long-range influence from the solid surface – would nonetheless stay on as multilayers on the solid surface rather than evaporate at any relative vapor pressure less than 100% saturation (see Cassie 1945). Second, it cannot explain the intense 'extra' uptake of polar molecules like water at near saturation vapor pressure. Thus, the BET theory could adequately explain adsorption only at below 50% vapor saturation. Higher than that, theoretical predictions and experimental data sharply diverge (see Figure 26). Third, condensed as *normal liquid water*, it cannot explain the extensive experimental demonstrations of altered physico-chemical attributes of the deep layers of water collecting on suitable solid surfaces (Henniker 1944).

That leaves only Bradley's general theory for gas molecules with permanent dipole moments. Yet, despite an endorsement by Brunauer, Emmett and Teller (1938, p 311), it is also full of holes. The fact that it shares a formal equation (Equation 1) with two other theories no longer tenable is already bad. By Bradley's own admission, that data fitting his equation does not prove that his theory is confirmed weakens it further.

Perhaps the most damaging to the Bradley isotherm for gas molecules with permanent dipole moments is this. It gives not an inkling as to how deep a layer of water a water-polarizing surface can produce. And as you notice above, it is the lacking of the theory that can give us a definitive answer to that question that would leave the legions of exciting and wonderful experimental data like so many orphans.

Indeed, without the support of a sound theoretical foundation even these spectacular observations could run the risk of sharing the same fate that had overtaken the already long list of truly exciting observations of long range interaction (see Henniker 1944, also Figures 29 and 30) in being treated like curios and relics rather than sound scientific knowledge. When these clear-cut scientific facts are looked upon with uncertainty, the PM theory of cell water, which has its foundation on nothing better than the Bradley isotherm, also suffers.

For all these reasons, I felt jubilant when I discovered a short cut and through it, developed a new theoretical foundation for the long-range dynamic structuring of water molecules and other suitable polar gases.

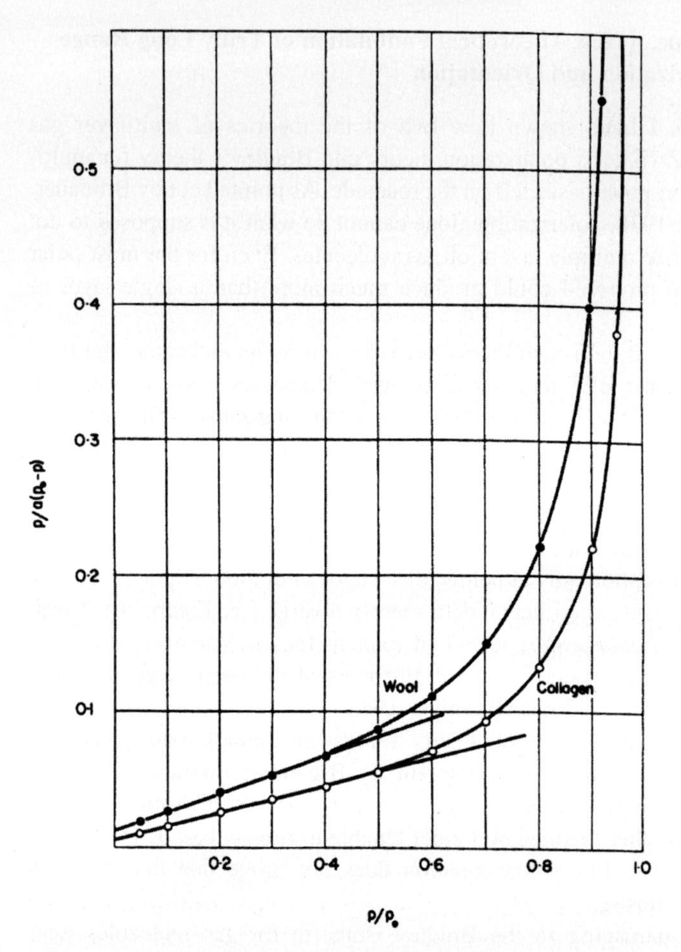

Figure 26. Water vapor sorption data from Bull (1944) on collagen and on wool were plotted according to the BET isotherm. Data fit the isotherm well up to about 40% relative vapor pressure. From this point and higher, the data and isotherm are far, far apart. In other words, the BET theory cannot explain the high uptake of water by these proteins at the high end of the vapor pressure values (from Ling 1965, by permission of the Annals of NY Academy of Sciences)

For details, the reader must consult my recent paper (Ling 2003). Suffice it to say here that even the best of human minds have limitations. Thus, when dealing with highly complex physiological problems, methods that had proved so powerful in dealing with simple systems may not yield equally rewarding results.

More specifically, the central problem in dealing with long range dynamic structuring of water lies in coping with the thermally agitated *permanent dipole moment* of the gas molecules near the polar surface. To overcome this difficulty, de Boer and Zwikker simply ignored the permanent dipole moment of the gas molecules. In his general theory for all gases on salt crystal surfaces, Bradley did not ignore the

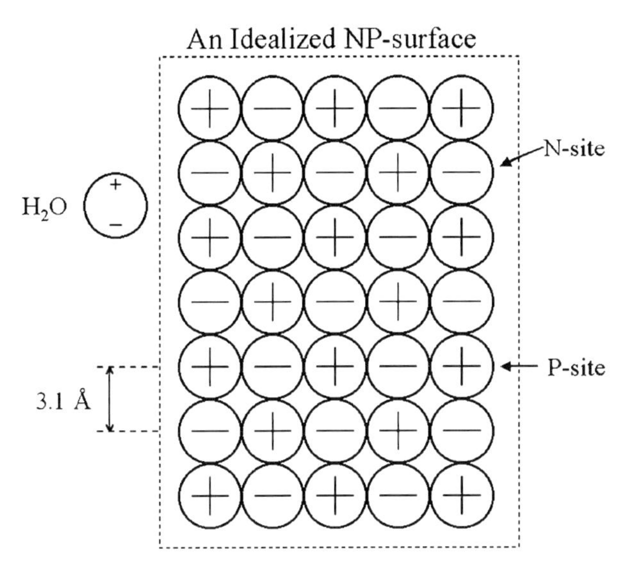

Figure 27. An idealized NP surface. The distance between a pair of the nearest-neighboring N and P site is equal to the distance, r, between neighboring water molecules in the normal liquid state and approximately 3.1 Å (from Ling 2003, by permission of the Pacific Press, Melville, NY)

permanent dipole moment. Rather, he ignored the fixed negatively charged sites or N sites of the salt crystal surface. So, instead of a study of what I would call a NP system or even an OP system, he in fact studied what I would call a PP system. As pointed out in section 4.1, such a uniformly charged surface cannot polarize-orient deep layers of polar gas molecules like water.

The short cut that I took was to get around the thermal agitation of the permanent dipole moments by bringing the temperature to just a little above absolute zero. I then introduced what I call an *idealized NP surface* (Figure 27) and added conditions that would eliminate extraneous factors like gravitation, border effects etc. The net result was that I have greatly simplified the problem by making electrostatics the sole determining factor in the polarization and orientation of water dipoles. Under this condition, each water molecule is fully characterized by three parameters: its diameter (r), its permanent dipole moment (μ) and its polarizability (α). Note also that the distance between the nearest neighboring N and P site is made to equal that of the water diameter, r, chosen.

The result of the computation is illustrated in Figure 28. The most striking feature is that the (negative) energy of water-to-water interaction, E^n, at the nth layer of water molecules, does not taper off with increasing distance from the idealized NP surface. Instead, as the distance increases toward infinity, E^n assumes a constant value described by the following equation:

$$E^n = (4\mu^2 r^3)/(r^3 - 8\alpha)^2 \quad (3)$$

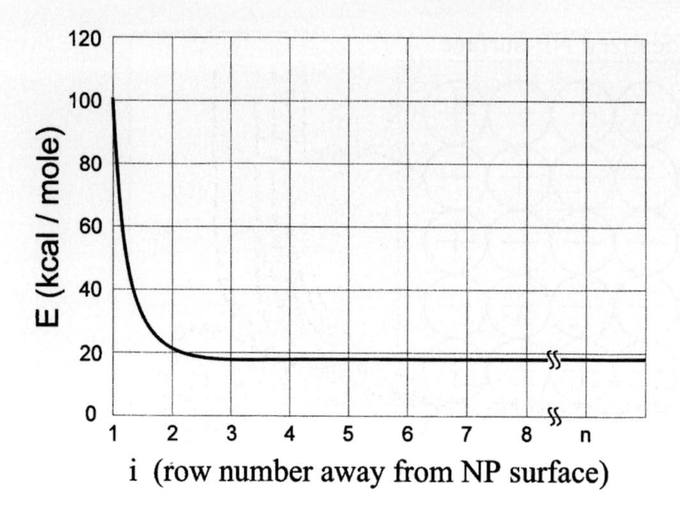

Figure 28. The adsorption energy of a water molecule in successive layers from an idealized NP surface at a temperature very near absolute zero. The theoretically computed adsorption energy per water molecule (E) at successive rows of water molecules away from an idealized NP surface. Note that as the distance between the water molecule and idealized NP surface increases, the adsorption energy does not taper off to zero. Rather, it continues at a constant value described by Equation 3. For detailed on the idealized NP surface, see Figure 27 (from Ling 2003, by permission of the Pacific Press, Melville, NY)

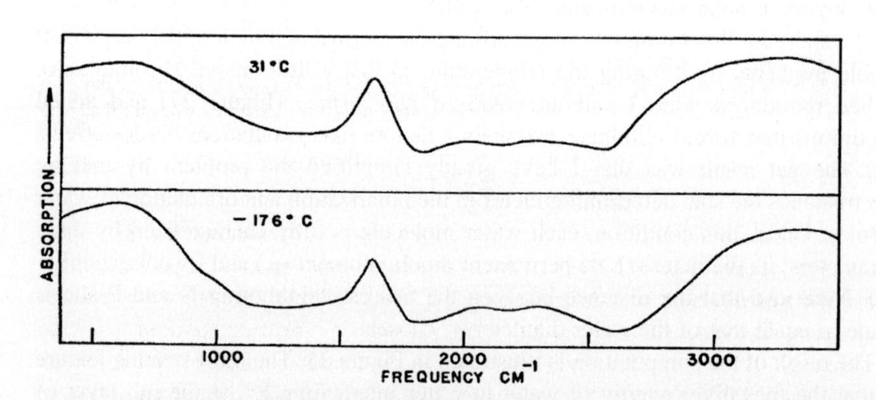

Figure 29. Infrared absorption spectrum of 10-micron-thick water film held between polished AgCl crystal plates. (For the near-ideal geometry of the AgCl NP surface, see Figure 9 in Ling 2003). The two (indistinguishable) spectra were observed respectively at ambiant temperature (top, at 31 °C) and at liquid air temperature (bottom, at −176 °). I am indebted to Dr. Rod Sovoie, the former associate of Dr. Giguère for the information on the thickness of the water film Prof. Giguère and Harvey studied (from Giguère and Harvey 1956, by permission of Canad. J. Chemistry)

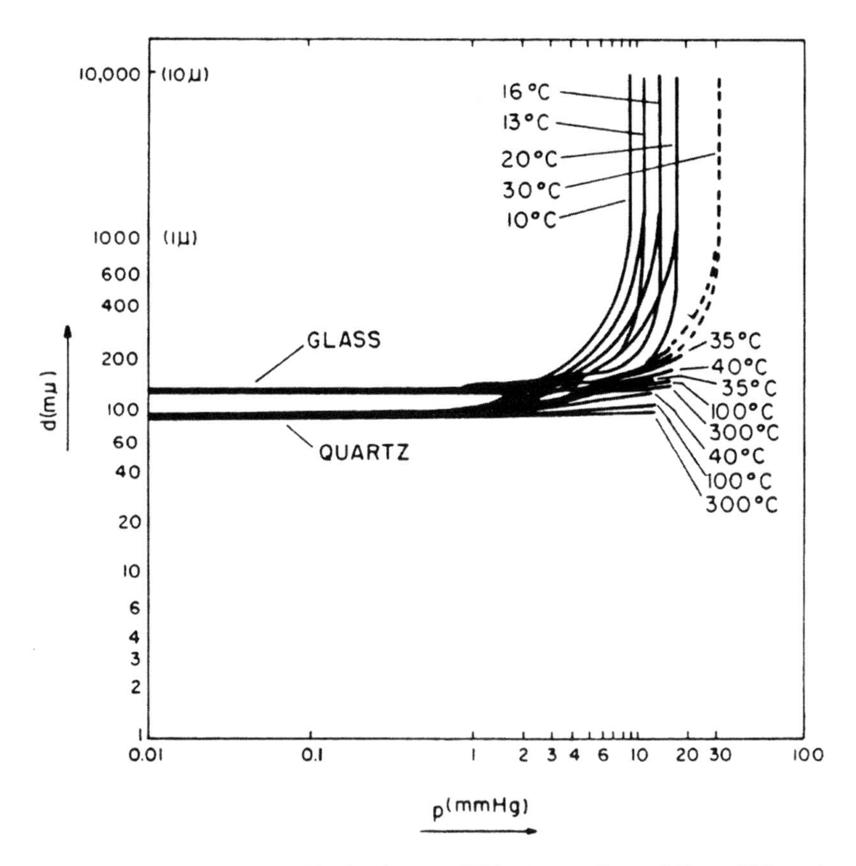

Figure 30. The vapor pressure (abscissa in mm of Hg) of water film at different thickness in mμ (or 10^{-7} cm or 10 Å) (ordinate). Water film of varying thickness was produced by placing water between one flat glass (or quartz) plate and one curved plate with radius of curvature of 35 m. The thickness mentioned above refers to the water film at the periphery of doughnut-like ring. When the thickness is 1 μ or thicker, the vapor pressure is not different from normal liquid water. However, when the thickness falls to 90 mμ (or 900 Å or some 300 water molecules thick), the vapor pressure becomes zero even at temperature as high as 300 °C (from Ling 1972, redrawn after Hori 1956)

Set μ equal to zero, E^n vanishes. This is why, as pointed out by Brunauer, Emmett and Teller that without the participation of the permanent dipole moment, polarization alone cannot produce multilayer adsorption. Also important is the value of r chosen. The value of 3.1 Å adopted was obtained by dividing the molar volume of water (18.016 cc) by the Avogadro number and taking its cube root as equal to r. For values of r shorter than 3.1 Å, the calculated E^n would be even stronger. For values much lower than 3.1 Å, the *ad infinitum* propagation of polarization-orientation would no longer be predicted. Ice crystals of different phases (e.g., ice I, ice X) may then emerge.

The most important contribution of this work is that a good theoretical foundation for the PM theory of cell water and for the long-range dynamic structuring of water and other polar molecules is now on hand. And this includes the spectacular recent finding of Zheng and Pollack just described. A more detailed analysis of their findings will be forthcoming based at once upon the PM theory of long-range dynamic water structuring (Ling 2003) and the PM theory of solute exclusion (Ling 1993).

Two additional predictions from what is summarized in Equation 3 are: (1) that under proper conditions, water layers held between two surfaces with charge distribution approaching that of an idealized NP surface, would not freeze under any attainable low temperature; (2) that water under similar conditions would not boil at temperature as high as 400 °C. Both predictions have been confirmed retroactively by observations of Giguère and Harvey (1956) and of Hori (1956) reported almost half a century ago. Their key figures are respectively reproduced here as Figures 29 and 30.

ACKNOWLEDGEMENTS

I thank Dr. Raymond Damadian, and his Fonar Corporation and its many friendly and helpful members for their continued support, Margaret Ochsenfeld and Dr. Zhen-dong Chen for their dedicated and skillful cooperation, and librarian Tony Colella and Michael Guarino, director of Media Services and Internet Services for their patience and tireless assistance.

APPENDIX 1

$$\text{(A3)} \qquad q = \exp\left\{\frac{1.23v\Delta E_s\left[1-(1-b)\frac{(kv)^n}{1+(kv)^n}\right]-(\Delta E_v+1.23\Delta e^*)v}{RT}\right\},$$

where q is the equilibrium distribution coefficient of the solute in question. v is the molecular volume (molar volume) of the solute and it is in cm^3. b is a small fractional number describing the probability of (very large) molecules in finding adsorbing sites on the water lattice. k and n are parameters describing the steepness of the declining probability of finding adsorbing sites with increase of molecular volume. ΔE_s is the *specific surface (or solute) polarization energy* per cm^2 in units of $cal.mol^{-1}\,(cm^2)^{-1}$, when the solute is moved from normal liquid water to the polarized cell water. ΔE_v is the *specific solvent polarization energy*, equal to the difference between the energy spent in excavating a hole 1 cm^3 in size in the polarized (cell) water and the energy recovered in filling up a 1 cm^3 hole left behind in the surrounding normal liquid water; it is in units of $cal.mole^{-1}(cm^3)^{-1}$. Δe^* is the *increment of the activation energy* for overcoming the greater rotational restriction per unit surface area in units of $cal.mole^{-1}(cm^2)^{-1}$, when a solute is transferred from normal liquid water phase to the polarized water phase. R and T are the gas constant and absolute temperature respectively.

REFERENCES

Bamford CH, Elliott A, Hanby WC (1956) Synthetic Polypeptides: Preparation, Structure and Properties. New York: Academic Press, p 310

Beatley EH, Klotz IM (1951) Biol Bull 101:215

Benson SW, Ellis DA, Zwanzig RW (1950) J Amer Chem Soc 72:2102

Best CH, Tatlor NB (1945) The Physical Basis of Medical Practice. 4th edn., Baltimore: The Williams & Wilkens Co., p 7, column 2

de Boer JH, Zwikker C (1929) Zeitschr Phyisk Chem B3:407

de Boer JH, Dippel CJ (1933) Rec Trav Chem Pays-Bas 52:214

Boyle PJ, Conway EJ (1941) J Physiol 100:1

Bradley RS (1936a) J Chem Soc:1467–1474

Bradley RS (1936b) J Chem Soc:1799–1804

Brunauer S, Emmett PH, Teller E (1938) J Amer Chem Soc 60:369

Bull H (1944) J Amer Chem Soc 66:1499

Cameron IL (1988) Physiol Chem Phys & Med NMR 20:221

Cassie ABD (1945) Trans Farad Soc 41:450

Conway EJ, Creuss-Callaghan G (1937) Biochem J 31:828

Doty P, Gratzer WB (1962) Polyamino Acids, Polypeptides and Proteins. Stahmann MA (ed) Madison: Univ. Wisconsin Press, pp 116–117

Dujardin F (1835) Annales des sciences naturelles; partie zoologique, 2d sér. 4:364

Durant W (1926) The Story of Philosophy. New York: Pocket Books, A Division of Simon and Schuster, reprinted all the way to at least 1961

Eastoe JE (1955) Biochem J 61:58.9

Edelmann L (1977) Physiol Chem Phys 9:313

Edelmann L (1984) Scanning Electron Microscopy. II:875

Edelmann L (1986) Science of Biological Specimen Preparation. Müller M, Becker RP, Boyde A, Wolosewick JJ (eds) Chicago: SEM Inc, AMF O'Hare, p 33

Freedman JC (1976) Biochem Biophys Acta 455:989

Gary-Bobo CM, Lindenberg AB (1969) J Coll Interf Sci 29:702

Giguére PA, Harvey KB (1956) Canad J Chemistry 34:798

Glasstone S (1946) Textbook of Physical Chemistry. 2nd edn., New York: D.van Nostrand

Graham T (1861) Phil Trans R Soc. London, vol 151, p 183

Grove A (1996) Only the Paranoid Survive. New York: Doubleday Dell Publ

Hall TS (1969) Ideas of Life and Matter. Chicago: Univ Chicago Press, p 310

Henniker JC (1949) Rev Modern Phys 21:322

Hodgkin AL (1971) The Conduction of the Nervous Impulse. Liverpool: Liverpool Univ. Press, p 21

Hoover SR, Mellon EF (1950) J Amer Chem Soc 72:2562

Hori T (1956) Low Temperature Science. A15:34 (Teion Kagaku, Butsuri Hen) (English translation: No. 62, US Army Snow, Ice and Permafrost Res. Establishment, Corps of Engineers, Wilmette, Ill. USA)

Huang HW, Hunter SH, Warburton VK, Mos SC (1979) Science 204:191

Katchman BJ, McLaren AD (1951) J Amer Chem Soc 73:2124

Katz JR (1919) Koloidchem Beihefte 9:1

Leeder JD, Watt IC (1974) J Coll Interf Sci 48:339

Ling GN (1952) Phosphorous Metabolism (Vol II). McElroy WD, Glass B (eds) Baltimore: The Johns Hopkins Univ. Press, p 748

Ling GN (1962) A Physical Theory of the Living State: The Association-Induction Hypothesis. Waltham MA: Blaisdell

Ling GN (1965) Ann NY Acad Sci 125:401

Ling GN (1969) Intern Rev Cytol 26:1

Ling GN (1970) Intern J Neurosci 1:129

Ling GN (1972) Water and Aqueous Solutions, Structure, Thermodynamics, and Transport Properties. Horne A (ed) New York: Wiley-Interscience, pp 663–699

Ling GN (1973) Physiol Chem Phys 5:295
Ling GN (1977) Physiol Chem Phys 9:319
Ling GN (1980) Cooperative Phenomena in Biology. Karreman G. (ed) NewYork: Pergamon Press, p 39
Ling GN (1984) In Search of the Physical Basis of Life. New York: Plenum Publ Corp
Ling GN (1992) A Revolution in the Physiology of the Living Cell. Malabar FL: Krieger Publ Co
Ling GN (1993) Physiol Chem Phys & Med NMR 25:145
Ling GN (1997) Physiol Chem Phys & Med NMR 29:123 Available via http://www.physiologicalchemistryandphysics.com/pdf/PCP29-2_ling.pdf
Ling GN (2001) Life at the Cell and Below-Cell Level: the Hidden History of a Fundamental Revolution in Biology. New York: Pacific Press
Ling GN (2003) Physiol Chem Phys & Med NMR 35:91 Available via http://www.physiologicalchemistryandphysics.com/pdf/PCP35-2_ling.pdf
Ling GN (2004) Physiol Chem Phys & Med NMR 36:1 Available via http://www.physiologicalchemistryandphysics.com/pdf/PCP36-1_ling.pdf
Ling GN, Hu WX (2004) Physiol Chem Phys & Med NMR 36:143 Available via http://www.physiologicalchemistryandphysics.com/pdf/PCP36-2_ling_hu.pdf
Ling GN, Ochsenfeld MM (1966) J Gen Physiol 49:819
Ling GN, Negendank W (1970) Physiol Chem Phys 2:15
Ling GN, Ochsenfeld MM (1973) Science 181:78
Ling GN, Walton C (1976) Science 191:293
Ling GN, Zhang ZL (1984) Physiol Chem Phys & Med NMR 16:221
Ling GN, Hu WX (1987) Physiol Chem. Phys. & Med. NMR 19:251
Ling GN, Hu WX (1988) Physiol Chem Phys & Med NMR 20:293
Ling GN, Ochsenfeld MM (1989) Physiol Chem Phys & Med NMR 21:19
Ling GN, Ochsenfeld MM, Walton C, Bersinger TJ (1980) Physiol Chem Phys 12:3
Ling GN, Zodda D, Sellers M (1984) Physiol Chem Phys & Med NMR 16:381
Ling GN, Niu Z, Ochsenfeld MM (1993) Physiol Chem Phys & Med NMR 25:177
Lloyd DJ (1933) Biol Rev Cambridge Phil Soc 8:463
Lloyd DJ, Phillips H (1933) Trans Farad Soc, vol 29, p 132
Macalum AB (1905) J Physiol (London) 32:95
McLaren AD, Rowen JW (1951) J Polymer Sci 7:289
Mellon EF, Korn AH, Hoover SR (1948) J Amer Chem Soc 70:3040
Mellon EF, Korn AH, Hoover SR (1949) J Amer Chem Soc 71:2761
Menten ML (1908) Trans Can Inst, vol 8, p 403
Perutz MF, Muirhead H, Cox JM, Goaman LCG, Mathews FS, McGandy EL, Webb LE (1968) Nature 219:29–32
Sponsler OL, Bath JD, Ellis JW (1940) J Phys Chem 44:996
Stirling AH (1912) James Hutchinson Stirling: His Life and Work. London, p 221
Tigyi J, Kallay N, Tigyi-Sebes A, Trombitas K (1980–81) International Cell Biology. Schweiger HG. (ed) Berlin: Springer, p 925
Walter H (1923) Jahrschr Wiss Bot 62:145
von Zglinicki T (1988) Gen Physiol Biophys 7:495
Zheng J, Pollack GH (2003) Phys Rev E68:031408

CHAPTER 2

MOLECULAR BASIS OF ARTICULAR DISK BIOMECHANICS: FLUID FLOW AND WATER CONTENT IN THE TEMPOROMANDIBULAR DISK AS RELATED TO DISTRIBUTION OF SULFUR

CHRISTINE L. HASKIN[1,*], GARY D. FULLERTON[2]
AND IVAN L. CAMERON[3]

[1,*] *University of Nevada Las Vegas, School of Dental Medicine*
[2] *University of Texas Health Science Center at San Antonio, Department of Radiology*
[3] *University of Texas Health Science Center at San Antonio, Graduate School of Biomedical Sciences, Cellular and Structural Biology*

Abstract: The temporomandibular articular disk was used to test the hypothesis that there is a positive relationship between the sulfur concentration and the amount of water held in the tissue, and an inverse relationship between sulfur concentration and the rate of fluid flow from the disk during compressive loading. Elemental concentrations were measured for sulfur, potassium, sodium, chlorine, phosphorus and calcium in each area of the disk by electron probe x-ray microanalysis. X-ray microanalysis showed high sulfur content coincident with histochemical localization of glycosaminoglycans. Further analysis of the elemental content revealed a strong correlation between sulfur and K^+, suggesting that the predominate counterion on fixed sulfates is a K^+ rather than Na^+. The resistance to fluid flow was measured by determining the cumulative grams of water forced from the tissue at multiple intervals during centrifugal loading. Values were expressed as grams water per gram dry mass and then plotted against time. Multiple regression analysis of sulfur content and water content values revealed a significant inverse, rather than a positive correlation between sulfur content and both the initial water content and the water content following centrifugal loading. Potassium content also had a strong negative correlation with water content. Curve analysis of flow rates revealed that there were two water compartments, an inner, more tightly held water compartment with a slower flow rate, and an outer compartment with a flow rate 2 to 3 times faster than that of

* Corresponding author. 1001 Shadow Lane, MS 7410, School of Dental Medicine, Las Vegas, NV 89106-4124, USA. Tel.: 1-702-774-2676; fax: 1-702-774-2552; *E-mail address:* Christine.haskin@unlv.edu.

G. Pollack et al. (eds.), Water and the Cell, 53–69.

the inner water compartment. The amount of water in the inner water compartment was negatively correlated with both K^+ and sulfur concentrations and the rate of flow was positively correlated with K^+ content. The amount of water in the outer water compartment was positively correlated with both Cl^- and Na^+ concentration, but the rate of flow in the outer compartment was not significantly correlated with the concentration of any of the elements measured. Therefore, the results lead us to rejection of the original hypothesis, but allowed development of a new hypothesis where the concentration of monovalent ions, such as Na^+ and K^+ interact with fixed charges in a way that significantly determines water content, resistance to fluid flow and the biomechanical properties of the disk. This new hypothesis was supported by the results from similar tests on a series of different ion-exchange resins

Keywords: Fluid Flow; Glycosaminoglycans; Hydration; Sulfation

1. INTRODUCTION

The composition and organization of the temporomandibular joint (TMJ) disk must provide an architecture to withstand the biomechanical demands of function. The articular disk is nearly acellular (2-5% cells by volume) and is composed of a crosslinked collagen fiber network with high resistance to stretch and tensile forces along the long axis of the fibers and flexibility in all other orientations (Mills et al., 1988; Detamore and Athanasiou, 2003a). Collagen fiber orientation has been described in both animal and human systems (Mills et al., 1988; Kino et al., 1989, Milam et al., 1991, Shengyi and Xu, 1991, Teng et al., 1991, Berkovitz et al., 1992, Minarelli and Liberti, 1997, Tong and Tideman 2001). In general, the collagen fibers run anteroposterior and interdigitate with circumferential fibers in the periphery of the bands (Scapino, 1983; Detamore and Athanasiou, 2003a; Detamore and Athanasiou, 2003b). Studies have demonstrated that the major collagen species is type I collagen (Milam et al., 1991; Kobayashi, 1992; Gage et al., 1990; Minarelli and Liberti, 1997) with type III collagen present in the posterior disk attachment (Gage et al., 1990).

While the arrangement of collagen fibers provides tensile strength and resistance to stretch, the type, concentration and distribution of proteoglycans in the temporomandibular joint disk are thought to reflect the type of mechanical loads placed on it (Tanaka et al., 2003; Tanaka and van Eijden, 2003). Proteoglycans (PG) are interspersed throughout the collagen network (Kuijer, 1986; Scott, 1988; Scott et al., 1989; Kempson, 1991) and contribute significantly to the biomechanical properties of the disk (Scott, 1989a; Scott, 1989b). Large aggregating PGs accelerate fibrillogenesis (Kuijer et al., 1988) and the small non-aggregating PGs retard fibrillogenesis (Kuijer et al., 1988). Small molecular weight PGs that alter the kinetics of collagen fibrillogenesis are typically found in areas with high resistance to tensile forces (Scott, 1988; Scott, 1989). High molecular weight PGs, similar to cartilage-type PGs, are present throughout the disk and can have a variety of glycosaminoglycan

(GAG) side chains, including chondroitin-4-sulfate, chondroitin-6-sulfate, dermatan sulfate and keratan sulfate (Mills et al., 1988; Nakano and Scott, 1989a; Kobayashi, 1992).

The glycosaminoglycans, the major component of the proteoglycans, have been reported to provide resistance to compression by forming highly charged and hydrated complexes within the extracellular matrix (Kempson et al., 1970; Kempson, 1980; Blaustein and Scapino, 1986; Klein-Nulend et al., 1987). Although immunohistochemical staining, which depends on epitope recognition, has shown a uniform distribution of PGs and GAGs (Nakano and Scott, 1989a, b; Kobayashi 1992), histological stains such as alcian blue at pH 1, which depend on the presence of sulfates, have shown that sulfated GAGs are not uniformly distributed but are concentrated primarily in areas thought to undergo compressive loading and are concentrated just beneath the inferior surface of the disk at the border between the thin intermediate zone and the anterior band of the disk (Kopp, 1976; Blaustein and Scapino, 1986; Mills et al., 1988). (Kopp, 1976; Kopp, 1978; Blaustein and Scapino, 1986; Mills et al., 1988; Kuc et al., 1989; Nakano and Scott, 1989a and 1989b; Shira, 1989; Tanaka et al., 2003).

The non-uniform distribution of sulfated GAGs in the cartilage results in localized variations in the density of fixed negative charges and in hydration properties of the tissue. Specifically, the polyanionic character of the sulfated GAGs is thought to attract a large hydration sphere and thereby provide the basis for resistance to compressive loading. This assumption has been supported by three major lines of research. First, the content of GAGs in articular cartilage increases when it is subjected to increased mechanical loading, and the topographical distribution of GAGs reflects the variation in loading from one area of the tissue to the next (Kopp, 1976; Blaustein and Scapino, 1986). Second, the stiffness and resistance to deformability of articular cartilage is primarily due to the PG and GAG content (Kempson et al., 1970; Myers and Mow, 1983). Third, measurements of the size of the hydration sphere of PG and GAGs in the collagen network indicate that they could hold four to five times the amount of water were it not for crosslinking in the collagen component of the cartilage (Ikada, Suzuki and Iwata, 1980). The PGs therefore develop a high Donnan osmotic pressure that retains water, provides resistance to fluid flow and develops a high hydrostatic pressure gradient that is the basis of the resistance to compression (Carney and Muir, 1988).

Thus, a model has developed that has three basic tenets: 1) the PG-collagen interactions maintain the organization and alignment of the collagen fibrils that provide tensile strength and resistance to stretch; 2) crosslinking of the collagen network immobilizes PG, and GAGs; and 3) the polyanionic character of the sulfated GAGs organizes a large sphere of hydration that provides the basis of resistance to compressive loading. However, the level of sulfation of the individual GAGs can vary greatly. Chondroitin sulfate can have 0.2 to 2.3 sulfates per disaccharide unit, dermatan sulfate can have 1.0 to 2.0 sulfates per disaccharide unit and keratan sulfate can have 0.9 to 1.8 sulfates per disaccharide unit (Alberts et al., 1984). Thus, the ability of the GAGs to organize water around fixed anionic groups may vary with the level of sulfation and significantly effect the hydration and biomechanical properties of the extracellular matrix. We therefore set out to test the hypothesis that a positive relationship exists between the concentration of sulfur and both the amount of water held in the tissue and the ability of the disk to resist the flow of fluid out of the disk under compressive loads.

2. MATERIALS AND METHOD

TMJ articular disks were harvested from juvenile (6-8 month old) pigs that were killed under general anaesthesia as part of a gastroenterology study. Animal care and procedures were performed in facilities approved by the American Association for the Accreditation of Laboratory Animal Care and according to institutional guidelines for the use of laboratory animals. Disk samples used in this study appeared normal without macroscopic evidence of degeneration, disease or joint damage.

2.1 Histological Study of Articular Cartilage

Following dissection, disks were tattooed to facilitate orientation and then fixed in 10% neutral buffered formalin. The specimens were dehydrated in ethanol (60%, 80%, 95%, 100%), cleared with xylene, impregnated with paraffin (57 °C) and blocked. Serial sagittal sections of the disk and articular surface of the condyles were stained with: Alcian blue at pH 1 for identification of areas with increased deposition of acidic sulfated GAGs; with the periodic acid-Schiff reaction (PAS) for localization of non-GAG carbohydrate; with Masson's trichrome for identification of collagen orientation in relation to cartilage matrix components; and with hematoxylin and eosin for evaluation of general morphology. Light micrographs were taken on a Zeiss photomicroscope.

2.2 X-ray Microanalysis

Based on the histochemical localization of the GAGs, the disk was divided into eight areas for x-ray microanalysis. The disks were removed from 6-8 month old pigs, and 2 × 2 × 2 mm pieces from the predetermined areas were cut out with a razor blade. The glenoid fossae side of the disk was placed down on a brass pin surface and the tissue immediately frozen by plunging the specimen into liquid propane cooled in liquid nitrogen. The specimens on frozen pins were stored in liquid nitrogen until cryosectioning. Beginning on the surface of the inferior aspect of the disk, 2-mm-thick cryosections were cut on a microtome at −30 °C for both specimen and knife. Because histological analysis had shown a concentration of sulfated GAGs 100-500 mm beneath the inferior surface of the disk, nine to twenty sections from representative areas were cut at a depth of 50-500 mm from the surface and then processed as previously reported (Smith et al., 1983). Grids were examined in a JEOL-35 scanning electron microscope at 25 kV accelerating voltage and imaged in the STEM mode. Each section was probed in ten different areas. Care was taken to probe only extracellular areas. The analysis was done with an Si(Li) x-ray detector and Tracor Northern NS-880 x-ray analysis system as previously reported. Multiple spectra were collected from eight different areas on the condylar surface of the TMJ disk. Area-specific data were averaged separately and peak-to-continuum values were converted to content (expressed as mmol/kg dry weight) on the basis of the 1988 Aminoplastic conversion standards (Roos and Barnar, 1984; Warley, 1990). Area-specific data were averaged separately and statistical analysis was performed for each element by location using a one-way analysis of variance (ANOVA). The Student-Newman-Keuls multiple range test was used to determine which means were significantly different. Multiple regression analysis was used to examine correlations between the different elements by area.

2.3 Resistance to Fluid Flow Under Centrifugal Loads

The resistance to loss of water under centrifugal loads was measured during centrifugation. The TMJs were divided into the same eight areas as for the x-ray microanalysis. Disks were obtained from 6-8 month old pigs, dissected from surrounding soft tissues and sectioned into 1 mm^3 pieces. The initial mass M_i was determined for each sample. Each sample was loaded into a microcentrifuge tube and placed over a bed of filter paper to keep the water separated from the disk during the braking period of the centrifuge. Samples were centrifuged for a total of 120 min (in 5-10 min intervals) on a 5 cm rotor at a total force of 13,000 g to provide stress of 4.0-5.0 MPa, assuming the individual sample masses represented the maximum tare subjected to the deepest part of the sample. Intermediate masses M_{ic} of the disk tissue were recorded at each centrifugation interval. Following centrifugation, all samples were dried to weight equilibrium at 100 °C in a vacuum oven and the final dry mass M_d was measured to allow the water content of each sample to be expressed as grams of water per gram dry mass ($M_{ic}/M_d = M_i\text{-}M_d/M_d$).

3. RESULTS

3.1 Histochemical Observation

The pig articular disk (Figure 1) was composed of a dense fibrous tissue composed of bundles of collagen fibers resembling a tendon. Fibroblast-like cells were dispersed throughout an extensive extracellular matrix consisting primarily of collagen. Chondrocyte-like cells were primarily found on the inferior surface of the disk just posterior and anterior to the thin band and were associated with an extensive extracellular matrix that had a high concentration of PGs with negatively charged sulfated GAGs as assayed by alcian blue staining at pH1. Periodic acid-Schiff stained sections (not shown) indicated carbohydrates were dispersed throughout the matrix and were closely associated with the collagen network.

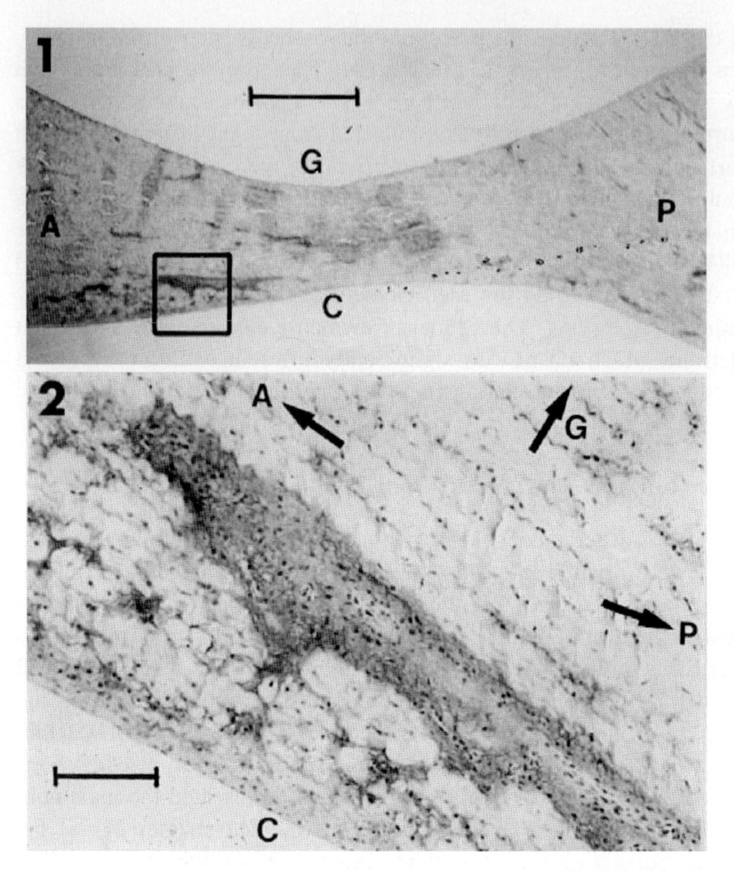

Figure 1. Localization of sulfated GAGs as shown by increased staining with Alcian blue in sagittal sections of the pig temporomandibular disk. Panel (2) is an enlargement of the area within the box in panel (1). Tissues were fixed as described in methods. Reference bar is 1 mm in panel 1 and 100 μm in panel 2. (G) glenoid fossa surface of the disk; (C) condylar surface; (A) anterior; (P) posterior

3.2 X-ray Microanalysis

The disk was divided into eight areas for x-ray microanalysis as diagrammed in Figure 2. The mean concentration of each element in each area, expressed as mmol/kg dry weight (Table 1) was used in an area-specific one-way analysis of variance for each element, in each area, as summarized in Table 2. The one-way analysis of variance indicated that the presumed load bearing areas of the disk (areas C, D, E, and F) were significantly different from the presumed non-load bearing areas of the disk (areas A, B, G, and H). The multiple regression analyses of elemental content of the pig TMJ disk, summarized in Table 3, showed that the distribution of K^+ is positively correlated with that of sulfur, explaining 56.6% of the variation in sulfur concentrations. Adding sodium to the equation did not significantly contribute to an explanation of sulfur content, suggesting that

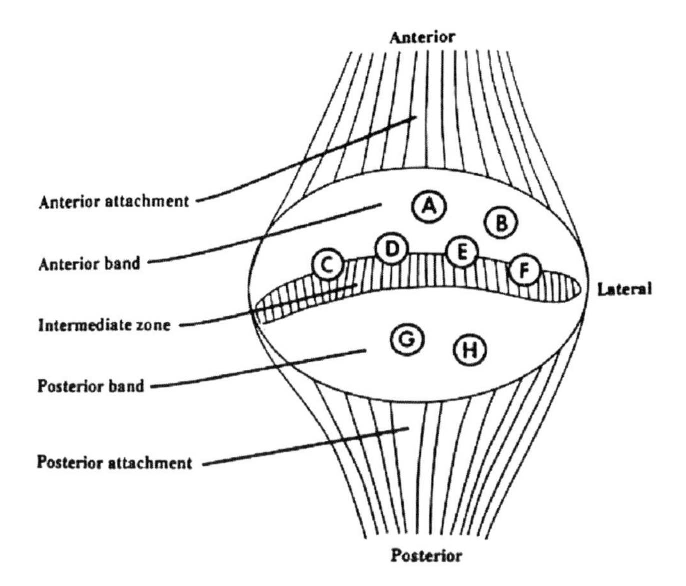

Figure 2. Designation of eight different areas on the condylar surface of the pig temporomandibular disk that were analyzed by x-ray microanalysis

K^+ counterions are preferentially associated with the fixed sulfates, i.e., sulfur concentration is the independent variable.

3.3 Resistance to Fluid Flow and the Water-holding Capacity under Compressive Load

Centrifugation experiments were done on whole disks sectioned into randomized, 1 mm^3 pieces. The total water in the sample was 1.77 g water/g dry mass, 0.88 g water/g dry mass was forced from the disk after 120 minutes centrifugation (in 5-10 minute intervals) at a stress of 4.0 MPa.

The resistance to fluid flow was also measured for the same locations on the disk as illustrated in Figure 2. The cumulative grams of water forced from each area of the disk was expressed as grams water per gram dry mass and then plotted against time under centrifugal load, such that the slopes of the curves defined the rate of water loss (i.e., flow rate through and out of the tissue). Thus, the slope for each location provided a numerical estimate of each area's resistance to fluid flow. Curve fit analysis (Figure 3) demonstrated that each area of the disk had two water compartments – an inner, tightly bound water compartment with a lower flow rate, and an outer, more loosely bound water compartment with a higher flow rate. The partitioning of water between inner and outer water compartments varied between non-load bearing areas and load bearing areas. In general, load bearing areas contained slightly less total water, partitioned a smaller percentage of that

Table 1. Mean concentration of elements by area in the pig temporomandibular joint articular disk as determined by x-ray microanalysis*

	Location							
	A	B	C	D	E	F	G	H
[S]	123.8 ± 2.8	118.0 ± 5.2	123.2 ± 2.6	126.1 ± 2.2	133.3 ± 3.8	154.4 ± 3.9	106.0 ± 5.4	115.1 ± 3.4
[K]	57.8 ± 1.6	41.5 ± 1.6	82.5 ± 2.2	78.1 ± 2.4	103.2 ± 5.0	113.0 ± 4.9	44.7 ± 3.0	48.8 ± 2.6
[Cl]	283.4 ± 4.2	284.8 ± 15.7	319.7 ± 6.3	281.2 ± 9.6	320.2 ± 11.5	355.4 ± 11.0	262.7 ± 3.8	192.2 ± 9.9
[Na]	172.4 ± 8.8	281.8 ± 20.9	287.1 ± 14.4	188.7 ± 10.8	250.8 ± 19.3	329.0 ± 14.7	201.2 ± 6.4	170.6 ± 9.7
[P]	29.3 ± 7.6	24.4 ± 4.1	22.0 ± 1.4	24.4 ± 2.0	59.8 ± 10.1	45.0 ± 4.5	17.8 ± 3.5	22.0 ± 2.1
[Ca]	3.8 ± 1.0	5.6 ± 0.9	1.3 ± 1.1	3.4 ± 0.8	2.4 ± 0.4	3.0 ± 0.7	5.0 ± 1.1	7.5 ± 0.9
[Mg]	7.7 ± 1.6	1.0 ± 2.3	0 ± 1.9	6.8 ± 1.1	4.9 ± 0.5	3.0 ± 1.0	2.4 ± 2.2	6.4 ± 1.0

* Values are expressed as mmol/kg dry weight (conversion based on the 1988 Aminoplastic Standards, Warley, 1990).

† see Figure 2 for the anatomical location refered to by letters in table; n = 10 samples for each element at each location.

Table 2. One-way analysis of variance in ion content between eight different areas on the condylar surface of the TMJ disk

	Area of disk GHBCADE		Area of disk BGHADCE		Area of disk HADGEBC
Sulfur		Potassium		Sodium	
($p = 0.001$)		($p = 0.001$)		($p = 0.001$)	
Area G		Area B		Area H	
Area H		Area G		Area A	
Area B		Area H		Area D	
Area C	*	Area A		Area G	
Area A		Area D	* * * *	Area E	* * * *
Area D	*	Area C	* * * *	Area B	* * * *
Area E	* *	Area E	* * * * * *	Area C	* * * *
Area F	* * * * * * *	Area F	* * * * * *	Area F	* * * * * *
	Area of Disk HGDABCE		**Area of Disk CEFDAGB**		**Area of Disk GHCDBAF**
Chlorine		Calcium		Phosphorus	
($p = 0.001$)		($p = 0.001$)		($p = 0.001$)	
Area H		Area C		Area G	
Area G	*	Area E		Area H	
Area D	*	Area F		Area C	
Area A		Area D		Area D	
Area B	*	Area A		Area B	
Area C	* *	Area G		Area A	
Area E	* *	Area B	*	Area F	* * * *
Area F	* * * * * * *	Area H	* * * * *	Area E	* * * * * * *

* Denotes pairs of groups significantly different at $p < 0.05$. S-N-K multiple range test was used to determine which means were significantly different.

† See Figure 2 for the anatomical locations referred to by letters in the table; $n = 10$ for each element at each location.

water into the inner, more tightly bound water compartment, and had lower flow rates than did non-load bearing areas.

As illustrated in Figure 3, curve fit analysis demonstrated two water compartments for each area. The data points from the outer water compartment (the least tightly bound water and therefore the first water to be removed from the tissue) fit logarithmic curves with r^2 values ranging from 0.976 to 0.993. The amount of water in the outer water compartment was positively correlated with Cl^- concentration ($r = 0.968$) and with Na^+ concentrations ($r = 0.782$) Areas F, C, E and D had low flow (high resistance). Multiple regression analysis indicated that 92.7% of the variation in the amount of water in the outer water compartment could be explained by the variation in Cl^- content. The rate at which water in the outer compartment leaves the tissue was not significantly correlated with any of the values measured.

The data points from the innermost water compartment (the most tightly bound water) fit simple linear lines with r^2 values ranging from 0.966 to 1.00. The amount

Table 3. Multiple regression analysis of selected elemental content of the condylar surface of the pig temporomandibular disk

Element	Stepwise entry of elements that make a significant ($p < 0.05$) contribution to explaining the dependent element content in pig articular disk				Cum % Constant
	First Factor	Second Factor	Third Factor	Fourth Factor	
[S]	[K] 56.6%	[Ca] 7.4%	[P] 1.6%		65.6%
	(0.485)*	(1.235)	(0.151)		81.13
[K]	[Cl] 65.6%	[S] 13.4%	[P] 1.9%	[Ca] 1.9%	81.9%
	(0.226)	(0.523)	(0.246)	(−0.980)	−59.60
[Na]	[Cl] 48.5%	[S] 2.6%	[Mg] 2.8%		53.9%
	(0.670)	(0.975)	(−2.736)		−76.68
[P]	[S] 40.4%	[K] 2.8%	[Ca] 4.4%		47.6%
	(0.535)	(0.228)	(−1.677)		−47.67

* The positive or negative slope of each linear equation is given in parentheses. The units of the slope are mmol/L independent element vs. mmol/L of dependent element. The constant to the equation is listed in the column to the far right. Next to each contributing element is listed the percent contribution to the explanation.

of water in the inner water compartment was negatively correlated with both K^+ and sulfur concentrations ($r = -0.848$ and -0.790 respectively). Multiple regression analysis indicated that 67.2% of the variation in the amount of water in the inner water compartment could be explained by the variation in K^+ content and that the relationship was negative, i.e., the higher the K^+ content, the lower the water content. If K^+ is eliminated as a variable, 65.0% of the variation in the amount of water in the inner water compartment could be explained by the variation in sulfur content. The rate at which water in the inner compartment leaves the tissue was positively correlated with the K^+ content ($r = 0.742$). Multiple regression analysis indicated that 47.5% of the variation in the rate of flow of water in this compartment was explained by the variation in K^+ content.

To evaluate the reproducibility of regional differences in the TMJ disks of different pigs of similar age, the flow rate, initial water content and final water content measurements were repeated on three different pigs. Although the absolute values vary from animal to animal, the relative ranking of each area tested (as diagrammed in Figure 2) are highly reproducible as determined by Spearman rank correlation coefficients. The Spearman's coefficient of rank correlation was chosen to test for association because it is the more appropriate test when there is less certainty about the reliability of close ranks. The rho values for the initial water content were 0.7619, 0.7904 and 0.9940 with $p < 0.01$. The rho values for the final water content were 0.9524, 0.9762 and 0.9762 with $p < 0.01$, and the rho values for flow rates were 0.6467, 0.9102 and 0.8976 with $p < 0.05$. These findings demonstrate that the methods used allow establishment of significant and reproducible differences that exist between the parameters measured in the TMJ disks from different animals.

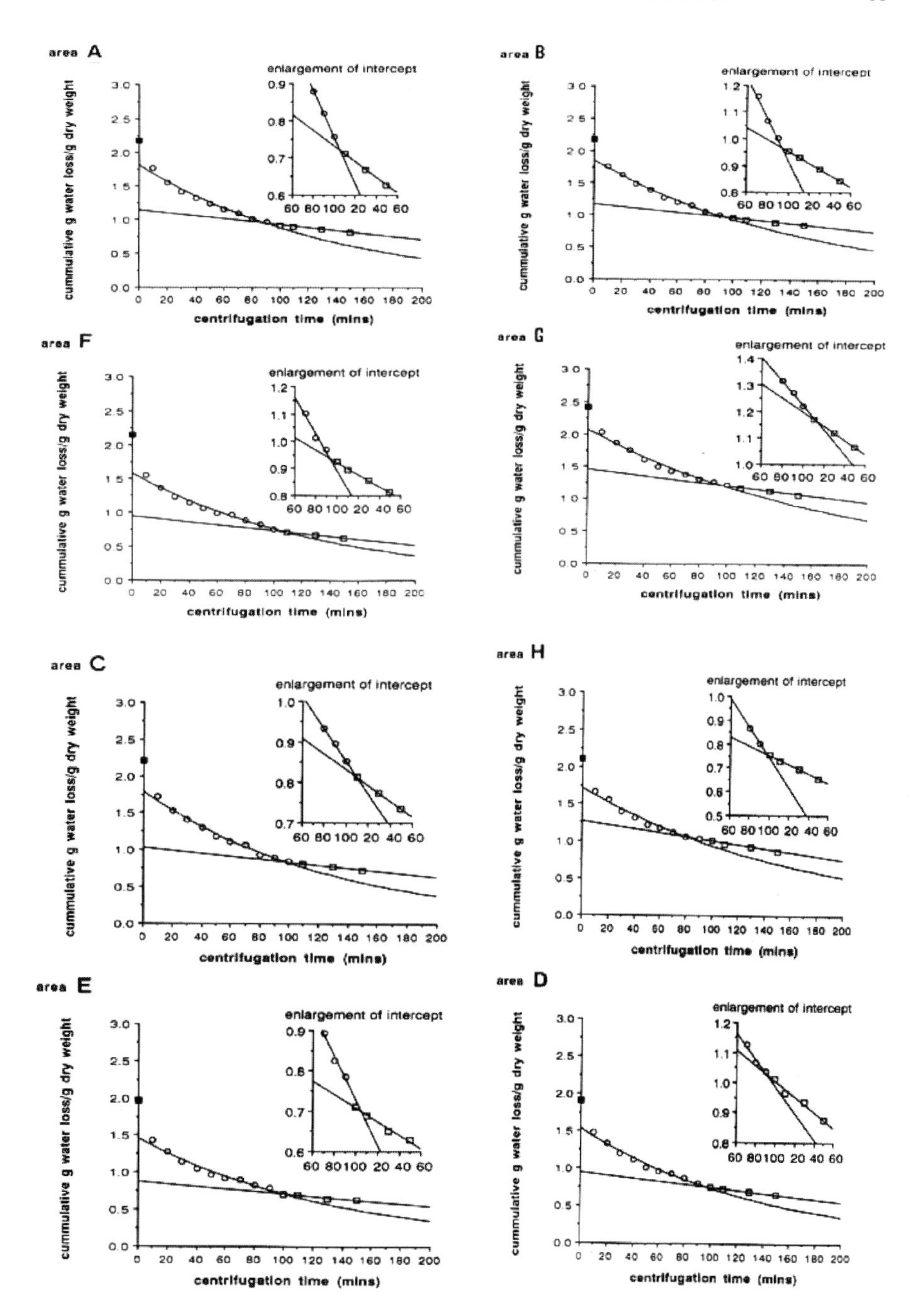

Figure 3. Resistance to fluid flow at eight different locations of the TMJ disk. The cummulative grams of water loss per gram dry weight are plotted against time under compressive load and describe a logarithmic line such that the slope is the log of water loss. The slope of the line derived for each location is therefore a numerical expression of the resistance to fluid flow under compressive load. The area letter refers to the location of tissue analyzed (see Figure 2)

4. DISCUSSION

The hypothesis being tested was that there was a positive relationship between the sulfur concentration and (1), the amount of water held in the tissue and (2), the ability of the disk to resist loss of fluid under compressive loads. The results of our investigation lead us to reject the hypothesis because there was a significant inverse, rather than a positive correlation, between both sulfur and potassium content and the amount of water in the inner water compartment. Secondly, there was a positive correlation between the K^+ and sulfur content and the rate of flow from the most tightly bound water compartment, i.e., the higher the K^+ and sulfur contents, the higher the flow rate. Although the ion content did not provide an explanation for the flow characteristics of the outer water compartment, the amount of water in the outer water compartment was positively correlated with the concentrations of Cl^- and Na^+.

It has been experimentally verified, in an artificial matrix, that both positive and negative fixed charges significantly increase both the initial water content and the final equilibrium water content of resins (unpublished results). This would seem to support the general model in the literature that the increased density of fixed charges would result in increased water content (Myers and Mow, 1983; Carney and Muir, 1988). However, since the extracellular matrix of the disk fibrocartilage can be expected to have substantial concentrations of each fixed charge groups, it is difficult to explain how the presence of sulfur, presumably in the form of fixed sulfates on the GAGs, would have an inverse correlation with water content, as we found in the TMJ disk. This indicates that some additional mechanism other than increased fixed charge density is altering the amount of water in the tissue. In vivo, the K^+ ion concentration was an even stronger negative predictor than sulfur concentrations of the amount of water in the tissue at force equilibrium. Moreover, through multiple regression analyses, we determined that sulfur appears to be preferentially associated with K^+ counterions. Thus, we propose that the role of the sulfated GAGs is not to organize a large sphere of hydration, but rather, due to their high density of fixed sulfates that preferentially accumulate K^+ counterions, to limit, as much as possible, the amount of water within a matrix that is undergoing compressive loading.

In addition to providing a more adequate explanation of observed water content, the new model is also consistent with the hydraulic conductivity data of Zamparo and Comper (1989), who demonstrated that the contribution of the fixed charge sites and their counterions to the hydraulic permeability of cartilage proteoglycans was minimal and that the shape and conformation of the macromolecule was the primary determinant of hydraulic conductivity. By measuring the sedimentation velocity of glycosaminoglycans in different salt solutions, Zamparo and Comper were able to quantitate the effect that the charge sites of glycsoaminoglycans had on the flow of water around the molecules. They demonstrated that when the effect of the charges was neutralized at high salt concentration, there were not significant differences in the sedimentation velocities of glycosaminoglycans with a high density of fixed charge sites and those with a low density of fixed charge sites. (Zampara and

Comper, 1989). Zamparo and Comper pointed out that sulfation was the last step in GAG synthesis and represented a high metabolic energy requirement–two molecules of ATP for every sulfate added and speculated that, in the absence of a significant contribution to the hydraulic properties of the molecule, the function of the sulfate groups could possibly be participation in electrostatic-binding interactions with other macromolecules in the extracellular matrix. We propose that the role of sulfated groups is primarily to provide the only biologically significant fixed charge that preferentially interacts with the K^+ counterion, rather than the Na^+ counterion, thereby reducing the total water around the ion pairs.

The distribution of counterions plays a dynamic role in modulating the water content and in determining the biomechanical properties of the extracellular matrix. We have experimentally verified, in an artificial matrix, that the presence of a K^+ counterion rather than a Na^+ counterion at the fixed charge site dramatically decreases the amount of water in a matrix (unpublished results). In addition, multiple regression analysis of the in vivo elemental content data suggests that sulfates are preferentially associated with K^+ rather than Na^+.

Why do fixed sulfates in the extracellular matrix preferentially accumulate K^+ ions? It has been shown that ion exchange resins preferentially accumulate a single counterion (Ling, 1984). For example, sulfate ion exchange resins selectively accumulate K^+ over Na^+ and the carboxylic and phosphoric resins selectively accumulate Na^+ over K^+ (Ling, 1984). If the basis of the selectivity were solely due to size of the hydrated counterion then there would be only one order of preference and all resins would selectively accumulate K^+ over Na^+. The basis of the selectivity can be accounted for by the differences in polarizability of phosphate, carboxylate and sulfate groups in comparison with that of water. Polarizability decreases in the order phosphate $>$ carboxylate $>$ water $>$ sulfate (Ling, 1984) and a consequence of the relative polarizability is that carboxylate and phosphate selectively accumulate Na^+ and sulfates selectively accumulate K^+. The consequence of selectively accumulating K^+ is less water of hydration initially and at equilibrium.

The assumption in the literature has consistently been that because the polyanionic GAGs can organize a large sphere of water over ionic and dipolar sites (Humzah and Soames 1988; Ikada et al., 1980) and because water is essentially non-compressible at physiological levels (MacDonald, 1975) that increased amounts of bound water (water restricted in its movement because of association with a dipolar or charge site) would be a mechanism by which tissues could resist the effects of compression.

Our in vivo data support the opposite conclusion., i.e., that one of the roles of the sulfated GAGs was to preferentially accumulate K^+ ions, thereby decreasing the relative amount of water present during compressive loading. Lower water content in areas of compressive loading should be an advantage. Areas with a larger fluid/solid ratio would undergo greater deformations in response to compressive loading and larger volumes of fluid would be forced in and out of the tissue during

Table 4. The resistance to fluid flow and water content by area in the pig temporomandibular disk

	Initial Water Content g H_20/g dry mass	Inner water Content g H_20/g dry mass	Inner Flow Rate*	Outer water Content g H_20/g dry mass	Outer Flow Rate
Intermediate Zone					
E	1.960	0.873 (44.5%)	-1.663^{-3x}	1.087 (55.5%)	−0.692* log(x)
C	2.208	1.024 (46.4%)	-1.925^{-3x}	1.184 (53.6%)	−0.893* log(x)
D	1.903	0.945 (49.7%)	-1.942^{-3x}	0.958 (50.3%)	−0.718* log(x)
F	2.134	0.941 (44.1%)	-2.100^{-3x}	1.193 (55.9%)	−0.806* log(x)
Anterior Band					
B	2.170	1.172 (54.0%)	-2.197^{-3x}	0.998 (46.0%)	−0.808* log(x)
A	2.178	1.147 (52.7%)	-2.236^{-3x}	1.031 (47.3%)	−0.839* log(x)
Posterior Band					
H	2.010	1.261 (62.7%)	-2.575^{-3x}	0.749 (37.3%)	−0.693* log(x)
G	2.415	1.460 (60.5%)	-2.625^{-3x}	0.955 (39.5%)	−0.847* log(x)

loading cycles. This would likely result in greater matrix destruction in areas with higher water content.

As seen in Table. 4, the higher the initial water content, the higher the flow rate. The most probable explanation for the increased flow rate in areas of high initial water content is that there is proportionately less solid matrix to resist compressive loads. In those areas with a decreased flow rate, the presence of K^+ with its smaller hydration sphere allows for a proportional increase in the solid matrix. One of the most significant effects of this decrease in initial water content may be a decrease in destructive shear forces in areas with a significant concentration of sulfates. Water may be incompressible under boundary conditions, but because of its viscous nature, it does not resist compressive loads under non-boundary conditions such as those in the TMJ disk and the frictional drag associated with its movement may have significant impact on matrix resistance to wear.

Finally, our data have shown that it may be possible to make significant alterations in the immediate biomechanical properties of a matrix system simply by altering the relative concentrations of specific counterions. In biological tissues such as the TMJ disk, the preferential accumulation of K^+ over Na^+ at fixed sulfates cannot be expected to be absolute, but rather a dynamic process dependent both on the ion content of the fluid bathing the tissues and the concentrations of the various fixed ions at specific sites within the tissue. Thus, dynamic changes can be made in the mechanical properties of a tissue without significant remodeling simply by changing the relative availability of counterions such as Na^+ and K^+.

5. SUMMARY AND CONCLUSIONS

1. Determination of water content and elemental analysis revealed that the amount of water present in specific areas of the temporomandibular disk is negatively correlated with increased potassium and sulfur concentrations.
2. Compressive loading of the TMJ disk revealed sequential loss of a faster flowing outer water compartment followed by loss of a slower flowing inner water compartment.
3. There was a positive correlation between the slower flow rate of the inner, more tightly bound water compartment and the potassium and the sulfur content of the disk.
4. K^+ was the preferred counterion of the negatively charged sulfated GAGs. Thus a statistically significant positive relationship was established between K and S content and the slower flow rate of the inner water compartment.
5. Monovalent ions, such as Na^+ and K^+, when bound to fixed charge sites, such as sulfated GAGs, appear to determine fluid flow rates and the biomechanical properties of the TMJ disk.

REFERENCES

Alberts B, Bray D, Lewis J, Raff M, Roberts K, Watson J (1984) Molecular Biology of the Cell. Garland Publishing, Inc., New York, NY, pp 702–707

Berkovitz BK, Robinson S, Moxham BJ, Patel D (1992) Ultrastructural quantification of collagen fibrils in the central region of the articular disc of the temporomandibular joint of the cat and the guinea pig. Arch Oral Biol 37:479–481

Blaustein DI, Scapino P (1986) Remodeling of the temporomandibular joint disk and posterior attachment in disk displacement specimens in relation to glycosaminoglycan content. Plat Recon Surg 78:756–764

Carney SL, Muir H (1988) The structure and function of cartilage proteoglycans. Physiol Re 68:858–910

Detamore MS, Athanasiou KA (2003a) Tensile properties of the porcine temporomandibular joint disc. J Biomech Eng 125:558–565

Detamore MS, Athanasiou KA (2003b) Motivation, characterization, and strategy for tissue engineering the temporomandibular joint disc. Tissue Eng 9:1065–1087

Fullerton GD, Cameron IL (1988) Relaxation of biological tissues. In: Wehrli FW, Shaw D, Kneeland JB (eds) Biomedical Magnetic Resonance Imaging–Principles, Methodology, and application. VCH Publisher, Inc., New York, pp 115–155

Fullerton GD, Ord VA, Cameron IL (1986) An evaluation of the hydration of lysozyme by an NMR titration method. Biochem Biophys Acta 869:230–246

Gage JP, Virdi AS, Triffitt JT, Howlett CR, Francis MJ (1990) Presence of type III collagen in disc attachments of human temporomandibular joints. Arch Oral Biol 35:283–288

Humzah MD, Soames RW (1988) Human intervertebral disk: Structure and Function. Anat Record 220:337–356

Ikada Y, Suzuki M, Iwata H (1980) Water in mucopolysaccharides. In: Rowland SP (ed) Water in Polymers. Washington, D.C., American Chemical Society, pp 287–305

Kempson GE (1980) The Mechanical Properties of Articular Cartilage. In: Sokoloff L (ed) The Joints and Synovial Fluid, vol 2. Academic, New York

Kempson GE (1991) Age-related changes in the tensile properties of human articular cartilage: A comparative study between the femoral head of the hip joint and the talus of the ankle joint. Biochim Biophy Acta 1075:223–230

Kempson GE, Muir H, Swanson SAV, Freeman MAR (1970) Correlations between stiffness and the chemical constituents of cartilage on the human femoral head. Biochim Biophysica Acta 215:70–77

Kino K, Ohmura Y, Kurokawa E, Shioda S (1989) Reconsideration of the bilaminar zone in the retrodiscal connective tissue of the TMJ. 2. Fibrous structure of the retrodiscal connective tissue and relation between those fibers and the disk. Nihon Ago Kansetsu Gakkai Zasshi 1:43–54 (English Abstract)

Kiyosawa K (1988) Precise expression of freezing-point depression in aqueous solutions. In: Lauger P, Packer L, Vasilescu V (eds) Water and Ions in Biological Systems. Birkhauser Verlag, Boston, pp 425–432

Klein-Nulend J, Veldhuijzen JP, van de Stadt RJ, van Kampen GPJ, Kuijer R, Burger EH (1987) Influence of intermittent compressive force on proteoglycan content in calcifying growth plate cartilage in vitro. J Biol Chem 262:15490–15495

Kobayashi J (1992) Studies on matrix components relevant to structure and function of the temporomandibular joint. Kokubyo Gakkai Zasshi 59:105–123. (English Abstract)

Kopp S (1976) Topographical distribution of sulphated GAGs in human temporomandibular joint disks. A histochemical study of an autopsy material. J Oral Path 5:265–276

Kopp S (1978) Topographical distribution of sulfated glycosmainoglycans in the surface layers of the human temporomandibular joint. A histochemical study of an autopsy material. J Oral Pathol 7:283–2294

Kuc TM, Nakano T, Scott PG (1989) The extracellular matrix of bovine temporomandibular joint disc. J Dent Res 68:229–242

Kuijer R, van de Stadt RJ, van Kampen GP, de Koning MH, van de Voorde-Vissers E, van der Korst JK (1986) Heterogeneity of proteoglycans extracted before and after collagenase treatment of human articular cartilage. II. Variations in composition with age and tissue source. Arthritis Rheum 29:1248–1255

Kuijer R, van de Stadt RJ, de Koning MH, van Kampen GP, van der Korst DK (1988) Influence of cartilage proteoglycans on type II collagen fibrillogenesis. Connec. Tissue Res 17:83–97

Ling GN (1984) The association-induction hypothesis I. Association of ions and water with macromolecules. In Search of the Physical Basis of Life. Plenum, New York and London, pp 145–180

MacDonald AG (1975) Physiological Aspects of Deep Sea Biology. Monographs of the Physiological Society No. 31. London, Cambridge University Press

Milam SB, Klebe RJ, Triplett RG, Herbert D (1991) Characterization of the extracellular matrix of the primate temporomandibular Joint. J Oral Maxillofac Surg 49:381–391

Mills DK, Daniel JC, Scapino R (1988) Histological features and in-vitro proteoglycan synthesis in the rabbit craniomandibular joint disk. Archs Oral Biol 33:195–202

Minarelli AM, Liberti EA (1997) A microscopic survey of the human temporomandibular joint disc. J Oral Rehabil 24:835–840

Myers ER, Mow VC (1983) Biomechanics of cartilage and its response to biomechanical stimuli. In: Hall BK (ed) Cartilage, vol 1. Academic, New York, 1983, pp 313–341

Nakano T, Scott PG (1989a) A quantitative chemical study of glycosaminoglycans in the articular disk of the bovine temporomandibular joint. Arch Oral Biol 34:749–757

Nakano T, Scott PG (1989b) Proteoglycans of the articular disc of the bovine temporomandibular joint. I. High molecular weight chondroitin sulfate proteoglycan. Matrix 9:277–283

Roos M, Barnar T (1984) Amionplastic standards for quantitative x-ray microanalysis of thin sections of plastic-embedded biological material. Ultramicroscopy 15:277–286

Scapino RP (1983) Histopathology associated with malposition of the human temporomandibular joint disc. Oral Surg 55:382–397

Scott JE (1988) Proteoglycan–fibrillar collagen interactions. Biochem J 252:313–323

Scott JE (1989) Ion binding: patterns of 'affinity' depending on types of acid groups. Symp Soc Exp Biol 43:111–115

Scott JE (1989) Secondary structures in hyaluronan solutions: Chemical and biological implications. Ciba Found Symp 143:6–15

Scott PG, Nakano T, Docc CM (1989) Proteoglycans of the articular disc of the bovine temporomandibular joint. II. Low molecular weight dermatan sulfate proteoglycans. Matrix 9:284–292

Shengyi T, Xu Y (1991) Biomechanical properties and collagen fiber orientation of TMJ discs in dogs: Part I. Gross anatomy and collagen fiber orientation of the discs. J Craniomandib Disord 5:28–34

Shira RB (1989) Histologic features of the temporomandibular joint disk and posterior disk attachment: Comparison of symptom-free persons with normally positioned disks and patients with internal derangement. Oral Surg Oral Med Oral Pathol 67:635–643

Smith NKR, Morris SS, Richter MR and Cameron IL (1983) Intracellular elemental content of cardiac and skeletal muscle of normal and dystrophic hamsters. Muscle Nerve 6:481–489.

Tanaka E, van Eijden T (2003) Biomechanical behavior of the temporomandibular joint disc. Crit Rev Oral Biol Med 14(2):138–150

Tanaka E, Aoyama J, Tanaka M, van Eijden T, Sugiyama M, Hanaoka K, Watanabe M, Tanne K (2003) The proteoglycans contents of the temporomandibular joint disc influence its dynamic viscoelastic properties. J Biomed Mater Res A 65(3):386–392

Teng S, Xu Y, Cheng M, Li Y (1991) Biomechanical properties and collagen fiber orientation of temporomandibular joint discs in dogs: 2. Tensile mechanical properties of the discs. J Craniomandib Disord 5:107–114

Tong AC, Tideman H (2001) The microanatomy of the rhesus monkey temporomandibular joint. J Oral Maxillofac Surg 59:46–52

Warley A Standards for x-ray microanalysis in biology. J Microsc 15:135–147

Zamparo O, Comper WD (1989) Hydraulic conductivity of chondroitin sulfate proteoglycan solutions. Arch Biochem Biophys 1990 274:259–269

CHAPTER 3

COHERENT DOMAINS IN THE STREAMING CYTOPLASM OF A GIANT ALGAL CELL

V.A. SHEPHERD*

* *Department of Biophysics, School of Physics, The University of NSW, NSW 2052, Sydney, Australia, Fax 61 2 9385 4484 and E-mail: vas@phys.unsw.edu.au*

Abstract: Giant internodal cells of the charophyte *Lamprothamnium* respond to hypotonic shock with an extended action potential and transient cessation of cytoplasmic streaming. The macro-structure of streaming cytoplasm was analysed before, during, and after hypotonic shock. Streaming cytoplasm contains coherent, cloud-like macroscopic domains, whose perimeter varies from hundreds to many thousands of micrometres. Some domains avidly associate with the fluorochrome 6-carboxyfluorescein (6CF), and others do not. The 6CF-labelled domains are recognisable through many cycles of streaming, despite constantly changing irregular edges. Domain perimeters were described by a fractal dimension of 4/3, the exponent of a power law fitted to a log-log plot of domain perimeter-area. Following hypotonic shock, the stable pattern of coherent domains enters an unstable phase of coalescence, and discrete domains subsequently amalgamate into stable, extended domains. Instability is associated with Ca^{2+} influx and Cl^{-} efflux, and a large increase in cell conductance. The electrophysiological K^{+} state, with greatly reduced conductance, is associated with the new, amalgamated stable state. The results support a concept of cytoplasm as a sponge-like percolation cluster, undergoing transition from discrete to extended domains. Results are discussed in terms of published theories concerning co-operative behaviour of supramolecular water-ion-protein complexes

Keywords: Cytoplasmic streaming, Fractal cytoplasm, Cell water, Characeae, Low density water, High density water, Gel phase transition, Water cluster

1. INTRODUCTION

Cytoplasmic streaming is one of the most mesmerising of cellular phenomena, and it has fascinated biologists since at least 1774, when Corti described streaming in giant internodal cells of *Chara*. No less fascinating is the excitability of these ‘green nerves’ or ‘green muscles’, which fire action potentials in response to touch,

G. Pollack et al. (eds.), Water and the Cell, 71–92.

electrical stimulation, osmotic shock, or sudden changes in temperature or light (Wayne, 1994).

Giant internodal cells of the Characeae, including *Chara*, are the workhorses of electrophysiology and studies of cytoplasmic streaming (Shimmen and Yokota, 1994; Tazawa and Shimmen, 2001). In an internodal cell a single band of streaming cytoplasm, or endoplasm, is split into two oppositely directed streams, separated by a helical 'neutral line'. The stationary cortical or 'gel' cytoplasm contains helically arrayed files of chloroplasts. Beneath these, at the interface with the endoplasm, are sets of 3-6 F-actin bundles, each containing ~100 actin microfilaments, with each filament 5-6 nm wide (Shimmen and Yokota, 1994; Grolig and Pierson, 2000). These bundles are responsible for rapid streaming (Shimmen and Yokota, 1994). Opposed polarity of actin filaments determines the opposed directions of streaming (Williamson, 1975). Myosin-associated endoplasmic reticulum slides along actin bundles (Kachar and Reese, 1988).

Action potentials have multiple signalling functions in plants (reviewed, Davies, 1987). Action potentials are interpreted as the result of Ca^{2+} influx to the cytoplasm, Cl^- efflux via Ca^{2+}-activated Cl^- channels, and K^+ efflux through voltage-gated K^+ channels (Wayne, 1994; Kikuyama, 2001). Acto-myosin driven cytoplasmic streaming stops transiently, due to Ca^{2+}-sensitivity of the myosin calmodulin light chain (Yokota et al., 2000). An action potential is accompanied by a transient drop in turgor pressure (Zimmermann and Beckers, 1978), and transient contraction of the cell (Oda and Linstead, 1975). The process of transient Ca^{2+} influx, Cl^- and K^+ efflux, and change in turgor pressure is also a fundamental motif used by plants to operate 'osmotic motors', during leaf movements, controlling stomatal aperture, trap closure in insectivorous plants, and turgor-pressure regulation in response to osmotic shock (Hill and Findlay, 1981).

Our understanding of both electrophysiology and streaming in plant cells is underpinned by concepts that many researchers have argued are outdated. First is the 'cytosol' concept (see Clegg, 1984), in which the aqueous cytoplasm is envisaged as a solution, equivalent to 100 mM KCl, in which other ions and macromolecules are dissolved or suspended. Second is the accompanying view that intra and intercellular transport are governed by diffusion (see Wheatley, 2003). Electrophysiology and cytoplasmic morphology are usually imagined as disparate phenomena, uncoupled except during the transient Ca^{2+} increase associated with cessation of cytoplasmic streaming.

However, enzymes of many major metabolic pathways are spatially linked, and organised into supramolecular complexes called "metabolons". Intermediates are channeled from one enzyme to the next without entering a bulk phase (reviewed, Al-Habori, 1995; Hochachka, 1999). Coherence in the glycolytic pathway and other multi-enzyme systems is related in turn to larger-scale structural organisation of the cytoskeleton (reviewed, Aon and Cortassa, 2002). Many metabolic pathways in plants are organised into metabolons, associated with the ER or endomembranes (reviewed, Winkel, 2004). Furthermore, intracellular K^+ is associated with proteins, and does not freely diffuse in cytoplasm (Edelmann, 1988). Intracellular ions have a far reduced chemical activity compared to dilute solutions of them (Cameron et al., 1988). Although it is often referred to as 'cytosolic free calcium', calcium does not diffuse in the cytoplasm (Trewavas, 1999).

Metabolism and signalling take place within a highly organised and synchronised cytoplasmic medium, which can even function as an intelligent machine, capable of data-processing, an '...intelligent, giant multienzyme complex...' (Albrecht-Buehler, 1985, p 4).

The streaming cytoplasm in Characean cells does not behave as a 'cytosol'. Pickard (2003) reasons that intercellular transport of macromolecules and vesicles is actin-mediated. Solutes smaller than the ~1 kDa molecular size exclusion limit for plasmodesmata could still be transported by diffusion (Pickard, 2003). However, even small fluorochromes (e.g., 6-carboxyfluorescein, 374 Da) microinjected into *Chara* endoplasm travel in the direction of the injected stream, moving intercellularly before changing direction at the ends of the cell (Shepherd and Goodwin, 1992a, 1992b). They do not move in the opposite direction to flow, or across streams. The endoplasm of a single cell is divided into essentially separate upwardly and downwardly directed transport streams, and this polarity has developmental significance, since antheridia develop only from node-cells on the downstream side (Shepherd and Goodwin, 1992a, 1992b). Furthermore, injected fluorochromes are not homogenously distributed in the cytoplasm, but are concentrated instead in cytoplasmic domains, which retain a recognisable and coherent structure (Shepherd and Goodwin, 1992a).

In this paper, I take a closer look at the morphology of cytoplasmic domains, and whether this changes during an action potential. Fluorochromes move quickly from the cytoplasm into the vacuole of *Chara* cells, and the action potential is rapid (Shepherd and Goodwin, 1992a), making it difficult to observe the domains for very long. I have instead looked at domain behaviour in young salt-tolerant *Lamprothamnium* cells, which retain 6-carboxyfluorescein (6CF) in the cytoplasm for long periods (Beilby et al., 1999). These young cells respond to hypotonic shock with transient Ca^{2+} influx, efflux of Cl^- and K^+, and turgor pressure regulation (Beilby and Shepherd, 1996); essentially a long, slow action potential.

The perimeter-area relationship of 6CF-labelled cytoplasmic domains shows scaling behaviour, with a fractal dimension. Following hypotonic shock, dramatic changes in domain morphology are paralleled by equally dramatic electrophysiological changes. These results cannot be explained by electrophysiological modelling. A vast literature demonstrates that cell-associated water has altered properties, including different density and solvent properties (reviewed by Clegg, 1984; Cameron et al., 1988; Wiggins, 1990; Clegg and Drost-Hansen, 1991; Drost-Hansen and Singleton, 1995; Pollack, 2001). Ion channel function and operation of molecular 'motors' can be interpreted in terms of the co-operative behaviour of supramolecular water-ion-protein complexes (Watterson, 1988, 1991, 1993, 1997; Wiggins, 1990, 1995a, 1995b, 1996, 2001;

Chaplin, 1999; Chaplin, http://www.lsbu.ac.uk/water, Pollack, 2001, 2003; Pollack and Reitz, 2001).

The results are discussed with reference to theories of organised cell water.

2. MATERIALS AND METHODS

2.1 Cells

Lamprothamnium sp. plants were collected from a creek adjacent to Lake Munmorah, Central Coast, NSW. Salinity of the creek was ~1/3 that of seawater. Young glassy internodal cells were cut from the apex of the plants and allowed to recover in 1/3 seawater for a week before experiments.

2.2 Fluorescence-labelling and Imaging

The cytoplasm of twelve cells was fluorescence-labelled with 6-carboxyfluorescein (6CF) by diluting 6-carboxyfluorescein diacetate (Molecular Probes, Eugene, OR, USA) stock solution (Shepherd et al., 1993), in 1/3 seawater to a final concentration of 40 μg/ml. Cells were labelled for two hours, and then mounted in fresh, Millipore filtered 1/3 seawater, in a grooved perspex slide whose base was a size zero coverslip. Hypotonic shock was induced by exchanging 1/3 for 1/6 seawater.

Cells were imaged with a sensitive Zeiss ZVS47EC cool CCD video camera coupled with a Zeiss Axiovert inverted fluorescence microscope (Carl Zeiss Pty Ltd., Oberkochen, Germany), using the FITC excitation/barrier filter combination to excite 6CF fluorescence and eliminate chloroplast autofluorescence.

2.3 Image-analysis of Cytoplasmic Domains Before and After Hypotonic Shock

Actin bundles involved in generating streaming are located at the interface between the endoplasm and cortical gel cytoplasm (Shimmen and Yokota, 1994). This region was located by focussing on the underside of the chloroplasts.

Still images of cytoplasmic domains in sharp focus at the interface were captured from digitised video sequences and analysed using NIH Image J software. Areas and perimeters of cytoplasmic domains were calculated from fifteen still images captured from a seven-minute sequence of streaming cytoplasm. All calculations were done at the same zoom/magnification scale, 1 μm = 0.2 pixels. This was selected given the limitations of domain size and image resolution. Twenty-five cytoplasmic domains were analysed from a cell in 1/3 seawater. Very small domains were excluded because of the large error in pixel counting, and very large domains did not fit into the field of view. The cytoplasm coalesced following hypotonic shock, and was too large for analysis.

Images of the cell in hypotonic solution were captured at critical times defined by electrophysiological analysis (Beilby and Shepherd, 1996; Beilby et al., 1999).

2.4 Scaling Behaviour of Cytoplasmic Domains

An average fractal dimension of domains was estimated by fitting a power law to the log-log plot of perimeter versus area. $P = KA^{D/2}$, where P is the domain perimeter, A is the domain area, K is a scaling constant, and D is the fractal dimension (Kenkel and Walker, 1996). The validity of the method was checked by fitting a power law to the log-log plot of the area/perimeter of 25 circles placed over domains of different sizes.

3. RESULTS

3.1 Fluorescence-labelling and Cytoplasmic Domains

The 6CF appeared to be non-toxic, since branches labelled with 6CF continued to grow. Amitotic fragmentation of the nucleus was still taking place in young cells. They contained a series of large and small spherical nuclei, as well as multiple kidney-shaped nuclei. All nuclei retained intense 6CF fluorescence throughout. Chloroplasts did not accumulate 6CF. The gel cytoplasm was only faintly fluorescent.

Coherent streaming cytoplasmic domains were visible in all twelve cells. 6CF was retained in the cytoplasm of the young cells. Older cells compartmented 6CF in vacuoles, as occurred in *Chara* (Shepherd and Goodwin, 1992a). However, 6CF-labelled domains were visible before this compartmentation occurred. Domains were also visible with phase-contrast microscopy of unlabelled cells.

Figure 1A-B shows a region of a cell in 1/3 seawater, as domains stream across it. 6CF was localised in irregularly shaped domains, streaming beneath helically arrayed chloroplast files. Domain morphology was complex yet coherent, reminiscent of clouds. Some domains were warped in the direction of flow, with a parabolic leading front. Morphology was malleable, with constantly changing irregular edges, deformations, transient connections and disconnections, yet individual domains were recognisable through complete end-to-end cycles of streaming. Coherence is demonstrated by the bending of an entire domain (asterix, Figure 1A). This occurred when an organelle in the domain centre temporarily paused in its streaming, whilst those at the edges continued to stream. Non-fluorescent motile regions of equally complex morphology, ('anti-domains') were interposed between labelled domains. These regions changed shape in concert with the streaming domains, as if displacing or displaced by them. Deeper-lying 6CF-labelled cytoplasmic aggregations spanned the central, non-labelled vacuole.

3.2 Changes in Cytoplasmic Structure During Hypotonic Shock

Changes in cytoplasmic structure following hypotonic shock are shown in Figure 1A-H. Key events in cytoplasmic structural change were coupled with key electrophysiological events (Beilby and Shepherd, 1996; Beilby et al., 1999; Shepherd et al., 2002), summarised below.

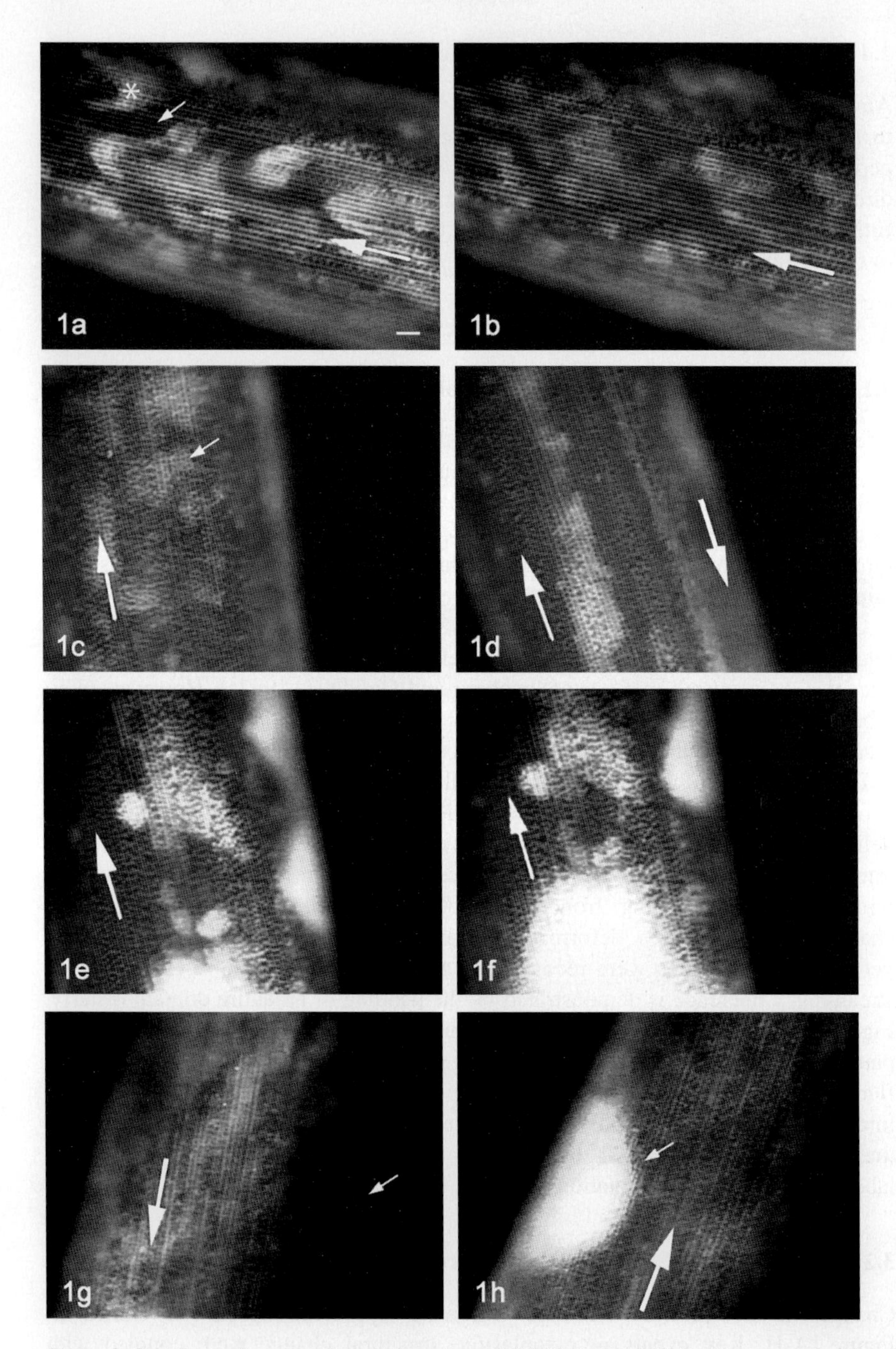

Figure 1. Cytoplasm of a young *Lamprothamnium* cell labelled with 6CF, in 1/3 seawater (A-B) and during hypotonic shock (C-H). Different regions of the cell were viewed, to reduce photobleaching and cell damage. Large arrows show direction of streaming. Scale bar = 100 μm. (A) and (B) Streaming

Key event 1. Electrophysiology and cytoplasmic macrostructure in the first few minutes after hypotonic shock.

An initial depolarisation has been attributed to stretch-activated ion channels. Subsequently, Ca^{2+} influxes to the cytoplasm, at first increasing from 80 nM to 300 nM (Okazaki et al., 2002), which is believed to activate Ca^{2+}–activated Cl^- channels. This is associated with a ten-fold increase in cell conductance. Subsequently, Ca^{2+} increases to ~400-600 nM (Okazaki et al., 2002) and cytoplasmic streaming stops.

Cytoplasmic streaming stopped after 40 sec. in hypotonic solution. Cytoplasmic domains (Figure 1A-B) stopped streaming but were still recognisable (Figure 1C). Chloroplasts were rearranged into alternate clumps and files. Laterally distributed 6CF-labelled endoplasm formed a labyrinthine pattern between chloroplasts.

Key event 2. Electrophysiology and cytoplasmic macrostructure 4-5 minutes after hypotonic shock.

Cytoplasmic Ca^{2+} declines (to ~ 300 nM; Okazaki et al., 2002) but Ca^{2+}–activated Cl^- channels remain in force.

Cytoplasmic streaming restarted in some regions only. Cloud-like domains elongated in the direction of streaming. Cytoplasmic 'curds', with increased fluorescence intensity and 'V-shaped' streaming fronts, appeared (Figure 1D).

Figure 1. cloud-like cytoplasmic domains (yellow/green fluorescence, rendered here in greyscale) interpenetrated by non-labelled, streaming domains (small arrow in 1A). Asterix in 1A shows bending of a labelled domain, brought about when an organelle in the domain centre paused in its streaming, whilst those at the edges continued to advance. (C) Cloud-like domains (small arrow) stop streaming and transiently retain their morphology. Labelled cytoplasm moves laterally between chloroplasts to form a labyrinthine pattern. Chloroplasts are rearranged into alternate files and clumps. Image captured 2 min. after introduction of hypotonic solution. (D) Cytoplasmic streaming restarts in some regions only, and labelled domains elongate in the direction of streaming. This image shows the two opposed streaming directions (large arrows) separated by the 'neutral line'. Image captured after 4 min. 03 sec in hypotonic solution. (E) and (F) Coalesced cytoplasmic 'curds', with increased 6CF fluorescence intensity, stream very slowly, collide and fuse with the few remaining cloud-shaped domains, which move more slowly or are stationary. Coalesced regions are interspersed with regions of dispersed fluorescence. Figure 1F was captured 10 min. 50 sec., and Figure 1G, 19 min. 30 sec., after introduction of hypotonic solution. (G) Domains coalesce into a large convoluted entity with diffuse fluorescence, displacing or displaced by the large equally complex 'antidomains'. Small arrow shows edge of the cell. The rate of cytoplasmic streaming recovered to normal at 25 min. 51 sec. Image captured after 30 min. in hypotonic solution. (H) The 6CF-labelled cytoplasmic entity retained diffuse fluorescence, but some thick coalesced regions, surrounding the large nucleus (small arrow), were present. Coalesced streaming cytoplasm moved in concert with unlabelled 'antidomains' of similar morphology, alternating in position. Cytoplasmic 'fingers' connected to the convoluted fluorescent cytoplasmic mass showed peristaltic movements at the gel-endoplasm interface. Image captured after 50 min. in hypotonic solution

Key event 3. Electrophysiology and cytoplasmic macrostructure 15-20 minutes after hypotonic shock.

Ca^{2+}–activated Cl^- channel activity is greatly reduced between ~15-19 min, paralleled by reduction in cytoplasmic Ca^{2+} (Okazaki et al., 2002). Large conductance K^+ channels begin to dominate the cell conductance as the conductance attributed to Cl^- channels declines. The cell potential recovers to $\sim E_{K+}$, the Nernst potential for K^+ ($\sim -80\,mV$). Cell conductance is reduced to ~ 4 times the resting level. The cell enters a 'K^+ state', dominated by large conductance K^+ channels.

Streaming cytoplasmic 'curds' collided and fused with remaining stationary cloud-like domains, forming large coalesced cytoplasmic masses, interpenetrated by vacuolar counterparts of similar complex morphology (Figure 1E, 1F).

Key event 4. Electrophysiology and cytoplasmic macrostructure 30 minutes after hypotonic shock.

The cell potential stabilises at $\sim -80\,mV$. Cell conductance is reduced to ~ twice the resting level.

Streaming recovered its initial rate. The entire 6CF-labelled coalesced cytoplasm streamed in concert with its equally complex, unlabelled counterpart (Figure 1G). The coalesced cytoplasm extended 'fingers' to the cell periphery, and these showed peristaltic movements.

Key event 5. Electrophysiology and cytoplasmic macrostructure 50+ minutes after hypotonic shock.

The cell potential hyperpolarises (to $\sim -120\,mV$). The cell conductance declines to less than the resting value. This state remains stable for hours.

The 6CF-labelled cytoplasm remained as a complex coalesced entity, interpenetrated by unlabelled domains (Fig. 1H). Peristaltic movements at the cell periphery pinched off some 6CF-labelled domains, and some labelled, 'cloud-like' domains, reappeared.

3.3 Scaling Behaviour/Fractal Dimension of Cytoplasmic Domains

The log-log plot of domain perimeter/area is shown in Figure 2. Minimum domain perimeter was $5.3 \times 10^2\,\mu m$, with area $6.2 \times 10^3\,\mu m^2$. Maximum domain perimeter was $5.7 \times 10^3\,\mu m$, with area $1.9 \times 10^5\,\mu m^2$. Larger and smaller domains were present, but were not analysed (see Methods). A reasonably straight line was fitted, over two orders of perimeter magnitude. The power law describing this plot is $P = 1.65A^{0.7}$. The average fractal dimension, D, estimated from these data, is 1.4, or $\sim 4/3$.

A log-log plot of perimeter/area calculated for a series of circles superimposed on the domains yielded the power law, $P = 3.45A^{0.5}$, over two orders of magnitude (not shown) with $D = 1.0$.

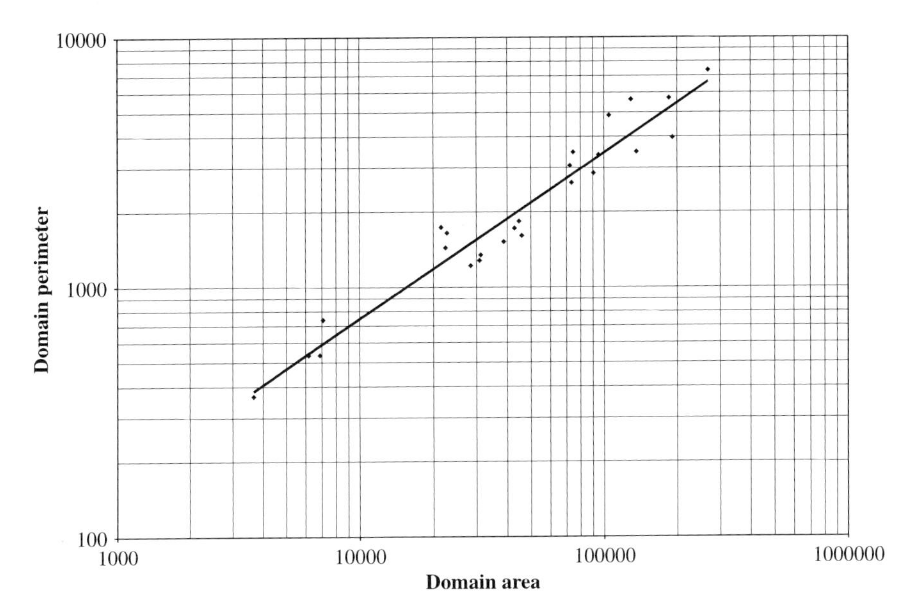

Figure 2. Scaling behaviour of 6CF-labelled cytoplasmic domains. A log-log plot of perimeter versus area for 25 cloud-like cytoplasmic domains, fitted with a power law, $P = 1.65\,A^{0.7}$, giving D (fractal dimension) = 1.4 or $\sim 4/3$

4. DISCUSSION

The streaming cytoplasm is not an homogenous phase. 6CF labels persistent 'cloud-like' domains (Figure 1A-B), akin to domains in *Chara* internodal cells (Shepherd and Goodwin, 1992a). Domains formed complex patterns in time and space, with constantly changing irregular edges, yet were coherent, and recognisable through complete cycles of streaming. Domains moved in association with streaming organelles. Many were curved in the direction of streaming (Figure 1A). Changes in rates of organelle movement within domains induced bending of the entire structure, sometimes in the opposite direction to streaming (asterix, Figure 1A). Domains were surrounded by streaming cytoplasm, which did not label with 6CF ('anti-domains').

There are two co-existent phases in the streaming endoplasm. One avidly associates with 6CF, and another does not.

Hypotonic treatment produced drastic changes in the distribution of cloud-like domains (Figure 1C-H). The 'key events' in electrophysiology were coupled with 'key events' in cytoplasmic structural change.

The perimeter-area relationship of the cloud-like domains showed scaling behaviour, with irregular perimeters (Figure 2) of estimated average fractal dimension of 4/3 (1.33).

Electrophysiological modelling (Beilby and Shepherd, 1996) does not predict or explain either the presence of domains, or the structural changes following hypotonic treatment. Some possible explanations will be discussed, based on ideas from the vast literature concerning structured water in cells, and the following conclusions.

4.1 Cytoplasmic Domains are Associated with the Actin Cytoskeleton

Domains were imaged at the gel/endoplasm interface, where actin bundles responsible for streaming are located (Shimmen and Yokota, 1994). Actin filaments synchronise the movements (and morphologies) of myosin-associated endoplasmic reticulum and endoplasmic organelles (Lichtscheidl and Baluska, 2000). Within the gel cytoplasm is situated an array of fine actin strands (Foissner and Wasteneys, 2000a) and peripheral cytoskeleton, which immobilises chloroplasts and actin bundles (Williamson, 1985).

Irregularly shaped, fluorescent extensions of some domains traversed the cell, passing through a non-fluorescent region interpreted as the vacuole. Transvacuolar cytoplasmic strands containing actin traverse the vacuole in *Chara* cells at a comparable developmental stage (Foissner and Wasteneys, 2000a).

The polarity of domain streaming, their associations with the gel-endoplasm interface, and their continuity with deeper-lying transvacuolar cytoplasm, all indicate that domains move in association with the actin cytoskeleton.

4.2 6CF Labels Charged Structures within Domains

Domains in endoplasm squeezed out of permeabilised cells contain clusters of 6CF-labelled micronuclei, fluorescent ER/actin bundle ensembles, unidentified fluorescent spherical organelles, and irregular aggregations of fluorescent cytoplasm (see Figure 1C of Shepherd and Goodwin, 1989). The latter might contain protein complexes for which 6CF has high affinity. Large and small spherical nuclei retained intense 6CF fluorescence (Figure 1H).

The cytoplasmic domains are essentially motile ‘islands’, containing charged structures labelled by 6CF.

4.3 Changes in Cytoplasmic Structure are Coupled with Changes in Electrophysiology Following Hypotonic Shock

We have intensively analysed the time-course of electrophysiological changes following hypotonic shock (Beilby and Shepherd, 1996; Beilby et al., 1999; Shepherd et al., 2002). The full response to hypotonic shock, including cessation of cytoplasmic streaming, takes place in young cells, like those in this study, which retain 6CF fluorescence in the cytoplasm. Older cells secrete sulphated polysaccharide mucilage, do not stop streaming, and accumulate the 6CF in their vacuoles (Shepherd et al., 2002).

In 1/3 seawater, the 6CF-labelled domains and unlabelled ‘antidomains’ cycled repetitively, with constantly changing but coherent morphology- a stable state (Figure 1A-B). Chloroplasts were arranged in files.

Key events 1-2 (Figure 1C-D) reflect structural and electrophysiological instability following hypotonic treatment. The dramatically depolarised cell experiences a large increase in conductance, attributed to Ca^{2+} increase in the cytoplasm, and Ca^{2+} activated-Cl^- efflux (Beilby and Shepherd, 1996; Okazaki et al., 2002). The domains transiently remain coherent when streaming stops (Figure C). Re-arrangement of chloroplasts suggests re-arrangement of the peripheral actin/protein cytoskeleton, which immobilises chloroplasts and actin bundles (Williamson, 1985).

Streaming restarts in key event 2, but Ca^{2+} remains sufficiently elevated for the Ca^{2+} activated-Cl^- conductance to continue (Okazaki et al., 2002). Streaming restarts in only some regions, where cloud-like domains progressively coalesce into thick cytoplasmic 'curds'. The dramatically increased 6CF fluorescence of 'curds' could be due to cytoplasmic condensation, concentrating the 6CF, or a pH change in curds only.

During key events 3-5 (Figure 1E-H), cytoplasmic structure and electrophysiology both shift to a new stable state. The Ca^{2+} activated-Cl^- channels are inactivated (Beilby and Shepherd 1996; Okazaki et al., 2002), Ca^{2+} declines in the cytoplasm, and large conductance K^+ channels dominate the progressively declining cell conductance. Thick cytoplasmic curds fuse with remaining domains and form a coalesced entity with complex topology, streaming in concert with an equally complex unlabelled entity. These entities, and the electrophysiological K^+ state, remain stable for hours. After one hour, the cell hyperpolarises, its conductance drops to below the initial level, and some 'cloud-like' 6CF-labelled domains reappear.

4.4 The 6-CF Labelled Domains have a Fractal Dimension of 4/3

The gross morphology of the domains is statistically self-similar across several orders of magnitude, shown by scaling behaviour and fractal dimension of $\sim$4/3. Since domains and anti-domains mutually displace one another, antidomains also have fractal perimeters. The domains, treated here as flat islands, are actually three-dimensional, and the area/volume relationship probably also scales. Such scaling behaviour suggests domain construction involves iteration of units. That is, it may be quantised.

4.5 The Lattice Concept of Cytoplasmic Construction

A lattice structure could underlie scaling behaviour. Modules of hydrated proteins of a basic size can form collectives, a cytoplasmic lattice interpenetrated by water (Clegg, 1984; Albrecht-Buehler, 1990), which may be the fundamental pattern of cytoplasmic organisation (Clegg, 1984). Macromolecules of critical size (1-5 kDa) induce vicinal water structuring. Vicinal water has enhanced H-bonding, lower density, greater viscosity, and is a poorer solvent (Drost-Hansen and Singleton, 1995). Vicinal water is distinct from bound water. It extends for at least 3-5 nm from a surface and possibly much further (Clegg, 1984; Drost-Hansen and Singleton,

1995), with the influence of the surface decaying with distance. There are at least three phases of water within the cytoplasm; bound, vicinal and more fluid or 'normal'.

4.6 The Lattice as a Percolation Cluster

Some time ago, Aon and Cortassa (1994) proposed that the cytoskeletal lattice is a percolation cluster, a kind of random fractal. Their hypothesis was based on dehydrated structures, including dried ovomucin, which formed branching dendritic structures (Aon et al., 2000), and electron micrographs of cytoskeletal proteins.

In percolation theory, the probability, P that an injected fluid will percolate through an infinite number of pores is low if there are few connected pores. A sudden transition occurs at a critical probability, P_{crit}. Below P_{crit}, the pores are connected in discrete clusters. Above P_{crit} all pores are suddenly globally connected. The percolation cluster is an attractive concept. It may explain the shift from discrete domains to extended domains of complex topology (Figure 1A-H).

Bruni et al., (1989) found an abrupt percolative transition in the rehydration behaviour of *Artemia* cysts, and suggested it involved proton conduction. Basic metabolism began, and the cysts suddenly became conductive, at only 35 g protein/g water hydration. At this hydration level, water had to be protein-associated rather than 'bulk phase' (Fulton, 1985). If percolation takes place, it involves water. What controls P_{crit}, creating 'global' or 'discrete' domain patterns?

The plant cell experiences turgor pressure, essentially an interplay between the pressure-generating vacuole and the tensile cell wall. The vacuole is osmotically active, gaining or losing water in hypo- or hypertonic solutions, and expelling K^+ and Cl^- during turgor regulation following hypotonic shock (Bisson, 1995). The vacuole forms connections with the myosin-associated ER, and this interpenetrates the endoplasm. Streaming organelles impact on the constantly changing ER morphology (Lichtscheidl and Baluska, 2000). The ER itself has fractal structure (Mandelbrot, 1983).

A cytoskeletal lattice in streaming endoplasm is intimately associated with endomembranes.

Such a lattice may be more like a sponge.

4.7 A Mental Picture of the Cytoplasm/Vacuole as a Motile Sponge

The cytoplasm of another giant but non-streaming algal 'cell', *Ventricaria ventricosa*, is sponge-like, and quantised into domains, each containing a nucleus, each interconnected to its neighbours by perinuclear microtubules ensheathed in thin cytoplasm (Shepherd et al., 2004). Each nucleus retains a constant cytoplasmic volume. The cytoplasm is interpenetrated by a topologically complex vacuole, the membrane-lined 'holes' in the cytoplasmic 'sponge'. Quantised cytoplasm enables regeneration of new organisms from cytoplasmic fragments. Following wounding a wide-meshed actin reticulum appears, and contracts the entire cytoplasm into 'islands', which then contract into hundreds of regenerative protoplasts (La Claire, 1989).

The cytoplasm/vacuole structure in *Lampothamnium* is also akin to a sponge, albeit a motile one. The multiple nuclei probably have perinuclear microtubules (Foissner and Wasteneys, 2000b), and presumably an associated conserved cytoplasmic volume. Actin bundles, along which they stream, associate with a delicate peripheral actin meshwork, and there are indirect associations with microtubules (Foissner and Wasteneys, 2000a). The endomembrane system/vacuole can be imagined as the 'holes' in the sponge, whose solid parts are a cytoskeletal and proteinaceous meshwork. The porosity of the cytoplasmic sponge increases towards the cell centre. In mature cells, it enlarges to form a central vacuole. In young cells, there remains a wide meshwork of actin-associated transvacuolar cytoplasm. The vacuole is linked to an endomembrane continuum of finer interconnected 'holes', including actin-associated ER at the gel/endoplasm interface.

P_{crit}, and the shift from discrete to coalesced domains, could involve a critical change in a fundamental dimension of the meshwork. This would involve a co-operative change in the dimensions of the 'holes' (endomembranes). What then, is the finest dimension, or 'endpoint dimension', of the proteinaceous meshwork of the cytoplasmic 'sponge'?

4.8 Endpoint Dimensions of a Motile Sponge-like Cytoplasm

The 'passive' molecular size exclusion limit (~1 kDa) for fluorescent probes travelling through the cytoplasmic channel in plasmodesmata indicates a baseline channel dimension of 3 to 5 nm (Shepherd and Goodwin, 1992a). Actin, the ER, and myosin are continuous from cell to cell through plasmodesmata (White et al., 1994; Blackman and Overall, 1998).

The cytoplasm of permeabilised *Chara* cells has the same molecular size exclusion limit as plasmodesmata, and a 'channel' dimension of ~3.4 nm (Shepherd and Goodwin, 1989). This may be associated with the protein meshwork cytoskeleton (Williamson, 1985), or delicate actin meshwork in the gel cytoplasm (Foissner and Wasteneys, 2000a).

The 'endpoint' dimensions of actin-containing plasmodesmata and protein-containing peripheral cytoplasm are the same, ~3.4 nm.

Myosin moves processively along actin with 35 nm steps (Tominaga et al., 2003), a ten-fold factor of the endpoint dimension. The ~17 nm dimension of putative 'microtrabeculae' in plant cells, whilst larger than the 3-6 nm dimension in animal cells (Wardrop, 1983), is still a five-fold factor of the endpoint dimension. The endpoint dimension is the same as the domain size of numerous diverse proteins, including aquaporin channel dimension (Watterson, 1991, 1997). Molecules far larger than the baseline ~1 kDa molecular size exclusion limit can move from cell-to-cell via the ER (Cantrill et al., 1999).

The existence of similar endpoint dimensions in the peripheral cytoplasm and the plasmodesmata could fulfill Mandelbrot's condition for percolation, where '... a shape drawn on a square or cube is said to percolate if it includes a connected curve joining opposite sides of the square or the cube...' (Mandelbrot, 1983).

4.9 Why 3-5 nm? Watterson's Wave-Cluster Theory

The end-point dimensions of the cytoplasm are the same as those of a water cluster or pressure pixel in wave-cluster theory (Watterson, 1988, 1991, 1993, 1997, 2005, http://www.lsbu.ac.uk/water). The supracellular cytoplasmic continuum thins down to an ultimate dimension in plasmodesmata and in the peripheral cytoplasm; the same dimension as a water cluster.

In wave-cluster theory, protein domains and water clusters form co-operative supramolecular aggregates. Water is quantised into clusters which have an edge ~3.5 nm long at RT, containing about 1400 molecules (Watterson, 1997). Clusters are constantly forming and collapsing in water, and so cluster organisation travels as a structure wave (oscillations in the H-bonded network). Protein domains are the same size as pre-existing water clusters (Watterson, 1991) and fit together with them, stabilising the clusters, and creating networks of protein domain/cluster structures, whose continuity is gated by the lengths of the travelling structure waves. That is, domain/clusters are macroscopic entities with defined vibrational modes.

Macroscopic pressure operates only down to the ~3.5 nm scale of a cluster, the 'pressure pixel', and tension operates below this dimension (Watterson, 1997). Switching between tension and pressure can explain transitions from gelled to fluid states in the cytoplasm.

The cytoplasmic endpoint dimensions sit at the edge of pressure/tension switching.

If P_{crit} varies with this dimension, 3.5 nm, then a change in cluster size could induce tension/pressure switching, coalescing the discrete domains. This would involve co-operative changes in cluster size in both the ER/vacuole and the cytoplasmic protein meshwork.

The 6CF-labelled domains may be macroscopic 'superclusters', ensembles of protein domain/water clusters held together by unifying structure waves, and separated from 'antidomain superclusters' by nodes in the structure waves. Co-operative fluctuations between protein domain/clusters, of basic size 3.5 nm, and domain 'superclusters', thousands of microns in size, could give rise to coherent patterns and fractal structure.

Following hypotonic treatment, an increase in the length of the structure wave (increase in cluster size) could unite the discrete domains into a single coalesced entity (see Figure 4; Watterson, 1997), thereby reducing their surface area and energy. Regions of different tension and pressure could be balanced by different cluster sizes (structure wave-length) travelling across the ER/vacuole membrane, changing wavelength and amplitude, but not energy. Solutes form nodes in the wave-cluster structure, and so solute concentration equals cluster concentration (Watterson, 1991). Clusters change concentration by changing size. Turgor pressure regulation could reflect progressive rebalancing of cluster sizes, with larger clusters forming as KCl exits the vacuole. The wave-cluster theory shows that changes in water structure can be propagated, and provides a reason for the dimensions of cytoplasmic 'endpoints'.

4.10 High and Low Density Water in Gels and Cells

Wiggin's experiments and theories (see Wiggins, 1990, 1995a, 1995b, 1996, 2001) show that there are two forms of water in cells, of lower (LDW) and higher (HDW) density. Water reduces its density (LDW) at a hydrophobic surface, with a corresponding increase in density (HDW) elsewhere (Wiggins, 1995a). A highly charged surface, with high counterion concentration and diffuse double layer, produces HDW in the double layer and LDW beyond it (Wiggins, 1995a).

The forms are in a continuum, and micro-osmosis, the shifting equilibrium between them, is an alternative mechanism for understanding ion channels, voltage gating, and operation of molecular 'motors'. Micro-osmosis is a '... vectorial, non-linear, sometimes self-limiting process, which can oscillate in time, can generate a force or create highly ordered spatial distributions of solutes...' (Wiggins, 1995b). At the extremes of the continuum, LDW resembles vicinal water, and is more strongly H bonded, more viscous, and accumulates weakly hydrated ions (e.g. $NO3^-$, HCO_3^-, Cl^-, K^+, NH_4^+; Wiggins, 1990, 1995a). It excludes strongly hydrated ions (Mg^{2+}, Ca^{2+}, H^+, and Na^+). HDW is more fluid, and accumulates Mg^{2+}, Ca^{2+} and Na^+. However, neither water nor solutes ever reach equilibrium, and local changes in density continue.

Put another way, the transition between weak/strong ion hydration depends on whether an ion binds more strongly to water than water does to itself (Collins, 1995). Weakly hydrated K^+ and Cl^- are 'sticky', adsorbing to weakly hydrated gels by partially dehydrating, whilst strongly hydrated Na^+ can flow through (Collins, 1995).

4.11 Is 6-CF a Probe for HDW/LDW?

6CF-labelling potentially distinguishes HDW from LDW in living cells. Initially, the 6CF-labelled domains, containing charged structures (DNA/protein, actin), could be regions of predominantly HDW. Preserving electroneutrality, ions accumulate into the water zone surrounding charged patches, and water equilibrates by increasing its density near the charged patches and decreasing it in hydrophobic regions (Wiggins, 1990). Most of the water between actin filaments would be LDW, whilst that nearest to charged patches would be HDW (more normal). The 6CF may be forced into the HDW adjacent to charged patches.

At higher resolution, the domains treated as whole 'islands' are probably complexes of far smaller domains (e.g., Figure 1C, Shepherd and Goodwin, 1989). At the low resolution shown here, the 6-CF labelled domains may include '...proteins, surrounded by high density, fluid, reactive water, containing most of the ions of the cytoplasm, clustering around actin filaments in such a way that their surface area is minimised…' (Wiggins, 1996). The 'anti-domains' would then represent travelling waves of LDW structuring.

The partitioning behaviour of 6CF could be determined by experiment, using gel bead/film systems used by Wiggins (1995a) to define HDW and LDW.

In Chaplin's cluster/density model (Chaplin, 1999; Chaplin, 2005; http://www.lsbu.ac.uk/water, http://www.lsbu.ac.uk.water.cell.html), expanded (less dense) icosahedral cluster structures are in dynamic equilibrium with denser, collapsed dodecahedral clusters. Ionic chaotropes (HCO_3^-, Cl^-, NO_3^-, NH_4^+, K^+) form clathrates in low-density icosahedral clusters, whilst ionic kosmotropes (SO_4^{2-}, HPO_4^{2-}, Mg^{2+}, Ca^{2+}, Na^+, H^+) prefer collapsed cluster structures. The basic icosahedral cluster size is ~3 nm, compatible with a pressure pixel in wave-cluster theory.

A shift in the HDW/LDW balance changes H bond strength, viscosity and solvent properties. The postulated P_{crit} could be determined by a critical transition between LDW/HDW.

Wiggin's and Chaplin's concepts emphasise the dynamic relationship between HDW/LDW. HDW in a region means LDW elsewhere. Plant cytoplasm and vacuoles would contain a dynamic balance between forms.

Vacuoles accumulate SO_4^{2-}, PO_4^{3-}, H^+, Ca^{2+}, Mg^{2+}, Na^+ (Table 5.2, 5.4; Hope and Walker, 1975); ions rejected by LDW. Na^+ and Ca^{2+} are at low levels in cytoplasm (Table 5.5, Hope and Walker, 1975). $K^+$and Cl^- also accumulate in vacuoles. Vacuoles of maize roots contain large amounts of Ca^{2+} and K^+, usually more K^+ than Ca^{2+} (~5:1 ratio), but the reverse is true in some cells (Canny and Huang, 1993). The K^+ / Ca^{2+} ratio might indicate the position of the LDW/HDW equilibrium. This could change with cell age. Older charophyte cells have more calcium in their vacuoles (Kirst et al., 1988), an indicator of age (Winter et al., 1987).

An 'average' cytoplasm is shifted in the direction of LDW; an 'average' vacuole/ER in the direction of HDW.

During hypotonic treatment, Ca^{2+} shifts transiently from ER/vacuole to cytoplasm. KCl subsequently exits the vacuole/ER, as Ca^{2+} progressively re-enters it. This suggests dynamic shifts in the LDW/HDW balance within, and between, these compartments.

Hydration of isolated cytoplasmic droplets causes Ca^{2+} to increase, and it possibly moves from the ER (Kikuyama, 2001). The increased Ca^{2+} in the cytoplasm is expected to the destroy LDW structuring associated with F-actin filaments, and shift the balance towards HDW. If K^+ binds to carboxylate (aspartic and glutamic acid) groups of F-actin, forming clathrates that induce icosahedral cluster structuring (LDW), then increased Ca^{2+} could displace the bound K^+, destroying the LDW clathrates and co-operatively converting associated water to HDW (see Chaplin, 2005; http://www.lsbu.ac.uk/water). Thus, coalescence of 6CF-labelled domains could reflect growth of HDW on a macroscopic scale.

Several early papers considered *Chara* cytoplasm as essentially a cation-exchanger. Negative charges changed sign during the action potential, accompanied by transient appearance of protons (Coster et al., 1968; Coster and Rendle, 1974). The mobility of Cl^- in cytoplasm simultaneously increased (Coster and Rendle, 1974). This is consistent with transient increase in HDW, changed carboxylate pK_a, and increased flexibility of the actin cytoskeleton (see Chaplin, http://www.lsbu.ac.uk.water.cell.html).

The LDW/HDW theories suggest microvariations in dynamic viscosity (viscosity/density) could occur during streaming. The single band of motile cytoplasm is functionally separated into upwardly and downwardly directed streams (Shepherd and Goodwin, 1992a). In upright cells, the downstream flows faster, is thinner, and less viscous than the upstream (Wayne et al., 1990). Shifts in the LDW/HDW balance could explain this asymmetry, with HDW dominating the downstream, and LDW, the upstream.

Trewavas (1999) argued that plant cell cytoplasm can behave like a neural net, with Ca^{2+} channels as nodes, gating 'on' or 'off'. Ca^{2+} does not diffuse, but inositol trisphosphate (IP_3) does, activating Ca^{2+} channels and producing calcium waves. Calcium waves could conceivably represent travelling waves of LDW/HDW structure.

4.12 Pollack's Phase-gel Transition Concept

The coupled electrophysiological/structural changes resemble polymer-gel phase changes that Pollack (2001, 2003; Pollack and Reitz, 2001) proposes as a unifying mechanism for explaining ion distributions, cell electric potentials, and other aspects of cell behaviour. These phase changes involve propagated changes in water structure. Pollack (2001) proposes an alternative mechanism for understanding streaming. Actin is envisaged as undergoing propagated structural changes (undulations) accompanied by changes in water structuring. Depending on levels of ATP and K^+, actin alternates between two phases or conformations, one linear, and another shorter, kinked, and more flexible. The linear conformation is associated with structured water. The kinked conformation is associated with unstructured water. Changes in surface charge (from more hydrophilic to more hydrophobic), due to Ca^{2+} bridging anionic sites, or protein-protein interactions, result in a propagated phase change.

Structured water, which rejects solutes, alternates with less structured water, which carries solutes along the propagated phase transition.

The 6CF-labelled domains could be the 'moving windows' of unstructured water predicted by the theory (Pollack, 2001, p 171–174). The unlabelled 'antidomains' would then represent moving regions of structured water, which reject the 6CF. The propagated phase transition '...melts local water into bulk water, creating a window within which solutes or organelles can be suspended. As the window moves, so do the solutes...' (Pollack, 2001, p 174). The theory predicts the observed alternation between streaming domains and 'antidomains'. Domain bending (Figure 1A), is explained, as an organelle in the centre pauses, having missed its 'window', until it is captured '... by a subsequent moving window...' (Pollack, 2001, p 174).

Ca^{2+} can cross-link negative sites on actin, bringing them together, destructuring and squeezing out water, resulting in a condensed phase (Pollack, 2001). This could explain the transient cell contraction and water loss accompanying an action potential (Oda and Linstead, 1975; Zimmermann and Beckers, 1978). Re-arrangement of chloroplasts suggests that the peripheral cytoskeleton and possibly underlying actin bundles had indeed been re-arranged during the extended Ca^{2+} release following hypotonic treatment. The increased fluorescence intensity of coalesced 'curds' (Figure 1D) suggests condensation. Protein or Ca^{2+}-induced crosslinking may result in collisions between 'windows' of unstructured water, as streaming stops, and restarts, in different regions at different times. Expansion of the condensed polymer gel involves expulsion of Ca^{2+}, by restructuring of protein-associated water, which rejects strongly hydrated Ca^{2+} (Figure 9.5, Pollack, 2001). The patterns of streaming domains could be a sol-gel dissipative structure, which bifurcates following Ca^{2+} change, entering a different stable state.

5. CONCLUSION

The streaming endoplasm of *Lamprothamnium* cells contains large coherent domains with irregular, fractal morphologies. There appear to be two phases in the streaming cytoplasm; coherent domains, which associate with 6CF, and 'antidomains', which do not. A hypotonic treatment induces coupled structural and electrophysiological instability, including Ca^{2+} influx and Cl^- efflux, a large increase in cell conductance, and probable re-arrangement of the peripheral and underlying actin cytoskeletons. The 6CF-labelled domains (and possibly their unlabelled counterparts) subsequently coalesce. A new stable state is reached, where the electrophysiological K^+ state, with greatly reduced conductance, is associated with the new, stable state of amalgamated domains.

Large-scale cytoplasmic morphology is thus coupled to cell electrophysiology.

These data, and the fractal perimeter of domains, support a percolation cluster concept of fractally organised cytoplasm (Aon and Cortassa, 1994). The streaming cytoplasm is imagined as a motile lattice with a critical dimensionality, a sponge-like protein meshwork containing 'holes' occupied by the ER and vacuole. The finest dimension of the sponge-like cytoplasmic lattice is equated with the $\sim$3.5 nm 'channel' dimension of actin-containing plasmodesmata and peripheral cytoskeleton.

Recent theories of water structuring in proximity to proteins and membranes offer possible explanations for the presence of cytoplasmic domains and for a critical transition between stable states. These theories are not mutually exclusive.

It is suggested here that the 6CF-labelled domains correspond to the 'moving windows' of unstructured water, which transport solutes in Pollack's theory of streaming (Pollack, 2001). They also correspond to the fluid, reactive HDW in Wiggin's and Chaplin's theories (Wiggins, 1990, 1995a, 1995b,

1996, 2001; Chaplin, 1999, 2005, http://www.lsbu.ac.uk/water). The equally complex unlabelled domains correspond to Pollack's solute-excluding 'structured water', and to the more strongly H-bonded LDW in Wiggin's and Chaplin's theories.

The ~ 3.4 nm dimension of the 'pressure pixel' in Watterson's wave-cluster theory is compatible with the finest 'channel' dimension in the postulated sponge-like cytoplasm. This dimension is thus poised at a threshold between cluster tension and pressure. It is poised at a threshold between LDW/HDW, since the expanded icosahedral cluster dimension (LDW) in Chaplin's theory is also ~3 nm. Changes in this critical dimension equate with changes in water structuring. This would take place within and between the ER/vacuole ('holes') and cytoplasmic meshwork ('sponge'). A Ca^{2+}–induced change in this dimension could signify percolative transition (or phase-change) through coupled changes in water structuring.

REFERENCES

Albrecht-Buehler G (1985) Is cytoplasm intelligent too? In: Shay JW (ed) Cell and Muscle Motility. 6. New York, London: Plenum Press, pp 1–21

Albrecht-Buehler G (1990) In defense of non-molecular biology. Int Rev Cytol 120:191–241

Al-Habori M (1995) Microcompartmentation, metabolic channeling and carbohydrate metabolism. Int J Biochem Cell Biol 27:123–132

Aon MA, Cortassa S (1994) On the fractal nature of cytoplasm. FEBS Lett 344:1–4

Aon MA, Cortassa S (2002) Coherent and robust modulation of a metabolic network by cytoskeletal organisation and dynamics. Biophys Chem 97:213–231

Aon MA, Cortassa S, Lloyd D (2000) Chaotic dynamics and fractal space in biochemistry: simplicity underlies complexity. Cell Biol Internat 24:581–587

Beilby MJ, Shepherd VA (1996) Turgor regulation in *Lamprothamnium papulosum*. 1. I/V analysis and pharmacological dissection of the hypotonic effect. Plant, Cell Environ 19:837–847

Beilby MJ, Cherry CA, Shepherd VA (1999) Dual turgor regulation response to hypotonic stress in *Lamprothamnium papulosum*. Plant, Cell, Environ 22:347–359

Bisson MA (1995) Osmotic acclimation and turgor pressure regulation in algae. Naturwissenschaften 82:461–471

Blackman LM, Overall RL (1998) Immunolocalisation of the cytoskeleton to plasmodesmata of *Chara corallina*. Plant J 14:733–742

Bruni F, Careri G, Clegg JS (1989) Dielectric properties of *Artemia* cysts at low water contents. Evidence for a percolative transition. Biophys J 55:331–338

Cameron IL, Fullerton GD, Smith NKR (1988) Influence of cytomatrix proteins on water and on ions in cells. Scanning Microsc 2:275–288

Canny MJ, Huang CX (1993) What is in the intercellular spaces of roots? Evidence from the cryo-analytical-scanning microscope. Physiol Plant 87:561–568

Cantrill LC, Overall RL, Goodwin PB (1999) Cell-to-cell communication via plant endomembranes. Cell Biol Int 23:653–661

Chaplin MF (1999) A proposal for the structuring of water. Biophys Chem 83:211–221

Clegg JS (1984) Properties and metabolism of the aqueous cytoplasm and its boundaries. Amer J Physiol 246:R133–R151

Clegg JS, Drost-Hansen W (1975) On the biochemistry and cell physiology of water. In: Hochachka PW, Mommsen TP (eds) Biochemistry and Molecular Biology of Fishes 1. Phylogenetic and Biochemical Perspectives. New York: Elsevier Science Publishers, pp 1–23

Collins K (1984) Sticky ions in biological systems. Proc Natl Acad Sci USA 26:12233–12239

Coster HGL, Syriatowicz JC, Vorobiev LN (1968) Cytoplasmic ion exchange during rest and excitation in *Chara australis*. Aust J Biol Sci 21:1069–1073

Coster HGL, George EP, Rendle VA (1974) Potentials developed at a solution cytoplasm interface in *Chara corallina* during rest and excitation. Aust J Plant Physiol 1:459–471

Davies E (1987) Action potentials as multifunctional signals in plants: a unifying hypothesis to explain apparently disparate wound responses. Plant Cell Environ 10:623–631

Drost-Hansen W, Singleton JL (1995) Our aqueous heritage: Role of vicinal water in cells. In: Bittar EE (ed) Principles of Medical Biology. 4. Cell Chemistry and Physiology. Greenwich Connecticut: JAI Press Inc, pp 195–215

Edelmann L (1988) The cell water problem posed by electron microscopic studies of ion binding in muscle. Scanning Microsc 2:851–865

Foissner I, Wasteneys GO (2000) Actin in characean internodal cells. In: Staiger CJ, Baluska F, Volkmann D, Barlow PW (eds) Actin: A Dynamic Framework for Multiple Plant Cell Functions. Developments in Plant and Soil Sciences. 89. Dordrecht: Kluwer Academic Publishers, pp 259–426

Foissner I, Wasteneys GO (2000) Nuclear crystals, lampbrush-chromosome-like structures, and perinuclear cytoskeletal elements associated with nuclear fragmentation in characean internodal cells. Protoplasma 212:146–161

Fulton AB (1982) How crowded is the cytoplasm? Cell 30:345–347

Grolig F, Pierson ES (2000) Cytoplasmic streaming: from flow to track. In: Staiger CJ, Baluska F, Volkmann D, Barlow PW (eds) Actin: A Dynamic Framework for Multiple Plant Cell Functions. Developments in Plant and Soil Sciences. 89. Dordrecht: Kluwer Academic Publishers, pp 165–190

Hill BS, Findlay GP (1981) The power of movement in plants: the role of osmotic machines. Q Rev Biophys 14:173–222

Hochachka PW (1999) The metabolic implications of intracellular circulation. Proc Natl Acad Sci USA 26:12233–12239

Hope AB, Walker NA (1975) The Physiology of Giant Algal Cells. London, New York: Cambridge University Press

Kachar B, Reese TS (1988) The mechanism of cytoplasmic streaming in characean algal cells: sliding of the endoplasmic reticulum along actin filaments. J. Cell Biol 106:1545–1552

Kenkel NC, Walker DJ (1996) Fractals in the biological sciences. Coenoses 11:77–100

Kikuyama M (2001) Role of Ca^{2+} in membrane excitation and cell motility in characean cells as a model system. Int Rev Cytol 201:85–114

Kirst GO, Jansen MIB, Winter U (1988) Ecological investigations of Chara vulgaris L grown in a brackish water lake: ionic changes and accumulation of sucrose in the vacuolar sap during sexual reproduction. Plant, Cell, Environ 11:55–61

La Claire JW (1989) Actin cytoskeleton in intact and wounded coenocytic green algae. Planta 177:47–57

Lichtscheidl IK, Baluska F (2000) Motility of endoplasmic reticulum in plant cells. In: Staiger CJ, Baluska F, Volkmann D, Barlow PW (eds) Actin: A Dynamic Framework for Multiple Plant Cell Functions. Developments in Plant and Soil Sciences. 89. Dordrecht: Kluwer Academic Publishers, pp 191–201

Mandelbrot BB (1983) The Fractal Geometry of Nature. New York: WH Freeman and Company

Oda K, Linstead PJ (1975) Changes in cell length during action potentials in Chara. J. Exp. Bot 26:228–239

Okazaki Y, Ishigami M, Iwasaki N (2002) Temporal relationship between cytosolic free Ca^{2+} and membrane potential during hypotonic turgor regulation in brackish water charophyte Lamprothamnium succinctum. Plant Cell Physiol 43:1027–1035

Pickard WF (2003) The role of cytoplasmic streaming in symplastic transport. Plant, Cell, Environ 26:1–15

Pollack GH (2001) Cells Gels and the Engines of Life. Seattle, Washington: Ebner and Sons Publishers

Pollack GH (2004) The role of aqueous interfaces in the cell. Adv Coll Interfac Sci 103:173–196

Pollack GH, Reitz FB (2001) Phase transitions and molecular motion in the cell. Cell Mol. Biol 47:885–900

Shepherd VA, Goodwin PB (1989) The porosity of permeabilised *Chara* cells. Aust J Plant Physiol 16:231–239

Shepherd VA, Goodwin PB (1992a) Seasonal patterns of cell-to-cell communication in *Chara corallina* Klein ex Willd. 1. Cell-to-cell communication in vegetative lateral branches during winter and spring. Plant, Cell, Environ 15:137–150

Shepherd VA, Goodwin PB (1992b) Seasonal patterns of cell-to-cell communication in Chara corallina Klein ex Willd. 2. Cell-to-cell communication during the development of antheridia. Plant, Cell, Environ 15:151–162

Shepherd VA, Beilby MJ, Shimmen T (2002) Mechanosensory ion channels in charophytes: the response to touch and to salinity stress. Eur Biophys J 31:341–355

Shepherd VA, Beilby MJ, Bisson MA (2004) When is a cell not a cell? A theory relating coenocytic structure to the unusual electrophysiology of *Ventricaria ventricosa* (*Valonia ventricosa*). Protoplasma 223:79–91

Shimmen T, Yokota E (1994) Physiological and biochemical aspects of cytoplasmic streaming. Int Rev Cytol 155:97–139

Tazawa M, Shimmen T (2001) How Characean cells have contributed to the progress of plant membrane biophysics. Aust J Plant Physiol 28:523–539

Tominaga M, Kojima H, Yokota E, Orii H, Nakamori R, Katayama E, Anson M, Shimmen T, Oiwa K (2003) Higher plant myosin XI moves processively on actin with 35 nm steps at high velocity. EMBO J 22:1263–1272

Trewavas A (1999) Le calcium, c'est la vie: Calcium makes waves. Plant Physiol 120:1–6

Wardrop AB (1983) Evidence for the possible presence of a microtrabecular lattice in plant cells. Protoplasma 115:81–87

Watterson JG (1988) A model linking water and protein structures. Biosystems 22:51–54

Watterson JG (1991) The interactions of water and protein in cellular function. In: Jeanteur P, Kuchino Y, Muller WEG, Paine PL (eds) Progress in Molecular and Subcellular Biology 12. Berlin, Heidelberg, New York, London, Paris, Tokyo, Hong Kong, Barcelona, Budapest: Springer Verlag, pp 113–134

Watterson JG (1993) The wave-cluster model of water-protein interactions. In: Green DG, Bossmaier T (eds) Complex Systems: From Biology to Computation. Amsterdam, Oxford, Washington, Tokyo: IOS Press, pp 36–45

Watterson JG (1997) The pressure pixel-unit of life? Biosystems 41:141–152

Wayne R (1994) The excitability of plant cells: with a special emphasis on Characean internodal cells. The Bot Rev 60:265–367

Wayne R, Staves MP, Leopold AC (1990) Gravity-dependent polarity of cytoplasmic streaming in Nitellopsis. Protoplasma 155:43–57

Wheatley DN (2003) Diffusion, perfusion and the exclusion principles in the structural and functional organisation of the living cell: reappraisal of the properties of the "ground substance 99. J Exp Biol 206:1955–1961

White RG, Badelt K, Overall RL, Vesk M (1994) Actin associated with plasmodesmata. Protoplasma 180:169–184

Wiggins PM (1990) Role of water in some biological processes. Microbiol Rev 54:432–449

Wiggins PM (1995a) High and low density water in gels. Prog Polym Sci 20:1121–1163

Wiggins PM (1995b) Micro-osmosis in gels, cells and enzymes. Cell Biochem Funct 13:165–172

Wiggins PM (1996) High and low-density water and resting, active and transformed cells. Cell Biol Int 20:429–435

Wiggins PM (2001) High and low-density intracellular water. Cell Mol Biol 47:735–744

Williamson RE (1975) Cytoplasmic streaming in *Chara*: A cell model activated by ATP and inhibited by cytochalasin B. J Cell Sci 17:655–668

Williamson RE (1985) Immobilisation of organelles and actin bundles in the cortical cytoplasm of the alga *Chara corallina* Klein ex. Willd Protoplasma 163:1–8

Winkel BSJ (2004) Metabolic channeling in plants. Annu Rev Plant Biol 55:85–107

Winter U, Meyer MIB, Kirst GO (1987) Seasonal changes of ionic concentrations in the vacuolar sap of *Chara vulgaris* L. grown in a brackish water lake. Oecologia 74:122–127

Yokota E, Muto S, Shimmen T (2000) Calcium-calmodulin suppresses the filamentous actin-binding activity of a 135-kilodalton actin-binding protein isolated from lily pollen tubes. Plant Physiol 123:645–654

Zimmermann U, Beckers F (1978) Generation of action potentials in *Chara corallina* by turgor pressure changes. Planta 138:173–179

CHAPTER 4

THE GLASSY STATE OF WATER: A 'STOP AND GO' DEVICE FOR BIOLOGICAL PROCESSES

S.E. PAGNOTTA[1] AND F. BRUNI[2,*]

[1] *Dipartimento di Medicina Sperimentale e Scienze Biochimiche, Università degli Studi di Roma "Tor Vergata", Via Montpellier, 00100 Roma, Italy*

[2] *Dipartimento di Fisica "E. Amaldi", Università degli Studi di Roma Tre Via della Vasca Navale 84, 00146 Roma, Italy*

Abstract: What is unique about the properties of intracellular water that prevent its replacement by another compound? We tackle this question by combining experimental techniques as diverse as Electron Spin Resonance, Thermally Stimulated Depolarization Current, broadband dielectric spectroscopy, and neutron diffraction to a set of samples, namely a globular enzyme, intact plant seeds, and porous silica glasses, largely differing in terms of composition and complexity. Results indicate that interfacial and intracellular water is directly involved in the formation of amorphous matrices, with glass-like structural and dynamical properties. We propose that this glassiness of water, geometrically confined by the presence of solid intracellular surfaces, is a key characteristic that has been exploited by Nature in setting up a mechanism able to match the quite different time scales of protein and solvent dynamics, namely to slow down fast solvent dynamics to make it overlap with the much slower protein turnover times in order to sustain biological functions. Additionally and equally important, the same mechanism can be used to completely stop or slow down biological processes, as a protection against extreme conditions such as low temperature or dehydration

Keywords: Water; Glassy phase; Lysozyme; Anhydrobiosis; confined water

1. INTRODUCTION

Water is one of the essential components of life and its functions are manifold. Is that all? It is taken as axiomatic that life requires water, and so strong is the belief that water is essential, that it has become a defining parameter for life – if there is no water there can be no life. But, as cleverly pointed out by a recent

* Corresponding author, *E-mail address:* bruni@fis.uniroma3.it

G. Pollack et al. (eds.), Water and the Cell, 93–112.

meeting on the molecular basis of life (Daniel et al., 2004), if all life really depends on water, we ought to be able to say why. What is unique about the properties of water that prevent its replacement by another compound? What are the biological functions of water at the molecular level, and why is it necessary? Water is implicated in many biomolecular processes, and although much effort has gone into trying to understand the ways in which water is involved in these processes, there has been much less focus on trying to identify the specific molecular characteristics of water that 'Nature' exploits, and that evolution has capitalized upon (Kuntz and Kauzmann, 1974; Rupley and Careri, 1991; Finney, 2004). It is intriguing to notice that water molecules involved in such processes, namely those water molecules in the first hydration shells of proteins or adsorbed on macromolecular structures such as biomembranes, exhibit dynamical and structural properties whose features are closely reminiscent of those of simpler glass-forming liquids and polymer systems, as shown by experiments (Singh et al., 1981; Doster et al., 1986; Doster et al., 1989; Green et al., 1994; Gregory, 1995) and computer simulations as well (Bizzarri et al., 1996; Arcangeli et al., 1998; Bizzarri et al., 2000; Peyrard, 2001; Tarek and Tobias, 2002; Bizzarri and Cannistraro, 2002). Is this glassy behavior a nifty description or a functional strategy? Here we propose that this glassiness of intracellular water, geometrically confined by the presence of solid intracellular surfaces, is a key characteristic that has been exploited by Nature in setting up a mechanism able to match the quite different time scales of protein and solvent dynamics, namely to slow down fast solvent dynamics to make it overlap with the much slower protein turnover times in order to sustain biological functions. Additionally and equally important, in a manner similar to a car's brake, the same mechanism can be used to completely stop, or slow down dramatically, biological processes when needed, for example in protection under extreme conditions such as low temperature (Walters, 2004) or dehydration.

The search for a functional connection between the glassy dynamics of interfacial water and that of the biological system and, in turn, with biological activity has been the subject of a considerable number of studies (Iben et al., 1989; Frauenfelder, 1989; Cusack and Doster, 1990; Frauenfelder et al., 1991; Diehl et al., 1997; Nienhaus et al., 1997; Wilson et al., 1997; Daniel et al., 1999; Fitter, 1999; Fenmore et al., 2004), but the question relative to the uniqueness of water's role in biological processes has not been, in our opinion, completely answered. Here we review our approach to this issue, tackled by combining experimental techniques as diverse as Electron Spin Resonance (ESR), Thermally Stimulated Depolarization Current (TSDC), broadband dielectric spectroscopy, and neutron diffraction. These techniques have been applied to a set of samples, namely a globular protein, intact plant seeds, and porous silica glasses, largely differing in terms of composition and complexity. Obviously, we are aware that the results obtained might be of validity only for a restricted array of samples, and therefore the conclusions drawn be of limited relevance. Nevertheless, intracellular water is always confined to some extent, and its description as a glassy system should hold not only for the investigated samples, but for a much wider set of proteins and organisms, irrespective of their composition.

The following discussion will start with the extreme case of anhydrobiotic organisms, where reduced mobility of intracellular components is a must to stop metabolism and assure stability over time. Adopting the above analogy between the water ability to form amorphous matrices, or glassy phases, and that of the brake system of a car, we can say that for an organism to enter in a state of anhydroboisis is like the entering of a car in a parking lot. The ability of glassy water to control the speed of enzymatic reaction will be then discussed, and finally some conclusions will be drawn in the last section.

2. STOPPING MOTION

The ability to survive dehydration shown by anhydrobiotic organisms, such as plant seeds, spores, bacteria, *Artemia* cysts etc., seems to be especially remarkable as a departure from the structural, as well as functional, dependences on the presence of intracellular water.

Anhydrobiotic organisms can lose practically all of their water in a completely reversible way. This remarkable property makes them model systems for studies on cell-associated water. Furthermore, and equally important, knowledge of the properties of intracellular water can help us to shed light to the mechanisms underlying the ability of dry biosystems to survive to dehydration.

Anhydrous biosystems include plant seeds, spores, pollen grains, bacteria, cysts of desiccated forms of organisms such as the brine shrimp *Artemia*, and nematodes. Among this wide list of organisms, a detailed investigation has been carried, by one of us, on axes of soybean seeds (*Glycine max* L.). For this desiccation tolerant system the existence of a hydration dependent glass-like transition has been observed using the spin-probe ESR technique (Bruni and Leopold, 1991): the ESR signal due to an hydrophylic spin probe inserted in the cytoplasm of intact and desiccation tolerant axes was recorded as a function of hydration and temperature. Notably, the same measurements for desiccation intolerant and heat killed soybean seed axes indicated that the transition temperature is independent of sample water content and constantly equal to 273 K. This suggests that the aqueous cytoplasm of these intolerant samples was freezable and not glassy. These measurements provided evidence that the cytoplasm of desiccation tolerant seeds is in a vitrified state, while desiccation intolerant organisms did not reveal such a glassy cell interior.

In order to identify the cytoplasm component responsible for the glassy state, pools of water molecules have been identified by means of thermally stimulated depolarization current (TSDC) method (Van Turnhout, 1987; Mascarenhas, 1987; Bucci and Fieschi, 1966). Briefly, the TSDC method consists of measuring, during controlled heating, the current generated by release of a polarized state in a dielectric sandwiched between two electrodes. Samples at a given water content are polarized by an applied DC electric field (E_p) at a temperature T_p for a time t_p. This polarization is subsequently frozen in by cooling the sample down to a temperature sufficiently low to prevent depolarization by thermal energy. The field is then switched off. The sample is short-circuited through an electrometer, and warmed up at a constant rate while the depolarization current is measured. A current peak will thus be observed at a temperature where dipolar disorientation, ionic migration or carrier release from traps is activated. In the case of dipolar orientation with a thermally activated single relaxation time, the depolarization current $I(T)$ is expressed as:

$$(1) \qquad I(T) = \frac{Q}{\tau_0} \exp\left(-\frac{E_a}{kT}\right) \exp\left[-\frac{1}{\beta\tau_0}\int_{T_0}^{T} \exp\left(-\frac{E_a}{kT}\right) dT'\right]$$

where Q is the area under the peak, τ_0 is a pre-exponential factor, E_a is the activation energy of dipolar reorientation, k is the Boltzmann's constant, β is the heating rate and T_0 is the temperature at which depolarization current starts to appear.

Examining tolerant soybean axes samples in the temperature range 100–340 K, and over water contents ranging from $h = 0.05$ to $0.30 g/g$, three relaxation mechanisms can be detected (A, B and C), as shown in Figure 1 for a sample at an hydration level $h = 0.15 g/g$. Similar spectra have been obtained over the hydration range $0.05 \leq h \leq 0.30 g/g$.

The agreement between experimental data (black symbols) and the fit obtained using Eq. 1 (solid line) suggests that the three peaks are due to dipolar re-orientation.

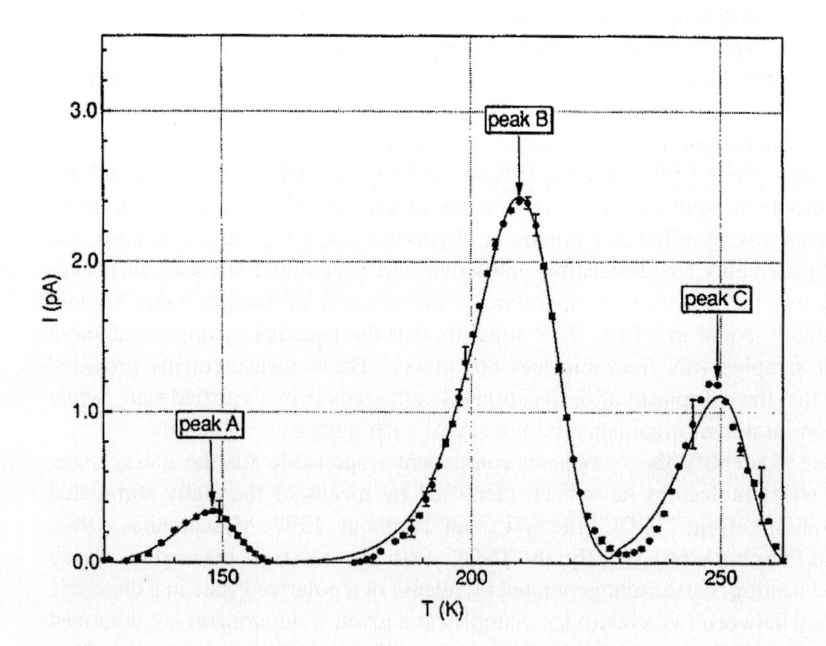

Figure 1. Typical TSDC spectrum obtained with a sample of soybean axes with $h = 0.15 g/g$. Solid line represent the fit of Eq. 1 to data

However, to exclude contributes arising from space charge relaxation of free-charge carriers cumulated at the electrode surfaces, and to get further informations on the nature of the three peaks it is necessary to study the dependance of peak temperature and amplitude on the polarizing field E_p. In fact, if we have a dipolar relaxation, the corresponding peak position should not depend on the field E_p, while its amplitude should be linearly dependent on it.

In Figure 2 such kind of test is performed on a soybean sample at $h = 0.096g/g$, with a polarization temperature $T_p = 297$ K kept constant in all the runs. While the temperature of the first and second peak (A and B, respectively) do not depend on the field, the third peak C, present in the first run, disappears from the spectrum in the following runs. In other words, temperature cycling, and five minutes isothermal temperature treatment at 297 K to polarize the sample, eliminated the relaxation process responsible for this current peak. The same behavior has been observed over the entire hydration range investigated. This important fact provide the first indication that peak C may be due to a glass-like transition, the rationale being that the glassy state is a metastable state, and the way in which a lower energy equilibrium state is reached depends solely on the thermal history of the sample. Another interesting indication arise when we compare the phase-diagram, obtained with ESR (already mentioned at the beginning of the paragraph), relating glass

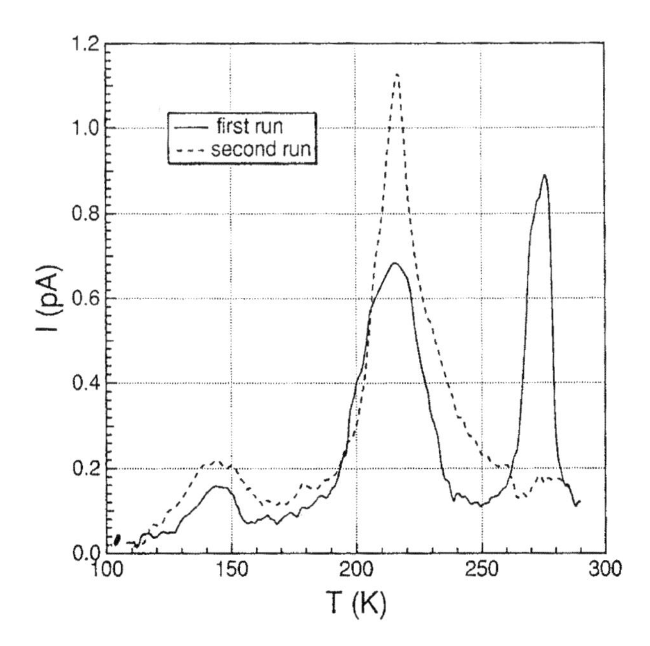

Figure 2. TSDC spectra for a soybean axes sample at $h = 0.096g/g$. Polarization temperature $T_p =$ 297 K. Polarizing field E_p was $1kV/cm$ in the first run, and $2kV/cm$ in the second run. Peak C disappears upon temperature cycling and isothermal treatment at $T = T_p$

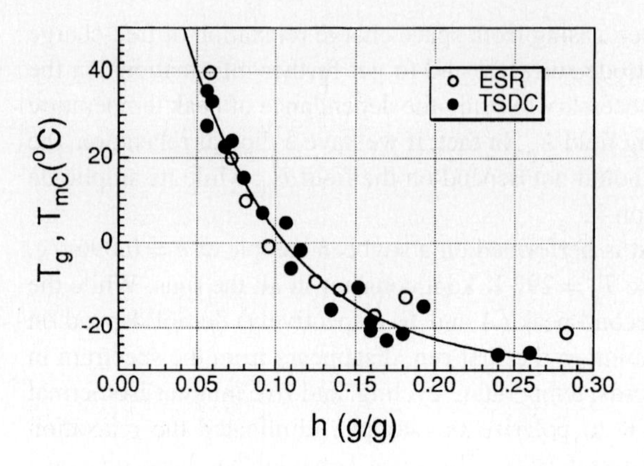

Figure 3. Peak C temperature T_m (filled symbols) and glass transition temperature T_g (open symbols) obtained with ESR, plotted as a function of water content *h* for soybean axes

transition temperature (T_g) to sample water content with the diagram showing the peak temperature of peak *C* (T_{mC}) again versus hydration *h* (Figure 3).

The good agreement between the two data sets provide further support to the assessment of the relaxation mechanism, responsible for peak *C*, as a glass transition. Actually, analogous results are observed also by studies on glass forming solutions (MacKenzie, 1977; Angell and Tucker, 1979; Takahashi and Hirsh, 1985) and on maize embryos (Williams and Leopold, 1989), using scanning calorimetry. The similarity between these studies and TSDC data make it reasonable to postulate that this current peak is due to relaxation of dipoles trapped in a glassy state.

As regards the other two peaks, analysis of peak *A* shows that it is characterized by low activation energy (in the range of reported values for the rotation of water molecule around a single hydrogen bond (Finney, 1982)) and by a small number of relaxing units. The dipolar relaxation responsible for it is attributable to reorientation of water molecules bound to other water molecules and/or polar groups on intracellular surfaces trough hydrogen binding. On the other hand, the dielectric dispersion corresponding to peak *B* has been attributed to rotation of CH_2OH groups, plasticized by water molecules.

The possibility that some of the water pools identified by TSDC may be associated with aqueous glasses in the cytoplasm of anhydrous systems, even at physiological temperatures, could be particularly relevant to anhydrous biology. Moreover, intracellular water is subjected to severe confining conditions (see for instance (Clegg and Drost-Hansen, 1991), resulting in significant changes of its structure and dynamics compared to bulk water. In oder to test the functional role of these aqueous glassy matrices, it is interesting to compare the results obtained on desiccation tolerant soybean seeds with analogous experiments performed on desiccation intolerant seeds, such as acorn (*Quercus rubra*) seeds.

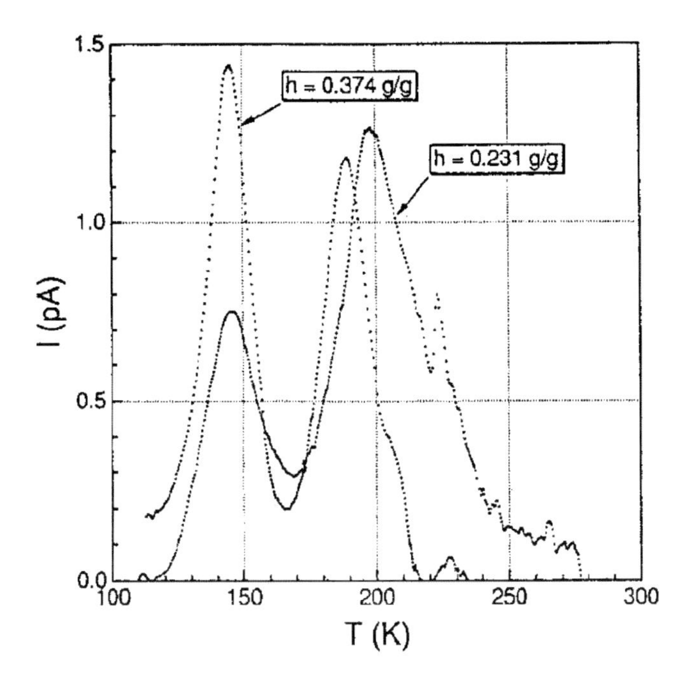

Figure 4. TSDC spectra for two samples of acorn axes at the water content indicated. Polarization temperature $T_p = 298$ K, and polarizing field $E_p = 4kV/cm$

Figure 4 shows TSDC spectra obtained for two samples of acorn seeds axes (1 and 2) at different hydration level ($h_1 = 0.374$ and $h_2 = 0.231g/g$). This kind of seed is tolerant to dehydration only up to about $h = 0.3g/g$, thus sample 1 is still germinable while sample 2 is not. What is immediately evident from these data is the absence of the high temperature C-peak, suggesting that, at the same water content, aqueous compartments of sensitive organisms are still liquids, with diffusion rates that do not preclude crystallization or chemical reactions, while aqueous compartments of dehydration insensitive organisms are in a glassy state. This important point strongly indicates the existence of cytoplasmic glassy domain as a key factor for the ability of anhydrobiotic organisms to tolerate desiccation. If water alone is responsible for the formation of a glassy cytoplasm, then the different behavior shown by tolerant and intolerant organisms is indeed puzzling, given the similar confining conditions and water content of the two kinds of organisms. Speculations as to the identity of the solutes that may be contributing to the observed glass formation focuses immediately upon the common sugars abundant in anhydrobiotic organisms. The ability of di- and oligosaccharides (Green and Angell, 1989) to protect membranes and proteins against dehydration can be thus connected with their ability to form or to induce glass formation at physiological temperatures. It has been also suggested that sugars are not the only components that participate in the intracellular glass formation, and intracellular protein might play an important role as well (Buitink et al., 2000). In general, the importance of the presence of glassy domains in the cytoplasm of biological organisms is in the fact that a glass is a liquid of high viscosity, such that it stops or slows down all chemical reactions requiring molecular diffusion (Franks, 1985). In doing so, a glassy state may assure quiescience and stability over time. Moreover, a glassy state could represent a useful mechanism to trap residual water molecules and to prevent damaging interactions between cell components (Vertucci, 1989). In addition, the resulting highly viscous phase can be readily melted upon addition of water, thus restoring the possibility for metabolic activity.

3. REDUCING SPEED

The biological relevance of the glassy state of cytoplasm in anhydrobiotic organisms, described in the previous paragraph, could be viewed as a phenomena confined only to particular systems able to survive in extreme condition like that induced by severe dehydration. Indeed this seems to be not the case. In particular, we will see how the glassy behavior of interfacial water might be of importance also for enzymatic activity for the globular protein lysozyme, and, in principle, it could be tentatively considered as a general mechanisms applicable to different protein systems to regulate their own functionality.

When we talk about the glassiness of water, independently of the systems to which it is eventually related, we can consider a static, structural phenomena or a dynamical one. In the first case we observe the three dimensional arrangement of H_2O molecules in the system, while in the second one we are interested in the dynamical transitions undergone by H_2O molecules. We can apply these considerations to interfacial water adsorbed on the protein surface and look first of all at its structure near the interface. To this end, the use of model systems for neutron diffraction experiments, can be an advantageous approach (Pertsemlidis et al., 1996).

Neutron diffraction is a powerful tool for studies on water, as neutrons can distinguish different isotopes of the same species and in particular hydrogen and deuterium (Bruni et al., 1998; Soper et al., 1998), thus allowing the extraction of site-site Radial Distribution Functions (RDF). In pure water we distinguish only two atomic sites, H and O, thus three site-site RDF, namely $g_{OO}(r)$, $g_{OH}(r)$ and $g_{HH}(r)$, can be obtained from the diffraction experiments. Porous silica glasses, such as Vycor, are good models for protein surfaces, due to the pore dimension ($\sim$40 Å), that is about the same diameter of a globular protein, and to the presence on the pore surface of oxygen and hydrogen atoms able to make hydrogen bonds with water molecules. Neutron diffraction experiments with isotopic substitution on water confined in Vycor can be performed to obtain the site-site atomic radial distribution function $g(r)$, that give us the probability, once fixed a given atom sites in the origin, to find another atom at a distance r. In Figure 5 such $g(r)$ are superimposed to the corresponding $g(r)$ for bulk water, to stress analogies and differences (Bruni et al., 1998; Soper et al., 1998). In particular, the water oxygen-oxygen radial distribution function has its second peak shifted to 4 Å as compared to 4.8 Å for bulk water; this is quite similar to what happens in bulk water when pressure is applied (Botti et al., 2004), and signals a substantial distortion of the hydrogen bond network. It is important to notice here that this reduced distance between water oxygens under confinement indicates that the density of confined water is on average larger than that of bulk water, as recently confirmed by molecular dynamics (MD) simulation studies (Merzel and Smith, 2002). As a consequence, water molecules interacting with a solid interface are subjected to a compressing action equivalent to an external pressure: this may bring the interfacial water molecules in a thermodynamic state where they remain liquid even at subzero temperatures, as experimentally observed for the first layers of H_2O around globular

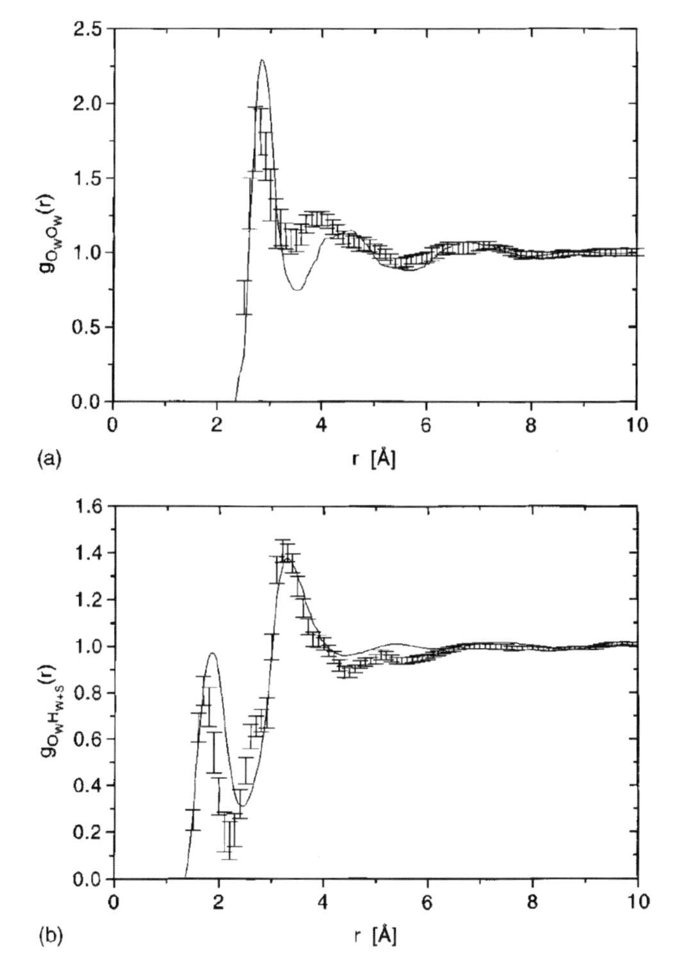

Figure 5. Calculated water-water radial distribution functions for water in Vycor, corrected for excluded volume effects, for (a) *OO*, (b) *OH*, and (c) *HH* correlations (lines with error bars) compared to the corresponding functions for bulk water (solid lines)

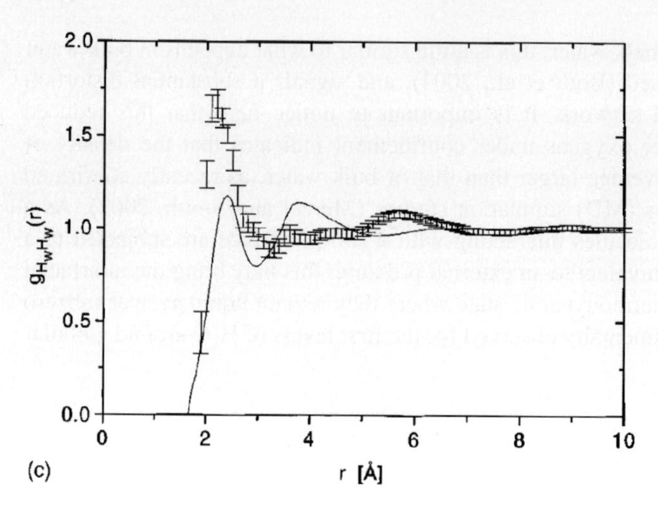

Figure 5. (Continued)

proteins (Sartor et al., 1995). The structure of the first layers of water molecules around an hydrophilic surface is therefore sensibly different from that of the bulk liquid; these water molecules can be easily supercooled and the amorphous (with respect to ice) matrix they make is often termed a glassy phase. Finally, it should be noted that the average number of hydrogen bonds per water molecule, as evaluated from the integral under the first peak of the intermolecular oxygen-hydrogen radilal distrubution function (see Figure 5) is reduced by 50% compared to bulk water. Once assessed the "structural" glassiness of water near interfaces, the following step is to look at its dynamical behavior in proteinic systems. This can be done for hydrated powders of lysozyme using broadband dielectric spectroscopy, a technique based on the measure of a complex quantity (*i.e.* admittance or impedance) as a function of angular frequency ω of a sample sandwiched between two electrodes. The measured admittance $Y_m(\omega)$ is directly related to the complex permittivity $\epsilon^*_m(\omega) = \epsilon'_m(\omega) - \imath\epsilon''_m(\omega)$ given that

$$\epsilon^*_m(\omega) = \frac{d}{\imath\omega\epsilon_o S} Y_m(\omega) \quad (2)$$

where $\imath = \sqrt{-1}$, ϵ_o is the permittivity of free space, S and d are respectively the electrode surface and gap. An example of measured complex permittivity for a sample of powdered lysozyme (water content $h = 0.26\ g/g$ and temperature $T = 270.4$ K) is depicted in Figure 6.

In general, a peak present in the permittivity spectrum (often termed as relaxation) reflects the existence of relaxing dipoles or moving charges in the sample, with a dynamics characterized by a relaxation time τ roughly corresponding to the inverse of the frequency at which the peak maximum occurs. For hydrated powders of lysozyme several studies observed the presence of a peak, ascribed to water assisted

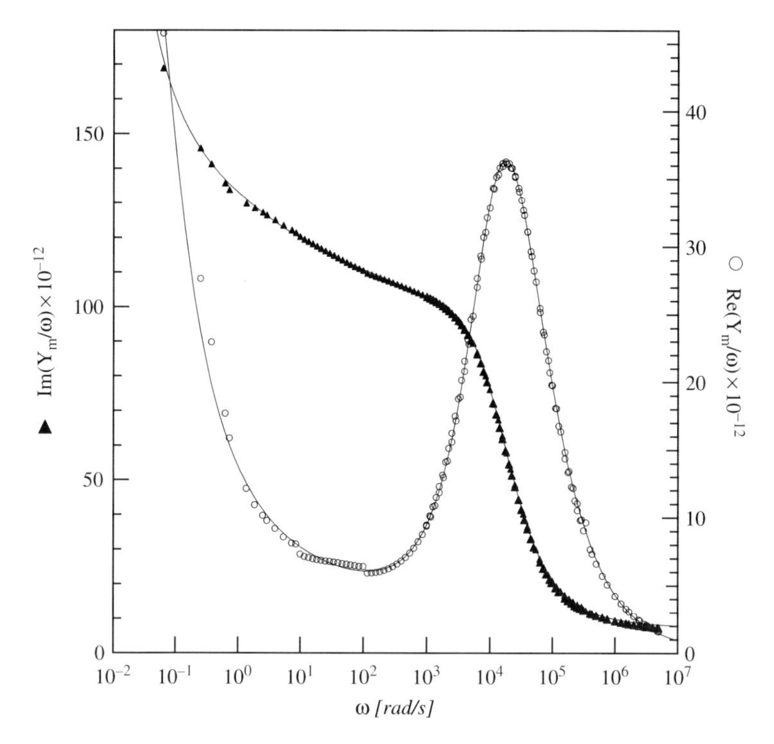

Figure 6. Angular frequency (ω) dependence of the imaginary (left axis) and real (right axis) components of the measured complex admittance divided by ω. This quantity is proportional to the complex permittivity. The dataset shown in this figure is made up by measurements performed on a lysozyme sample at $T = 261\text{K}$ and at a water content $h = 0.26g/g$. The solid lines through the data are the result of a fit procedure that takes into account the complex admittance of low frequency dispersion (LFD) and sample relaxations

proton migration over protein surface, in the frequency window between 10^3 and 10^6 Hz, at room temperature (Rupley and Careri, 1991). Measurements of the proton mobility are closely linked to the dynamics of the water molecules themselves. Early studies of proton mobility in ice (Kunst and Warman, 1980), as well as recent quantum mechanical calculations (Marx et al., 1999), showed that protons are not moving independently of the surrounding water molecules and that their dynamics is governed by solvent fluctuations.

To follow the thermodynamic of the system, looking for a possible glass transition, sample relaxation has to be followed as a function of temperature. To this end, we notice that one of the standard signature of a glassy system is the characteristic temperature dependence of the real and imaginary components of sample permittivity at constant frequency. Such dependence can be more easily seen in the $\epsilon''(T)$ data, as shown in Figure 7 for the same lysozyme sample previously described.

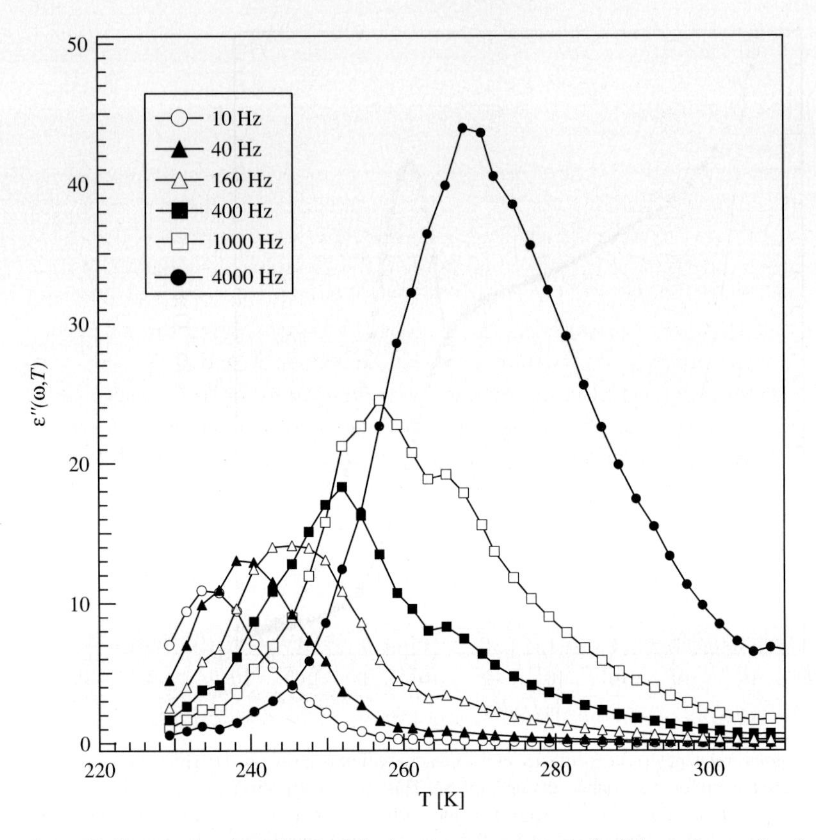

Figure 7. Temperature dependence of the imaginary component ϵ'' of the main sample relaxation. In order to show typical glasslike slowing down, $\epsilon''(\omega, T)$ is plotted at various angular frequencies ω

For a given measurement frequency, the ϵ'' data show a peak at a temperature that characteristically decreases as the frequency decreases. One can easily find that the behavior of our $\epsilon''(T)$ curves is very similar to that of proton glasses, such as random mixtures of ferroelectric and antiferroelectric crystals (Courtens, 1984, 1986). Moreover, the three canonical features of a glassy system (Ediger et al., 1996; Angell et al., 2000), such as non-Arrhenius temperature dependence of the dielectric relaxation time, non-exponential relaxation processes, along with non-ergodic behavior below a transition temperature, can be found, as we will discuss in the following.

First of all, it is known that the information about the behavior of the relaxation spectrum can be directly extracted by using a special representation for the real component of the complex permittivity $\epsilon'(\omega, T)$ in a so-called temperature-frequency plot (Kutnjak et al., 1993; Levstik et al., 1998). A detailed description of this method is given by (Kutnjak et al., 1993). Here we will only briefly summarize its essential steps to keep our focus on the results. In the first step a reduced dielectric permittivity is defined as

$$\delta = \frac{\epsilon'(\omega, T) - \epsilon_\infty}{\Delta\epsilon(T)} = \int_{z_1}^{z_2} \frac{g(z)dz}{1 + (\omega/\omega_a)^2 \exp(2z)} \tag{3}$$

The natural assumption is adopted that the distribution of relaxation times $g(\tau)$ is limited between lower and upper cutoffs z_1 and z_2, respectively, where $z_1 = \ln(\omega_a \tau)$ and ω_a is an arbitrary unit frequency. High frequency dielectric constant ϵ_∞ (found to be temperature independent) and the amplitude of the dielectric dispersion $\Delta\epsilon(T) = \epsilon_s - \epsilon_\infty$ were determined using the fit routine previously described (Bruni and Pagnotta, 2004). In the second step, by scanning δ, now playing the role of an experimentally adjustable parameter, between the values 1 and 0, ϵ' will vary between ϵ_s, the static dielectric constant, and ϵ_∞, and the filter in the second part of Eq. 3 will scan the distribution of relaxation times $g(z)$, thus probing various segments of the relaxation spectrum (Kutnjak et al., 1993). For each fixed value of δ a characteristic temperature-frequency profile was obtained in the (T, ω) plane. (T, ω) plot for a hydrated lysozyme sample is shown in Figure 8.

The characteristic bending of the data at low temperatures indicates a divergent behavior of all relaxation times in the relaxation spectrum. This implies that even at

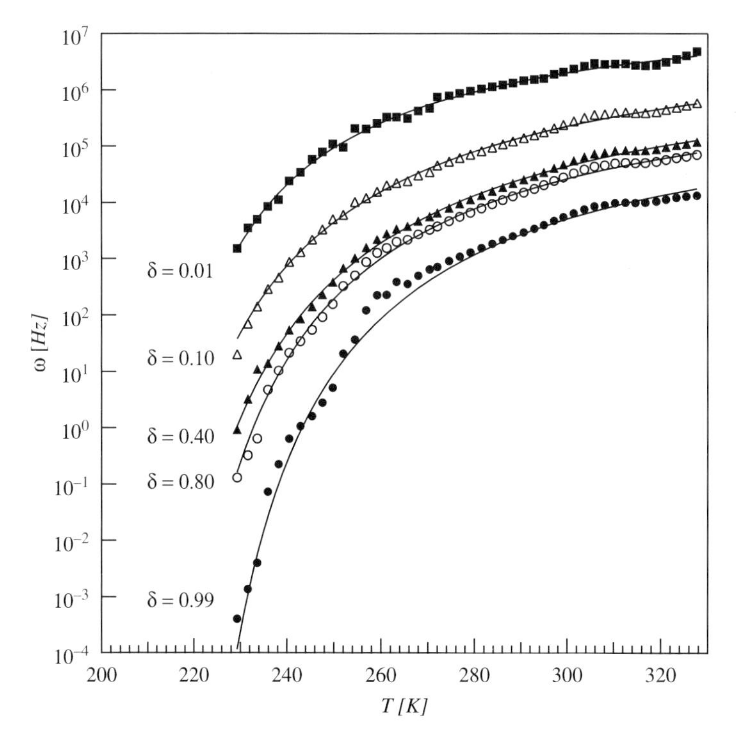

Figure 8. Temperature-frequency plots for several fixed values of the reduced dielectric constant δ (see Eq. 3). Solid lines are fits obtained with a VFT equation

small values of δ (probing the high frequency region of the relaxation spectrum) all relaxation angular frequencies in the (T, ω) plane have a non-Arrhenius temperature dependence, and can be effectively described by a Vogel-Fulcher-Tamman (VFT) equation:

$$\omega(T) = \omega_o \exp[-A/(T - T_o)] \tag{4}$$

as shown by the fit (solid lines) to each curve in Figure 8. Hence, the fit parameters ω_o, A, and T_o can be determined for each value of δ: their value, extrapolated at $\delta \to 1$ are $\omega_o = 4.96 \pm 0.48 \times 10^7 s^{-1}$, $A = 326 \pm 12$ K, and $T_o = 198.05 \pm 0.90$ K. It is important to notice that VFT temperature T_o coincides with the temperature indicating the onset of hydrogen bond network dynamics at the protein-water interface (see Figure 3 of (Bizzarri and Cannistraro, 2002)), thus confirming the role of the hydrogen bonded network on the protein surface in triggering the dynamics of the migrating protons, responsible for the measured dielectric relaxation. The interplay between these two dynamical behaviors, that of hydrogen bonds and that of interfacial water molecules, is quite interesting as it shows that the onset of a dynamical transition in proteins (we recall that the hydrogen bond network on the macromolecule surface is made by ionizable side chains, bound water, and nearby peptide backbones) mimics that of water probed through the H-bond dynamics (Bizzarri and Cannistraro, 2002; Tarek and Tobias, 2002). Another significant point regards the value of the parameter A, that now coincide with the energy ($\sim 0.028eV$) required to transfer an hydrated proton between adjacent H_2O molecules (Kuznetsov and Ulstrup, 1994; Lobaugh and Voth, 1996; Marx et al., 1999), confirming that the observed dielectric response is due to migrating protons along H-bonded water molecules adsorbed on the protein surface.

Even if noteworthy from a physical perspective, the existence of a glass transition around 200 K in the water-lysozyme system is apparently, with the exception of the economically important cryoogenic storage of tissues (Walters, 2004), useless from a biological point of view. However, remarkable biological conclusions can be reached from the previously described results simply with other few experiments and some further analysis. In fact, if we repeat the broadband dielectric measurements on the hydrated powders of lysozyme several times, changing each time the hydration level of the protein, we can build relaxation time versus hydration curves, at fixed temperature (Pizzitutti and Bruni, 2001). Among this family of curves the most interesting is obviously that one calculated at room temperature, where the protein is fully functional.

Figure 9 shows that relaxation times decreases with increasing hydration, and in this case it is possible to fit experimental data with a VFT modified function

$$\tau(h) = \tau_{oh} \exp\left(\frac{B_h}{h - h_o}\right) \tag{5}$$

where the variable temperature in Eq. 4 has been replaced with the variable h, indicating the water content of the sample. The possibility of such description of the

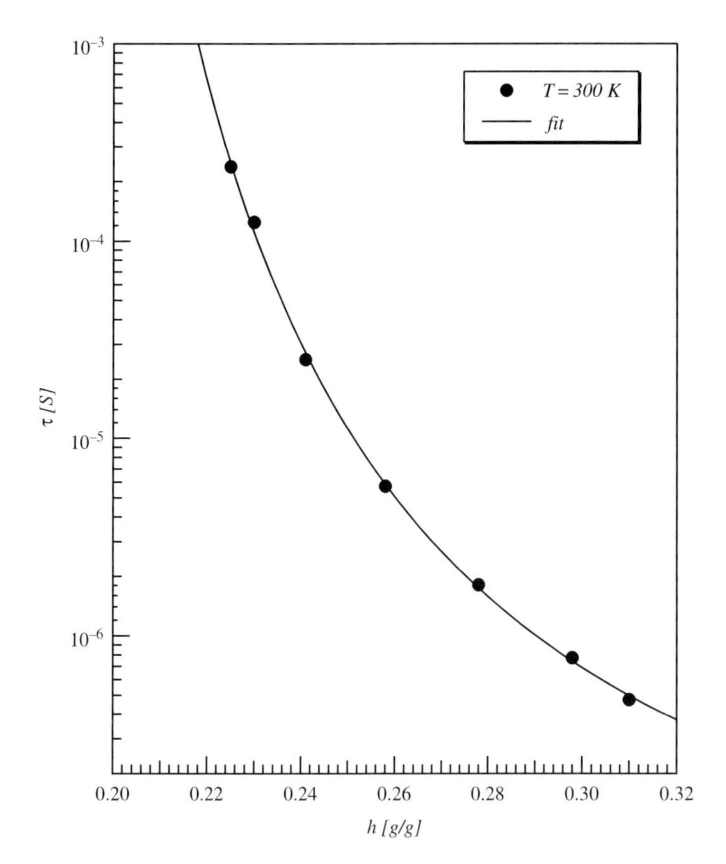

Figure 9. Hydration dependence of the dielectric relaxation time $\tau(h)$ at constant temperature (T = 300 K). Solid line is the fit with the modified VFT equation (see Eq. 5)

measured relaxation time, indicated by the goodness of the fit, is clearly unexpected, and the parameter h_o, playing here the same role of the glass transition temperature T_o, is found to be equal to 0.16 (g/g). Most importantly, the fact that the singular hydration h_o at which the relaxation time diverges is identical to the percolation threshold (see below) h_c of the dc protonic conductivity of the same system at the same temperature (Rupley and Careri, 1991) is an unpredictable finding relating a well established property of the protein with its functional glassiness. This indicates that the dynamics of the migrating protons is blocked below a critical hydration threshold that coincides with the water content required to trigger lysozyme enzymatic activity. The divergence of $\tau(h)$ for h approaching h_o suggests the absence of long range connectivity between hydrogen bonded water molecules. At hydration below h_o, the dynamics of the system of charges over the protein surface becomes non-ergodic, in analogy with the behavior of glasses below T_o. This remarkable coincidence underlines the link between the dynamic of interfacial water and biological function: enzymatic activity of lysozyme is known to be a hydration dependent phenomena, and this biological functionality is caused by the displacements of protons along hydrogen-bonded water molecules on the protein surface, with ionizable groups on the protein surface as sources/sink of migrating protons. In this contest, it is important to recall that percolation theory has been used to describe systems where spatially random events and topological disorder are of intrinsic importance (Bundle and Haulin, 1991; Stauffer, 1992). Briefly, the most interesting feature of a percolation process is the presence of a collective effect where long-range connectivity among the elements of the system suddenly appears at a critical concentration of such elements. In other words, at the percolation threshold an extended cluster of elements spans the system. In the case of hydrated lysozyme powders, and for hydrated biosamples in general, hydrophilic sites on the protein surface (or on intracellular proteins surface in the case of more complex systems) occupied by a water molecule can be considered as conducting elements, while empty sites as non-conducting ones. The percolation model has been successfully applied to several systems, and percolation thresholds for proton conduction have been observed in samples of purple membrane of *Halobacterium halobium* (Rupley et al., 1988), and anhydrobiotic organisms such as *Artemia* cysts (Bruni et al., 1989a) and in samples of maize seeds (Bruni et al., 1989b). This agrees with the accepted view that protein surfaces are quite similar in their ionizable side-chain distributions. In all cases studied so far the critical water content h_c for protonic conductivity was found to coincide with that observed for the onset of the sample-specific biological function. It is very interesting to notice that formation of spanning water networks on protein surfaces has been recently investigated for the first time by means of computer simulations on a model protein powder and on the surface of a simple protein molecule (Oleinikova et al., 2005; Smolin et al., 2005). The authors of these studies confirmed that formation of an infinite water network in the protein powder occurs as a two dimensional percolation transition at a critical hydration level, which is close to the values observed experimentally. The presence of a percolative transition indicates the occurrence of long range connectivity among water molecules in all the systems investigated. Noteworthy, theoretical work by Lemke and Campbell (Lemke and Campbell, 1996) and a recent study by our group (Pagnotta et al., 2005) has shown that it is possible to look at the glass transition in terms of a percolative transition.

How the presence of a glassy or percolative transition could be relevant to regulate enzymatic activity? The regulation of protein activity is often accomplished through the control of equilibrium among allosteric conformations. Differences in binding affinities of small molecules to specific regulatory sites among these conformations modulate this equilibrium. Looking at Fig. 9, we notice that the relaxation time of protons moving along hydrogen bonded water molecules on the lysozyme surface goes from $0.1\mu s$ to $1ms$ by reducing the number of water molecules from 0.32 to 0.20 g/g. The glassiness of migrating protons (slaved to that of the interfacial water molecules), along the hydrogen bond network at the protein surface, might have been envisaged to couple the intrinsically fast proton transfer process (occurring in the picosecond time scale (Drukker et al., 1998; Smedarchina et al., 2000)) with the much slower process governing enzymatic activity. In other words, the characteristic slowing down of water molecules dynamics in a glassy state can be considered as a "regulatory device" for lysozyme activity.

4. CONCLUSIONS

The present work has used only a small fraction of the data that already exist in the literature. Moreover, we have restricted our discussion to a very limited number of cases, but nevertheless, we feel we have reasonably tested our initial hypothesis and this allows us to draw conclusions that might be of general validity.We have shown that water is directly involved, either by itself or together with small intracellular solutes such as sugars and proteins, in the formation of an intracellular glassy matrix. This glassy matrix is essential to stop or to dramatically slow down metabolic reactions and this could be the mechanism that anhydrobiotic organisms have adopted to remain viable during periods of almost complete dehydration. With the reasonable assumption that all intracellular water is near, and therefore perturbed respect to bulk water, solid surfaces such as protein and macrostructures, we have looked at the structure and at the dynamics of these 'confined' water molecules. The microscopic structure of confined water molecules is sensibly different from that of the bulk liquid, and the dynamic of confined water molecules shows all the features of a glassy system.

At this stage we have assumed that hydration, instead of temperature, should be taken as the natural variable triggering catalytic processes as the number of water molecules available for binding to a single protein in intracellular solution can be varied by the presence of several natural solutes (Colombo et al., 1992; Bulone et al., 1993). In particular, we found that the presence of a glassy water matrix at the protein interface, can provide the essential reduction of speed required to match the intrinsically fast proton transfer with the much slower time scale of lysozyme activity. This mechanism, quite similar to a car's brake, could also be useful to proteins whose catalityic activity is not sustained by a proton transfer process. The standard view of the catalytic properties of enzymes focuses on the binding energy differences between the ground state and the transition state arising from arrangements of residues in the active site. There is an alternative view, however, that suggests that protein dynamics might play a role in catalysis (Kohen et al., 1999): given the interplay between solvent and protein dynamics (Pèrez et al., 1999; Caliskan et al., 2004; Fenmore, (2004)), the presence of a glassy matrix can therefore represent a general mechanism to modulate biological functionality. Thus, if biomolecular processes require a framework with an inbuilt ability to allow a modulation of time scales, the water molecule, and in particular its ability to form a glassy matrix, appears to be in pole position to do the job effectively. Are there any other suitable molecules?

REFERENCES

Angell CA, Tucker JC (1979) Heat capacity changes in glass-forming aqueous solutions and the glass transition in vitreous water. J Phys Chem 84:268–272

Angell CA, Ngai KL, McKenna GB, McMillan PF, Martin SW (2000) Relaxation in glassforming liquids and amorphous solids. J Appl Phys 88:3113–3157

Arcangeli C, Bizzarri AR, Cannistraro S (1998) Role of interfacial water in the molecular dynamics-simulated dynamical transition of plastocyanin. Chem Phys Letters 291:7–14

Bizzarri AR, Rocchi C, Cannistraro S (1996) Origin of the anomalous diffusion observed by MD simulation at the protein-water interface. Chem Phys Letters 263:559–566

Bizzarri AR, Paciaroni A, Cannistraro S (2000) Glasslike dynamical behavior of the plastocyanin hydration water. Phys Rev E 62:3991–3999

Bizzarri AR, Cannistraro S (2002) Molecular Dynamics of Water at the Protein-Solvent Interface. J Phys Chem B 106:6617–6633

Botti A, Bruni F, Imberti S, Ricci MA, Soper AK (2004) Ions in water: The microscopic structure of concentrated NaOH solutions. J Chem Phys 120:10154–10162

Bruni F, Careri G, Clegg JS (1989) Dielectric properties of Artemia cysts at low water contents. Evidence for a percolative transition. Biophys J 55:331–338

Bruni F, Careri G, Leopold AC (1989) Critical exponents of protonic percolation in maize seeds. Phys Rev A 40:2803–2805

Bruni F, Leopold AC (1991) Glass transitions in soybean seed. Relevance to anhydrous biology. Plant Physiol 96:660–663

Bruni F, Ricci MA, Soper AK (1998) Water confined in Vycor glass. I. A neutron diffraction study. J Chem Phys 109:1478–1485

Bruni F, Pagnotta SE (2004) Dielectric investigation of the temperature dependence of the dynamics of a hydrated protein. Phys Chem Chem Phys 6:1912–1919

Bucci C, Fieschi R (1966) Ionic thermocurrents in dielectrics. Phys Rev 148:816–823

Buitink J, van den Dries I, Hoeckstra FA, Alberda M, Hemminga MA (2000) High critical temperature above Tg may contribute to the stability of biological systems. Biophys J 79:1119–1128

Bulone D, San Biagio PL, Palma-Vittorelli MB, Palma MU (1993) The role of water in hemoglobin function and stability. Science 259:1335–1336

Bundle A, Haulin S (eds) (1991) Fractals and Disordered Systems. Springer, Berlin

Caliskan G, Mechtani D, Roh JH, Kisliuk A, Sokolov AP, Azzam S et al (2004) Protein and solvent dynamics: How strongly are they coupled? J Chem Phys 121:1978–1983

Clegg JS, Drost-Hansen W (1991) On the biochemistry and cell physiology of water. In: Hochachka A, Mommsen B (eds), Biochemistry and Molecular Biology of Fishes. Elsevier, pp 1–23

Colombo MF, Rau DC, Parsegian VA (1992) Protein solvation in allosteric regulation: A water effect on hemoglobin. Science 256:655–659

Courtens E (1984) Vogel-Fulcher scaling of the susceptibility in a mixed-crystal proton glass. Phys Rev Lett 52:69–72

Courtens E (1986) Scaling dielectric data on $Rb_{1-x}(NH_4)H_2PO_4$ structural glasses and their deuterated isomorphs. Phys Rev B 33:2975–2978

Cusack S, Doster W (1990) Temperature dependence of the low frequency dynamics of myoglobin. Measurement of the vibrational frequency distribution by inelastic neutron scattering. Biophys J 58:243–251

Daniel RM, Finney JL, Rèat V, Dunn R, Ferrand M, Smith JC (1999) Enzyme dynamics and activity: Time-scale dependence of dynamical transitions in glutamate dehydrogenase solution. Biophys J 77:2184–2190

Daniel RM, Finney JL, Stoneham M (2004) The molecular basis of life. Is life possible without water? Phil trans R Soc Lond B **359**:1143

Doster W, Bachleitner A, Dunau R, Hiebl M, Lüscher E (1986) Thermal properties of water in myoglobin crystals and solutions at subzero temperatures. Biophys J 50:213–219

Doster W, Cusack S, Petry W (1989) Dynamical transition of myoglobin revealed by inelastic neutron scattering. Nature 337:754–756

Diehl M, Doster W, Petry W, Schober H (1997) Water-coupled low-frequency modes of myoglobin and lysozyme observed by inelastic neutron scattering. Biophys J 73:2726–2732

Drukker K, de Leeuw SW, Hammes-Schiffer S (1998) Proton transport along water chains in an electric field. J Chem Phys 108:6799–6808

Ediger MD, Angell CA, Nagel SR (1996) Supercooled liquids and glasses. J Phys Chem 100:13200–13212

Fenmore PW, Frauenfelder H, McMahon BH, Young RD (2004) Bulk-solvent and hydration-shell fluctuations, similar to α- and β-fluctuations in glasses, control protein motions and functions. Proc Natl Acad Sci USA 101:14408–14413

Finney JL (1982) Solvent effects in biomolecular processes. In: Franks F, Mathias SF (eds), Biophysics of Water. John Wiley & Sons, New York, pp 55–58

Finney JL (2004) Water? What's so special about it? Phil Trans R Soc Lond B **359**:1145–1165

Fitter J (1999) The temperature dependence of internal molecular motions in hydrated and dry α Amylase: the role of hydration water in the dynamical transition of proteins. Biophys J 76:1034–1042

Franks F (1985) Biophysics and Biochemistry at Low Temperatures. Cambridge University Press, Cambridge

Frauenfelder H (1989) New looks at protein motions. Nature 338:623–624

Frauenfelder H, Sligar SG, Wolynes PG (1991) The energy landscapes and motions of proteins. Science 254:1598–1603

Green JL, Angell CA (1989) Phase relations and vitrification in saccharide-water solutions and the trehalose anomaly. J Phys Chem 93:2880–2882

Green JL, Fan J, Angell CA (1994) The protein-glass analogy: Some insights from homopeptide comparison. J Phys Chem 98:13780–13790

Gregory RB (1995) Protein hydration and glass transition behavior. In: Gregory RB (ed), Protein-Solvent Interactions. Marcel Dekker, Inc., New York, pp 191–264

Iben IET, Braunstein D, Doster W, Frauenfelder H, Hong MK, Johnson JB et al (1989) Glassy behavior of proteins. Phys Rev Lett 62:1916–1919

Kohen A, Cannio R, Bartolucci S, Klinman JP (1999) Enzyme dynamics and hydrogen tunneling in a thermophilic alcohol dehydrogenase. Nature 399:496–499

Kuntz ID Jr, Kauzmann W (1974) Hydration of proteins and polypeptides. Adv Protein Chem 28:239–345

Kunst M, Warman JM (1980) Proton mobility in ice. Nature 288:465–467

Kutnjak Z, Filipic C, Levstik A, Pirc R (1993) Glassy dynamics of $Rb_{0.4}(ND_4)_{0.6}OD_2PO_4$. Phys Rev Lett 70:4015–4018

Kuznetsov AM, Ulstrup J (1994) Dynamics of low-barrier proton transfer in polar solvents and protein media. Chem Phys 188:131–141

Lemke N, Campbell IA (1996) Random walks in a closed space. Physica A 230:554–562

Levstik A, Kutnjak Z, Filipic C, Pirc R Phys. Rev. B (1998), 57:11204.

Lobaugh J, Voth GA (1996) The quantum dynamics of an excess proton in water. J Chem Phys 104:2056–2069

MacKenzie AP (1977) Non-equilibrium freezing behavior of aqueous systems. Phil Trans R Soc London B **278**:167–189

Marx D, Tuckerman ME, Hutter J, Parrinello M (1999) The nature of the hydrated excess proton in water. Nature 397:601–604

Mascarenhas S (1987) Bioelectrets: Electrets in biomaterials and biopolymers. In: Sessler GM (ed), Electrets. Springer-Verlag, Berlin, pp 321–346

Merzel F, Smith JC (2002) Is the first hydration shell of lysozyme of higher density than bulk water? Proc Natl Acad Sci USA 99:5378–5383

Nienhaus GU, Müller JD, MacMahon BH, Frauenfelder H (1997) Exploring the conformational energy landscape of proteins. Physica D 107:297–311

Oleinikova A, Smolin N, Brovchenko I, Geiger A, Winter R (2005) Formation of spanning water networks on protein surfaces via 2D percolation transition. J Phys Chem B 109:1988–1998

Pagnotta SE, Gargana R, Bruni F, Bocedi A (2005) Glassy dynamics of a percolative water-protein system. Phys Rev E 71:031506

Pèrez J, Zanotti J-M, Durand D (1999) Evolution of internal dynamics of two globular proteins from dry powder to solution. Biophys J 77:454–469

Pertsemlidis A, Saxena AM, Soper AK, Head-Gordon T, Glaeser M (1996) Direct evidence for modified nsolvent structure within the hydration shell of a hydrophobic amino acid. Proc Natl Acad Sci USA 93:10769–10774

Peyrard M (2001) Glass transition in protein hydration water. Phys Rev E 64:011109

Pizzitutti F, Bruni F (2001) Glassy dynamics and enzymatic activity of lysozyme. Phys Rev E 64:052905

Rupley JA, Siemankowski L, Careri G, Bruni F (1988) Two-dimensional protonic percolation on lightly hydrated purple membrane. Proc Natl Acad Sci USA 85:9022–9025

Rupley JA, Careri G (1991) Protein hydration and function. Adv Protein Chem 41:37–172

Sartor G, Hallbrucker A, Mayer E (1995) Characterizing the secondary hydration shell on hydrated myoglobin, hemoglobin, and lysozyme powders by its vitrification behavior on cooling and its calorimetric glass-liquid transition and crystallization behavior on reheating. Biophys J 69:2679

Singh GP, Parak F, Hunklinger S, Dransfeld K (1981) Role of adsorbed water in the dynamics of metmyoglobin. Phys Rev Lett 47:685–688

Smedarchina Z, Slebrand W, Fernández-Ramos A (2000) A direct-dynamics study of proton transfer through water bridges in guanine and 7-azaindole. J Chem Phys 112:566–573

Smolin N, Oleinikova A, Brovchenko I, Geiger A, Winter R (2005) Properties of spanning water networks at protein surfaces. J Phys Chem B 109:10995–11005

Soper AK, Bruni F, Ricci MA (1998) Water confined in Vycor glass. II. Excluded volume effects on the radial distribution functions. J Chem Phys 109:1486–1494

Stauffer D, Aharony A (1992) Introduction to Percolation Theory, Taylor and Francis, London.

Tarek M, Tobias DJ (2002) Role of protein-water hydrogen bond dynamics in the protein dynamical transition. Phys Rev Lett 88:138101

Takahashi T, Hirsh A (1985) Calorimetric studies of the state of water in deeply frozen human monocytes. Biophys J 47:373–380

Van Turnhout J (1987) Thermally stimulated discharge of electrets. In: Sessler GN (ed), Electrets. Springer-Verlag, Berlin, pp 81–215

Vertucci CW (1989) Effects of cooling rate on seeds exposed to liquid nitrogen temperatures. Plant Physiol 90:1478–1485

Walters C (2004) Temperature dependency of molecular mobility in preserved seeds. Biophys J 86:1253–1258

Williams RJ, Leopold AC (1989) The glassy state in corn embryos. Plant Physiol 89:977–981

Wilson G, Hecht L, Barron LD (1997) Evidence for a new cooperative transition in native lysozyme from temperature-dependent Raman optical activity. J Phys Chem B 101:694–698

CHAPTER 5

INFORMATION EXCHANGE WITHIN INTRACELLULAR WATER

MARTIN F. CHAPLIN*
Department of Applied Science, London South Bank University, Borough Road, London SE1 0AA, UK

Abstract: A linkage between intracellular phenomena, involving the structuring of water, is described which associates the polarised multilayer theory with gel sol transitions. Intracellular K^+ ions are revealed to form ion pairs with acid rich domains on static proteins, particularly F-actin. Such structures then create low density water clustering by a cooperative process that is able to influence other sites and so transfer information within the cell

1. INTRODUCTION

We are rapidly assembling much information concerning the structure and biological rationale for the constituents of cells, particularly at a molecular level. However, there are many puzzles still to be solved, particularly concerning their holistic operation. At least partially to blame for our lack of understanding is the widespread ignorance concerning the role that water plays, together with the concentrated intracellular environments which are very different from the dilute solutions often used in research. It is clear to all that much useful biochemistry can and has been discovered using dilute preparations from homogenised dead cells. On the other hand, living cells are very different containing more concentrated solutes, more organised protein, more surface, more phases and much smaller aqueous compartments. They are also much more difficult to study. Although it clearly should not come as a surprise to find that living cells possess characteristics that

---

* Corresponding author. Tel: +44-207-815-7970; fax: +44-207-815-7699; E-mail address: martin.chaplin@lsbu.ac.uk.

G. Pollack et al. (eds.), Water and the Cell, 113–123.

are very much more than the sum of their parts, this is often seemingly ignored and in vitro experiments may well sometimes mislead.

Water is the one key material that has often been ignored, when looking at the operation of individual molecules and metabolic processes. Although researchers are starting to acknowledge its significance in particular reactions and processes, water's more general importance is less well recognized. Water's cellular significance becomes clearer when explanations for some puzzles are sought, such as 'How are potassium ions able to maintain a high concentration inside cells whereas sodium ions are found mainly outside?' and 'How do cells remain functional even when large holes are made in their surface membranes?'

These important questions stir up a hotbed of debate amongst cell physiologists. Some still hold to intracellular water being little different from extracellular water and its function merely consigned to that of 'space-filler'. This view seems to be the one most promoted, often by default, in current textbooks. This idea relies on water mostly acting as an uncomplicated environment for the cellular processes, which are determined by the structure and activity of the macromolecules alone. Uncritical support for this view of cells as bags of 'ordinary' water comes from the mechanism originally proposed for the ionic partitioning of K^+ and Na^+, with the cell membrane potential and K^+ ion membrane porosity being solely responsible for the high intracellular K^+ ion concentrations and ATP-driven ion pumps responsible for the low intracellular Na^+ concentrations. The original, and still often quoted, source for this view is Conway (1957). However, examination of his data shows much higher intracellular K^+ ion concentrations than can be explained by this mechanism alone. In answer to the second question above, the introduction of large holes in cell membranes does not initiate the rapid ionic exchange down the concentration gradients as expected between the extracellular and intracellular environments, even if membrane self-repair is not a possible mechanism (Pollack, 2003). For these and other reasons, described later, it is no longer generally accepted that cells can be treated as simply a bag of water, full of molecules, where the water simply acts as an inert medium (Cameron, 1997). Explanation of these and other natural intracellular phenomena involves the strange properties of water.

There are different views as to how the water inside the cells affects cellular function. Ling (2003) has proposed for many years that water forms polarised multi-layers against extended protein surfaces. There is much experimental support for the foundations of this theory but little, if any, experimental support for the required structural changes in the proteins or the involvement of extended protein surfaces as proposed. Pollack (2001) proposes that the water is involved in intracellular changes between sol and gel states. This is an interesting and useful idea but in search of a clear molecular mechanism.

Cell water has been found to possess reduced density consequent upon greater hydrogen bonding (Clegg, 1984) and this structuring changes with the metabolic state of the cell (Hazlewood, 2001); low density water (LDW) predominating in the resting cell converting to higher density water (HDW) in the active cell (Wiggins, 1996).

In this paper, we propose that the water actively changes its molar volume, due to its hydrogen bonded structuring. This enables diverse intracellular processes, in a manner compatible with the basic ideas of both Ling (2001), by forming polarised water layers, and Pollack (2001), by forming gel-like less mobile aqueous environments. In brief, the explanation involves changes in the density and clustering of intracellular water modulated by the mobility of key proteins, which in turn are controlled by the energy status and ionic content of the cell.

2. THE TWO-STATE NATURE OF WATER

Water possesses many properties that seem somewhat anomalous when compared to other liquids (Chaplin, 2005). Some of these, such as its high melting and boiling points can be simply explained as due to water's hydrogen bonded clustering but explanations for other properties are not so straightforward. Over the last ten years broad ranging evidence has accumulated concerning a two-state structuring

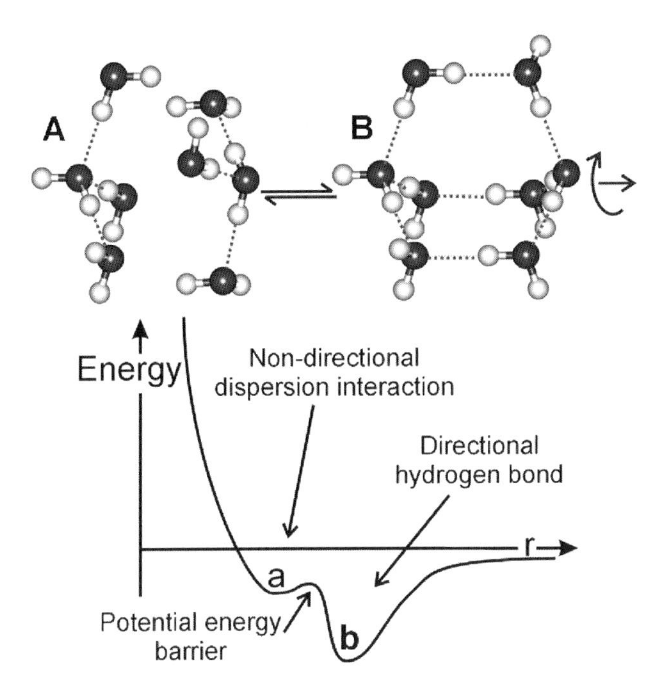

Figure 1. Two-state water clustering. The competition between maximizing van-der Waals interactions (**A**, yielding higher orientation entropy, higher density and individually weaker but more numerous water-water binding energies; high density water) and maximizing hydrogen bonding (**B**, yielding more ordered structuring, lower density and fewer but stronger water-water binding energies; low density water) is finely balanced, easily shifted with changing physical conditions, solutes and surfaces. The potential energy barrier between these states ensures that water molecules prefer either structure **A** or **B** with little time spent dwelling in intermediate structures. An individual water molecule may be in state **A** with respect to some neighbours whilst being in state **B** with respect to others (e.g., as in ice-VII). The shallow minimum (**a**), due to non-bonded interactions, lies up to 20% inside the deeper minimum (**b**) due to hydrogen bonding, even allowing for a 15% closer approach of individual hydrogen-bonded water molecules. In spatial terms, minimum (**a**) is far more extensive as the hydrogen-bonded minimum (**b**) is restricted in its geometry, being highly directional

within liquid water, which enables an uncomplicated explanation of many of the remaining anomalies (Robinson et al., 1999; Bartell, 1997; Compolat et al., 1998; Langford et al., 2001; Wiggins, 2001; Cho et al., 2002; Raichlin et al., 2004; Chaplin, 2000). This theory involves the presence of liquid aqueous environments with a higher molar volume and a similar density to that of solid ice. The water in such clusters flicker between partners as their hydrogen bonds constantly make and break. Observed over a long time scale, such clusters appear as favoured arrangements. These low-density water clusters lack long-range order and do not consist of ice-like crystals. They do contain water linked by hydrogen bonds in an open fully four-coordinated tetrahedral arrangement. At the smallest scale the water may be thought of as an equilibrium between two water tetramers (Figure 1); (A) held closely by non-bonded interactions forming a more dense structure and (B) held further away and linked by hydrogen bonds forming a less dense structure. There is little difference in energy between the structures so that the equilibrium is easily affected by the presence of solutes and surfaces. Both increased temperature and pressure shift the equilibrium to the left.

Although the natural structuring in ambient water is balanced in its clustering, given the right conditions the low density clusters can grow to form larger non-crystalline low-density clusters. Such clustering is based on dodecahedral water clusters, similar to those found in the crystalline clathrate hydrates. Under ideal conditions these may form extensive icosahedral $(H_2O)_{280}$ structures (Chaplin, 2000). The presence of ions with low surface charge density (e.g., K^+), surfaces and kosmotropic solutes all tend to increase the low-density nature of the intracellular aqueous environment (Wiggins, 2002).

3. WATER IN INTRACELLULAR SOLUTIONS

The different characteristics of the intracellular and extracellular environments manifest themselves particularly in terms of restricted diffusion inside cells and a high intracellular concentration of solutes that further promote the low-density clustering of water. This tendency towards a low density structuring is reinforced by the confined space within the cell stretching the hydrogen-bonded water. Additionally, the extensive surface effects of the membranes (e.g., liver cells contain ~100,000 μm^2 membrane surface area) help create the tendency towards low-density water inside cells as their lipids contain mainly hydrophilic kosmotropic head groups with structures that encourage this organization for the associated interfacial water.

The difference in concentration between intracellular and extracellular ions is particularly apparent between sodium (Na^+; intracellular, ~12 mM; extracellular, ~142 mM) and potassium (K^+; intracellular, ~140 mM; extracellular, ~4 mM) ions. The interactions between water and Na^+ ions are stronger than those between water molecules, which in turn are stronger than those between water and K^+ ions (Tongraar and Rode, 2004); all being explained by the differences in surface charge density; that of the smaller Na^+ ions being nearly twice that of K^+ ions. Thus

Na^+ ions are capable of breaking water's hydrogen bonds and impose its own ordered hydration water with raised density but K^+ ions are not. Entry of Na^+ ions into a more-ordered low density aqueous environment is energetically less favoured due to the consequential and additional water-water hydrogen bond disruption, only partly compensated by increased entropy. K^+ ions have positive entropy of hydration in bulk water due to their surrounding water molecules possessing increased freedom of movement. However this is also why K^+ ions to partition into low density aqueous environments (Wiggins, 2002), so releasing this entropic energetic cost. Ca^{2+} ions (intracellular, ~0.1 μM; extracellular, ~2.5 mM, with a surface charge density over twice that of Na^+ ions) have even stronger destructive effects on any low-density hydrogen bonding, than Na^+ ions. Other studies confirm this preference of K^+ ions towards, and Na^+ ions away from, low-density water (Collins, 1995). The ions partition according to their preferred aqueous environment; in particular, the K^+ ions are preferred within the intracellular environment and naturally accumulate inside the cells at the expense of Na^+ ions. This process will occur simply as a result of the water structuring and the machinations of the membrane ion-pumps and/or membrane potential are not required, although they speed the process.

It is worth noting that the cellular membrane ion-pumps cannot produce the large differences in ionic composition observed in the absence of other mechanisms (Conway, 1957), simply as the (ATP) energy required far exceeds the energy that is available to the cell (Ling, 1962; Ling, 1997; Hazlewood, 2001). Also, in contrast to that written in several undergraduate textbooks, many studies show that cells do not need an intact membrane or active energy (i.e., ATP) production to maintain their concentration gradients (Pollack, 2001; Ling, 2001). Ion pumps must thus be present for additional, perhaps fail-safe, purposes such as speeding up, or helping to prime the partition process, after metabolically linked changes in ionic concentration.

4. PROTEIN EFFECTS ON WATER STRUCTURING

The degree of density lowering of the intracellular water is determined by the solutes, their concentration and the state of motion of intracellular protein; mobile proteins creating more disorder in the clustering compared with more static proteins. Water has conflicting effects in the mixed environments around proteins due to the variety of amino acids making up their surfaces. Weak hydrogen bonding between and around the protein and surface water molecules allows greater protein flexibility. Strong hydrogen bonding endows the protein with greater stability and solubility. There is generally an ordered structure in the closest water molecules surrounding the protein, with both hydrophobic clathrate-like and hydrogen bonded water molecules each helping the other to optimise water's hydrogen-bonding network. Protein carboxylate groups are generally surrounded by strongly hydrogen-bonded water whereas the water surrounding the basic groups, arginine, histidine and lysine tends towards a more open clathrate structuring. The formation of partial clathrate cages over hydrophobic areas maximizes the non-bonded interactions without loss of hydrogen bonds. Carboxylate groups, however, usually only fit a collapsed water structure (see below) creating a reactive fluid zone. The diffusional rotation of the proteins will cause changes in the water structuring outside this closest hydration shell. At the breaking surface, hydrogen bonds are ruptured, creating a zone of higher density water. Protein diffusional rotation thus creates a surrounding high-density water zone with many broken hydrogen bonds (Halle and Davidovic, 2003).

Protein's two acidic amino acids, aspartic and glutamic acid, possess two oxygen atoms in their side chain carboxylate groups that are situated closer (~2.23 Å) than occurs between water molecules in bulk liquid water (~2.82 Å). Aqueous hydrogen bonding to these carboxylate oxygen atoms normally causes surrounding high-density water clustering due to the closeness of the held water molecules. Such hydrogen bonding induces increased negative charge on these oxygen atoms leading to a reduction in the acidity of the carboxylic acids (i.e., their pK_a is raised). If, however, the charges on the carboxylate oxygen groups are reduced, by an imposed negative electric field such as the formation of a surrounding clathrate cage (see later), the acidity is increased (i.e., their pK_a is reduced). It is found that Na^+ ions prefer binding to the weaker carboxylic acids whereas K^+ ions prefer the stronger acids (Ling, 2001).

Na^+ and K^+ ions also behave differently when close to the carboxylate groups; K^+ ions have a preference for forming ion pairs, where there is direct contact between the two ions, whereas Na^+ ions form solvent separated pairings (Fournier et al., 1998). This is due to the Na^+ ions holding on to their water strongly such that the carboxylate groups cannot compete to replace them. This solvent separated pairing is ideally situated for forming strong hydrogen bonds to the carboxylate oxygen atoms and so reducing the acidity of the carboxylate groups (Collins, 1997). The K^+ ions prefer to be within a clathrate water cage and this preference both reinforces its direct ion pairing to the carboxylate group and discourages aqueous hydrogen bonding to the associated carboxylate groups, so increasing the acidity of the carboxylate groups. Also confirming this, it has been experimentally proven that Na^+ ions cannot take part in dodecahedral clathrate structuring whereas K^+ ions prefer this environment (Khan, 2004).

Evidence for the association of K^+ ions with proteins' aspartate and glutamate groups has been presented elsewhere (Ling, 2001) where it is shown experimentally that (1) there is low intracellular electrical conductance, (2) intracellular K^+ ions possess strongly reduced mobility (3) there is a one to one stoichiometric absorption of K^+ ions to the carboxylate groups and (4) the K^+ ion absorption sites are identified as the aspartate and glutamate side chains of the intracellular proteins.

5. PROTEIN MOBILITY CONTROLS WATER STRUCTURING

Actin is a highly conserved and widespread eukaryotic protein (42–43 kDa) responsible for many functions in cells. Non-muscle cells contain actin in amounts 5–10% of all protein, whereas muscle cells contain about 20%. Actin is converted between

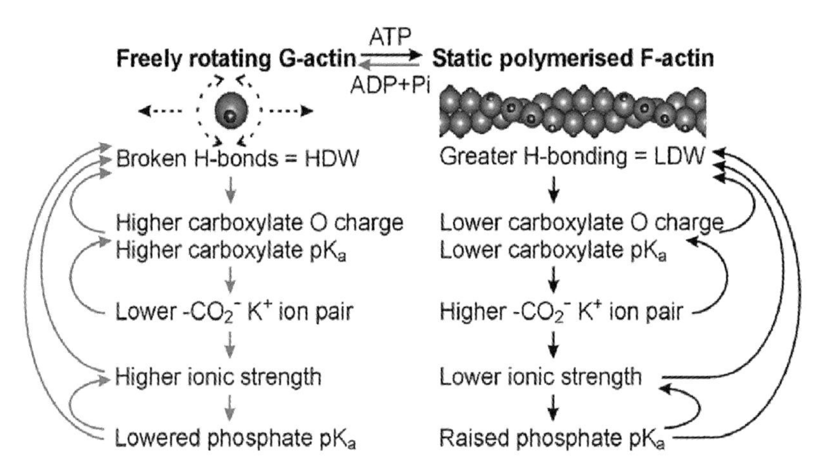

Figure 2. Actin converts between monomer (G-actin) and filament (F-actin). The N-terminal acidic cluster is shown projecting from the protein in the cartoons. Free monomer exhibits both translational and rotational diffusion, which gives rise to broken hydrogen-bonding in its surroundings (HDW). Cooperative events then increase this disorder. F-actin is surrounded by less mobile water due to its static surface and gives rise to cooperative events, detailed on the right, which increase the low density nature of its surrounding water (LDW). The presence of competing sodium ions tends to produce solvent-separated ion carboxylate – Na^+ ion pairs, which increase the charges on the carboxylate oxygen atoms so leading to further H-bond destruction, as shown on the left

a freely translating and rotating molecule (G-actin; about 4–6 nm diameter) and a static right-handed double helical protein filament (F-actin; up to several microns in length) by ATP (Figure 2); a process involving the conversion of an α-helix to a β-turn in one of its structural domains (Otterbein et al., 2001). Each molecule of the freely rotating G-actin can influence a large volume of water extending beyond its effective radius of gyration, causing a significant reduction in the intracellular low-density aqueous clustering. Filamentous actin (F-actin) has a much more ordered structure so creating more order in its surrounding water. At their surfaces the protein fibres trap water, which has consequentially decreased movement (i.e., lower entropy). In order to attempt to keep the water energy potential constant throughout, therefore, the water has to form stronger bonds with more negative bond energy (i.e., more negative enthalpy). This results in more directed bonds, causing greater structuring and lower density. Also, the enclosure of water in fibre-surrounded pools, involves capillary action that stretches the confined water and lowers its chemical potential (Trombetta et al., 2005), so ensuring that it is of lower density and hence more highly structured than the bulk water.

All actin molecules contain a conserved post-translation acetylated acidic N-terminus with several neighbouring aspartic and/or glutamic acids, for example the N-acetyl-aspartyl-glutamyl-aspartyl-glutamyl sequence in rabbit muscle α-actin. When actin polymerises in the cell under the action of ATP to form a filamentous structure, these highly negatively-charged antennae are placed on the exposed outer edge of the helix, where they may find further use as binding sites for other proteins, such as myosin. Tubulin, another intracellular structural protein that forms relatively immobile structures (using GTP rather than ATP) within cells is also a candidate for increasing the low-density nature of intracellular water (Mershin et al., 2004). It possesses an even more extensively negatively-charged acidic C-terminal conserved antenna, of about eight, usually glutamate, carboxylate groups, that serves similar functions. Although recognised for their biological importance, both the acidic N-terminus of actin and the C-terminus of tubulin have been previously somewhat overlooked in structural comparisons, due perhaps to their absence from crystallographic data and the lack of conserved structures due to the apparent interchangeability of the acid residues.

As overlapping fields from nearby groups enhance counter-ion association (Kern, 1948), F-actin's multiply negatively charged N-terminus will cause the attraction of cations into its vicinity. Under conditions when the carboxylic acids are weaker, both K^+ and Na^+ ions may form solvent separated species. This competition results in a preference for Na^+ ions and the formation of localised high-density water clustering. However, the natural rotational diffusion of the protein will tend to sweep such ions, and their associated water, away. If the protein stops rotating, Na^+ ions tend to destroy any low density structuring around carboxylate groups of the protein. However, the intracellular Na^+ ion concentration is generally far lower than that of K^+ ions and the K^+ ions will compete successfully for these sites. Given more acidic carboxylate groups, K^+ ions ion pair directly to such carboxylate groups, particularly when attracted into their microenvironment, due to the presence of the contiguous acidic amino acids.

6. COOPERATIVE WATER STRUCTURING PROCESSES

Binding of K^+ ions by the carboxylate groups lowers the ionic strength of the intracellular solution. As this ionic strength decreases, the acidity of phosphate groups decreases resulting in the conversion of the intracellular doubly charged HPO_4^{2-} ions to the singly charged $H_2PO_4^-$ ions, more favourable to low density water clustering (Ebner et al., 2005). All intracellular phosphate entities will behave similarly. Further support for this process is given by F-actin becoming more static in the presence of about 100 mM K^+ (Slosarek et al., 1994). Thus, the cooperative effects of the changes between the formation of static filaments and the freely diffusional proteins can be summarized in Figure 2.

Formation of K^+-carboxylate ion pairs leads to the formation of a surrounding clathrate water structuring that further guides icosahedral water structuring, so ensuring maximal hydrogen-bond formation and informing neighbouring carboxylate groups (Figure 3). This signalling cooperatively reinforces the tetrahedrality of the water structuring found between these groups. The clathrate cages allow rotational mobility (like a ball-and-socket joint) of the low density clusters, enabling the hydrogen bonding to search out cooperative partners. Cooperative interactions between such clathrated ion pair centres allow information to be passed around the cell. This has been previously described as quantum coherence and

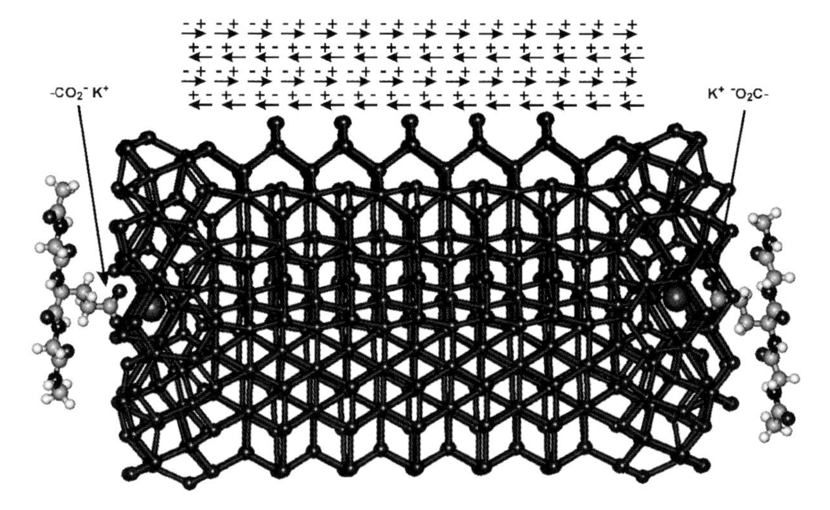

Figure 3. This cartoon shows the clustering around two K^+-carboxylate ion pairs as may be attached to part of two proteins' structures. In this diagram they are about 4 nm apart with 7–8 shells of water around each surface as is typically found between intracellular proteins. This distance is approximately as found between acid termini in actin and tubulin fibres. The water network is shown as linked (i.e., hydrogen bonded) oxygen atoms without showing their associated hydrogen atoms. The hydrogen bonding initially forms clathrate cages around the ion pairs, followed by a more extensive icosahedral arrangement (Chaplin, 2000). Extension of the hydrogen bonding along 'rays' connecting the neighbouring sites then develops. The water molecules show alignment both along the axis of the cylindrical cluster and in planes perpendicular to it. Once these 'rays' link, the hydrogen bonding of each reinforces the other in a cooperative manner, so strengthening the linkage and reinforcing the overall low density aqueous environment. As the aqueous clathrate cage possesses a more negative charge on its interior and a more positive charge on the outside, there is a marked polarization in the water molecules along the connecting axis that both allows these hydrogen bonding interactions and is in general agreement with the polarized multilayer theory of Ling (2003)

linked to information processing and brain function (Tuszynski et al., 1998). Such cooperative interactions also form the mechanistic link in the gel phase transitions of Pollack (2001).

Although the clustering involves a major drop in entropy, this is compensated by a more-negative enthalpy due to the stronger bonding of the fully tetrahedral hydrogen-bonded structure. It is consistent with Ling's association-induction polarised multilayer model (Ling, 2001), as can be seen from the net dipoles emanating from the clathrate arrangement, but offers a more realistic explanation. The initial icosahedral size (3 nm diameter; Chaplin, 2000), optimally surrounding each ion pair, also equals the water domain size proposed by Watterson (1997). Further support for this model is given by a number of studies concerning different types of intracellular water, such as 'normal' bulk and 'abnormal' osmotically inactive interfacial water (Garlid, 2000). This fully tetrahedral structuring possesses five-fold symmetry, which prevents easy freezing in line with the pronounced supercooling found for intracellular water.

Extension of the clathrate network, and its associated low density water, enables K^+ ion binding to all aspartic and glutamic acid groups, not just the key ones within the crucial N-terminal acidic centres. Thus, the sol-gel transition of Pollack (2001) may be interpreted as the conversion to low density water clustering (the gel state) due to clathrate clustering around K^+-carboxylate ion pairs.

In the presence of raised levels of Na^+ and/or Ca^{2+} ions, as occasionally occurs during some cell functions, these ions will compete and replace some of the bound K^+ ions. These newly formed solvent separated Na^+ and/or Ca^{2+} ion pairings destroy the low-density clathrate structures and initiate a cooperative conversion of the associated water towards a denser structuring.

7. CONCLUSION

In conclusion the aqueous information transfer within the cell involves the following. Intracellular water structuring is governed in part by the mobility of the proteins. Freely rotating proteins create zones of higher density water, which tend towards a lower density clustering if the rotation is prevented. K^+ ions partition into intracellular water in preference to Na^+ ions. Such K^+ ions ion pair to static charge-dense intracellular macromolecular structures. These ion paired K^+-carboxylate groupings prefer local clathrate water causing local low density water structuring. This low density water structuring cooperatively influences and reinforces the low-density character of neighbouring sites' water structuring. In this way information may be passed from site to site within the cell. Na^+ and Ca^{2+} ions can destroy this low density structuring in a cooperative manner.

REFERENCES

Bartell LS (1997) On possible interpretations of the anomalous properties of supercooled water. J Phys Chem 101:7573–7583

Cameron IL, Kanal KM, Keener CR, Fullerton GD (1997) A mechanistic view of the non-ideal osmotic and motional behavior of intracellular water. Cell Biol Int 21:99–113

Campolat M, Starr FW, Scala A, Sadr-Lahijany MR, Mishima O, Havlin S, Stanley HE (1998) Local structural heterogeneities in liquid water under pressure. Chem Phys Lett 294:9–12

Chaplin MF (2000) A proposal for the structuring of water. Biophys Chem 83:211–221

Chaplin MF (2005) Water Structure and Behavior. Available via http://www.lsbu.ac.uk/water

Cho CH, Urquidi J, Singh S, Park SC, Robinson GW (2002) Pressure effect on the density of water. J Phys Chem A 106:7557–7561

Clegg JS (1984) Interrelationships between water and cellular metabolism in Artemia cysts. XI. Density measurements. Cell Biophys 6:153–169

Collins KD (1995) Sticky ions in biological systems. Proc Natl Acad Sci USA 92:5553–5557

Collins KD (1997) Charge density-dependent strength of hydration and biological structure. Biophys J 72:65–76

Conway EJ (1957) Nature and significance of concentration relations of potassium and sodium ions in skeletal muscle. Physiol Rev 37:84–132

Ebner C, Onthong U, Probst M (2005) Computational study of hydrated phosphate anions. J Mol Liquids 118:15–25

Fournier P, Oelkers EH, Gout R, Pokrovski G (1998) Experimental determination of aqueous sodium-acetate dissociation constants at temperatures from 20 to 240 °C. Chem Geol 151:69–84

Garlid KD (2000) The state of water in biological systems. Int Rev Cytol 192:281–302

Halle B, Davidovic M (2003) Biomolecular hydration: From water dynamics to hydrodynamics. Proc Natl Acad Sci 100:12135–12140

Hazlewood CF (2001) Information forgotten or overlooked: Fundamental flaws in the conventional view of the living cell. Cell Mol Biol 47:959–970

Kern W (1948) Die aktivität der natriumionen in wäßrigen lösungen der salze mit polyvalenten säuren. Makromol Chemie 2:279–288

Khan A (2004) Theoretical studies of $Na(H_2O)^+{}_{19-21}$ and $K(H_2O)^+{}_{19-21}$ clusters: explaining the absence of magic peak for $Na(H_2O)^+{}_{20}$. Chem Phys Lett 388:342–347

Langford VS, McKinley AJ, Quickenden TI (2001) Temperature dependence of the visible-near-infrared absorption spectrum of liquid water. J Phys Chem A 105:8916–8921

Ling GN (1962) A physical theory of the living state: the Association-Induction Hypothesis, Waltham, Mass: Blaisdell

Ling GN (1997) Debunking the alleged resurrection of the sodium pump hypothesis. Physiol Chem Phys Med NMR 29:123–198

Ling GN (2001) Life at the cell and below-cell level. The hidden history of a functional revolution in Biology, New York: Pacific Press

Ling GN (2003) A new theoretical foundation for the polarized-oriented multiplayer theory of cell water and for inanimate systems demonstrating long-range dynamic structuring of water molecules. Physiol Chem Phys Med NMR 35:91–130

Mershin A, Kolomenski AA, Schuessler HA, Nanopoulos DV (2004) Tubulin dipole moment, dielectric constant and quantum behavior: Computer simulations, experimental results and suggestions. BioSystems 77:73–85

Otterbein LR, Graceffa P, Dominguez R (2001) The crystal structure of uncomplexed actin in the ADP state. Science 293:708–711

Pollack GH (2001) Cells, gels and the engines of life: A new unifying approach to cell function, Washington: Ebner and Sons Publishers

Pollack GH (2003) The role of aqueous interfaces in the cell. Adv Colloid Interface Sci 103:173–196

Raichlin Y, Millo A, Katzir A (2004) Investigation of the structure of water using mid-IR fiberoptic evanescent wave spectroscopy. Phys Rev Lett 93:185703

Robinson GW, Cho CH, Urquidi J (1999) Isobestic points in liquid water: Further strong evidence for the two-state mixture model. J Chem Phys 111:698–702

Slosarek G, Heintz D, Kalbitzer HR (1994) Mobile segments in rabbit skeletal-muscle F-actin detected by H-1 nuclear magnetic resonance spectroscopy. FEBS Lett 351:405–410

Tongraar A, Rode BM (2004) Dynamical properties of water molecules in the hydration shells of Na^+ and K^+: Ab initio QM/MM molecular dynamics simulations. Chem Phys Lett 385:378–383

Trombetta G, Di Bona C, Grazi E (2005) The transition of polymers into a network of polymers alters per se the water activity. Int J Biol Macromol 35:15–18

Tuszyński JA, Brown JA, Hawrylak P (1998) Dielectric polarization, electrical conduction, information processing and quantum computation in microtubules. Are they plausible? Phil Trans Math Phys Eng Sc pp 356:1897–1926

Watterson JG (1997) The pressure pixel – unit of life? BioSystems 41:141–152

Wiggins PM (1996) High and low density water and resting, active and transformed cells. Cell Biol Int 20:429–435

Wiggins PM (2001) High and low density intracellular water. Cell Mol Biol 47:735–744

Wiggins PM (2002) Water in complex environments such as living systems. Physica A 314:485–491

CHAPTER 6

BIOLOGY'S UNIQUE PHASE TRANSITION DRIVES CELL FUNCTION

DAN W. URRY
BioTechnology Institute, University of Minnesota, Twin Cities Campus, 1479 Gortner Avenue, Suite 240, St. Paul, MN 55108-6106, Bioelastics Inc., 2423 Vestavia Drive, Vestavia Hills, AL 35216-1333

Abstract: Systematic designs, physical characterizations and data analyses of elastic-contractile model proteins have given rise to a series of physical concepts associated with phase transitions of hydrophobic association and with the nature of elasticity that provide new insight into the function of a number of protein machines, namely, 1) Complex III of the electron transport chain wherein electron transfer pumps protons across the inner mitochondrial membrane, 2) the F_1-motor of ATP synthase that uses return of protons to produce the great majority of ATP in living organisms, 3) the myosin II motor of muscle contraction that uses ATP hydrolysis to produce movement, 4) the kinesin bipedal motor that walks along microtubules to transport cargo within the cell, and 5) the calcium-gated potassium channel. The physical processes utilize an understanding of the change in Gibbs free energy due to hydrophobic association, ΔG_{HA}, the water-mediated repulsion between hydrophobic domains and charged groups, ΔG_{ap}, and stretching of interconnecting chain segments that attends hydrophobic association

Keywords: phase transition, inverse temperature transition, apolar-polar repulsion, hydrophobic hydration, Gibbs free energy for hydrophobic association, protein machines, energy conversion, Complex III, Rieske Iron Protein, ATP synthase, myosin II motor, kinesin, calcium-gated potassium channel

Abbreviations: ECMP, elastic-contractile model proteins; ΔG_{HA}, the change in Gibbs free energy of hydrophobic association; ΔG_{ap}, apolar-polar repulsive free energy of hydration; RIP, Rieske Iron Protein

G. Pollack et al. (eds.), Water and the Cell, 125–149.

1. INTRODUCTION

1.1 Meeting the Challenge of Pollack's 'Cells, Gels and the Engines of Life.'

In the G. H. Pollack Book (2001), 'Cells, Gels and the Engines of Life' the final chapter calls for 'A New Paradigm for Cell Function.' It does so with four sections – Structured Water; Pumps, Channels, and Membranes; Phase Transitions, and Lessons from Biology. The objective of this paper is to bring functional detail to each section of the new paradigm. It does so by describing the distinctive and versatile *phase transition* of biology (Urry, 1992; 1997), which involves association/dissociation of hydrophobic domains in water. This requires consideration of two classes of *structured water* (Urry, 2006c) – water that hydrates hydrophobic groups and water that hydrates polar, especially charged, groups. Both classes of interfacial water decry the common assumption wherein water is treated as a uniform dielectric constant of approximately 80 right up to the surface of protein where the dielectric constant abruptly changes from 80 to 5 or less. While this approach has been useful in the past, in the present and future it is a fiction that needs to be abandoned.

Proteins perform the crucial energy conversion functions that sustain the cell. As examples of *lessons from biology*, we consider key classes of energy conversion. Protein crystal structures available from the Protein Data Bank allow analyses of three major classes of protein machines: 1) Complex III of the electron transport chain of the inner mitochondrial membrane that *pumps protons* (Urry, 2006a; 2006c) from the matrix side of the inner mitochondrial membrane to the space between the inner and outer mitochondrial *membranes*, 2) ATP synthase (Urry, 2006b; 2006c) (comprised of coupled intra-membranous and extra-membranous rotary motors) associated with the inner mitochondrial membrane that uses the flow of proton back across the inner mitochondrial membrane to form almost 90% of the ATP due to oxidation of glucose, and 3) the linear motors (Urry, 2005a; 2006a; 2006c), myosin II and kinesin, that perform the mechanical work of essential movements within the cell. These protein machines couple hydrophobic association and/or apolar polar repulsion with elastic deformation to achieve function. Also, continuing to flesh-out the new paradigm called for by Pollack, we note insights into the function of yet another membrane bound protein machine, a calcium-ion activated potassium ion *channel* (Mota and Teixeira, 2005).

1.2 Hydrophobic Hydration Exists, but with an Inverse Twist

As shown by Butler as early as 1937, *hydrophobic hydration exists but with a unique twist*. The system studied was the dissolution in water of a series of linear alcohols – starting with methanol, CH_3-OH, then addition of one CH_2 to give ethanol, CH_3-CH_2-OH, addition of a second CH_2 to give n-propanol, CH_3-CH_2-CH_2-OH, a third to give n-butanol, CH_3-CH_2-CH_2-CH_2-OH, and a fourth to give n-pentanol, CH_3-CH_2-CH_2-CH_2-CH_2-OH, commonly known as amyl alcohol. Butler's striking finding was that hydration of each added CH_2 group was favorable. Hydration of each added CH_2 group is exothermic. As hydration forms around each of the CH_2 groups, heat is given off due to formation of a favorable hydration shell, the pentagonal structure of which has subsequently been determined (Stackelberg and Müller, 1951; Teeter, 1984).

One, of course, then asks, Why is n-octanol with seven CH_2 groups insoluble in water? Herein lies the twist. Solubility is governed by the Gibbs free energy for solubility, ΔG (solubility) $= \Delta H - T\Delta S$, where ΔH is the heat released on dissolution and ΔS is the decrease in entropy as bulk water becomes the structured water around hydrophobic groups. When heat is given off, ΔH is negative, and for the above alcohol series Butler found that $\Delta H/CH_2 = -1.3\,kcal/mole$. But from the decrease in solubility resulting from the addition of each CH_2 group, Butler calculated for the series that $(-T\Delta S)/CH_2 = +1.7\,kcal/mole$. Accordingly, the mean ΔG(solubility) for each CH_2 group added would be $+0.4\,kcal/mole$.

Since the solubility of methanol starts with a significant negative ΔG(solubility), it takes seven CH_2 groups before ΔG(solubility) becomes sufficiently positive that solubility can be said to be lost. Thus, hydrophobic hydration increases for a hydrophobic domain until so much forms that ΔG(solubility) becomes positive and hydrophobic association (insolubility) results. *Therein lies the inverse twist.* Hydrophobic hydration increases until there is too much and then it disappears as the positive $(-T\Delta S)$ term overwhelms the negative ΔH term and hydrophobic association occurs.

1.3 Apolar-Polar Repulsion Controls Hydrophobic Association/Dissociation and Can Cause Charged States to Become Less Charged

So, whatever controls the actual or potential amount of hydrophobic hydration for a sufficiently hydrophobic domain determines whether hydrophobic association or hydrophobic dissociation occurs. In our view competition for hydration between apolar (hydrophobic) groups and polar (e.g., charged) groups controls whether hydrophobic association or dissociation results. The two disparate species move away from each other as they seek hydration unperturbed by the other. We call this an apolar-polar repulsive free energy of hydration and designate it as $\mathbf{\Delta G_{ap}}$ (Urry, 1992; 1997).

As will be seen below, not only does $\mathbf{\Delta G_{ap}}$ control hydrophobic association/dissociation, but in this way $\mathbf{\Delta G_{ap}}$ controls the structural changes that result in protein function. One example is for an external force to drive a hydrophobic domain into opposition through water with charge, as occurs in ATP synthase. In this case, we propose that the system reacts to lower the repulsion by decreasing the charge, that is, when the most charged state, ADP and P_i (inorganic phosphate) is faced-off through water with the most hydrophobic face of the γ-rotor, the result is to lower repulsion by formation of ATP (Urry, 2006b; 2006c).

1.4 Biology's Unique Phase Transition

Loss of protein function by cold denaturation occurs most notably due to hydrophobic dissociation of protein subunits comprising the functional protein. For example, on lowering the temperature of F_1-ATPase, the F_1-motor of ATP synthase, the subunits dissociate. On raising the temperature the subunits reassemble. Reassembly on raising the temperature occurs because solubility of the hydrophobic domains of the subunits is lost, that is $\Delta G(\text{solubility}) = \Delta H - T\Delta S$ becomes positive as the increasing temperature causes the positive $(-T\Delta S)$ term to become larger than the negative ΔH term.

1.5 Essential Equivalence of Intra-Molecular and Inter-Molecular Hydrophobic Association

In our model protein system hydrophobic association both intra-molecular and inter-molecular occurs, but so too are the hydrophobic associations in biology's protein machines. The change in thermodynamic properties for a hydrophobic domain going from hydrated to associated is essentially the same whether the second part of paired hydrophobic domains is intra-molecular or inter-molecular. Therefore, properly treated thermodynamics for the readily characterized inverse temperature transition, which involves both intra-molecular and inter-molecular hydrophobic associations, becomes applicable to hydrophobic association between globular protein subunits and also within protein subunits.

2. MATERIALS AND METHODS

2.1 The Elastic-Contractile Model Protein (ECMP) System

For the last several decades we have been studying a family of elastic-contractile model proteins (ECMP), wherein the parent repeating pentamer is $(\text{glycyl-valyl-glycyl-valyl-prolyl})_n$, abbreviated as $(\text{Gly-Val-Gly-Val-Pro})_n$ or simply $(\text{GVGVP})_n$. The most common designed modifications are designated as (GXGVP). The basic model protein, $(\text{GVGVP})_n$ with $n \approx 200$, is soluble in all proportions in water below 25 °C and phase separates on raising the temperature to 37 °C to form a more ordered viscoelastic state (Urry, 1992; 1997). Because an increase in temperature gives rise to an increase in order, we recognize a hydrophobic hydration that becomes less ordered bulk water as the model protein becomes more ordered by hydrophobic association. We have called this unique phase transition an *inverse temperature transition of hydrophobic association*. On cross-linking $(\text{GVGVP})_n$, an entropic elastic matrix is obtained that can perform thermo-mechanical and chemo-mechanical transduction, that is, the conversion of thermal energy into mechanical work and of chemical energy into mechanical work, respectively.

2.2 Definition of the Heat and Temperature of the Inverse Temperature Transition

Using the elastic-contractile model protein (ECMP) that exhibits an endothermic inverse temperature transition, differential scanning calorimetry determines the heat, $\mathbf{\Delta H_t}$ per mole of pentamer (GXGVP), and on heating the onset temperature, $\mathbf{T_t}$, of the inverse temperature transition is also referenced to a mole of pentamer for a guest amino acid residue, X, and for other biologically interesting chemical modifications of amino acid residues. This allows development of a hydrophobicity scale based on the change in Gibbs free energy for hydrophobic association, $\mathbf{\Delta G_{HA}}$, that results due to replacement of a Val residue by a guest residue (see Tables 1 and 2). Since $\mathbf{\Delta G_{HA}}$ is defined as the change in Gibbs free energy for hydrophobic association, which is for the water insolubility of hydrophobic groups, $\mathbf{\Delta G_{HA}} = -\Delta G$(solubility of hydrophobic groups).

Table 1. Hydrophobicity Scale for amino acid residues in terms of $\Delta G°_{HA}$, the change in Gibbs free energy of hydrophobic association

Residue X	T_t°C	$\Delta G°_{HA}$ (GXGVP) kcal/mol-pentamer
W: Trp	−105	−7.00
F: Phe	−45	−6.15
Y: Tyr	−75	−5.85
H° : His°	−10($\mathbf{T_t}$)	−4.80 (from graph)
L: Leu	5	−4.05
I: Ile	10	−3.65
V: Val	26	−2.50
M: Met	15	−1.50
H^+: His^+	30 ($\mathbf{T_t}$)	−1.90 (from graph)
C: Cys	30 ($\mathbf{T_t}$)	−1.90 (from graph)
E°: Glu(COOH)	20 (2)	−1.30 (−1.50)
P: Pro	40	−1.10
A: Ala	50	−0.75
T: Thr	60	−0.60
D°: Asp(COOH)	40	−0.40
K°: Lys(NH_2)	40 (38)	−0.05 (−0.60)
N: Asn	50	−0.05
G: Gly	55	0.00
S: Ser	60	+0.55
R: Arg	60 ($\mathbf{T_t}$)	+0.80 (from graph)
Q: Gln	70	+0.75
Y^-: Tyr($\phi - O^-$)	140	+1.95
D^-: Asp(COO^-)	170 ($\mathbf{T_t}$)	≈ +3.4 (from graph)
K^+: Lys(NH_3^+)	(104)	(+2.94)
E^-: Glu(COO^-)	(218)	(+3.72)
Ser ($PO_4^=$)	860 ($\mathbf{T_t}$)	≈ +8.0 (from graph)

Data within parentheses utilized microbial preparations of poly(30 mers), e.g., $(GVGVP\ GVGVP\ GXGVP\ GVGVP\ GVGVP\ GVGVP)_n$, with n ≈ 40. The notation (from graph) indicates that the value of $\mathbf{T_t}$ was used to obtain $\mathbf{\Delta G°_{HA}}$(GXGVP) from the experimental sigmoid curve of $\mathbf{T_t}$ versus $\mathbf{\Delta G°_{HA}}$ from Urry (2004, 2006). $\mathbf{T_t}$-values adapted from Urry (2004).

Table 2. Hydrophobicity Scale (preliminary $\mathbf{T_t}$ and $\mathbf{G_{HA}}$ values) for Chemical Modifications and Prosthetic Groups of Proteins.[a] $\mathbf{T_t}$ = Temperature of Inverse Temperature Transition for poly[f_V (VPGVG),f_X(VPGXG)]

Residue X	$\mathbf{G_{HA}}$ (kcal/mol)[g]	$\mathbf{T_t}$, linearly extrapolated to $f_X = 1$
Lys (dihydro NMeN) [b,d]	−7.0	−130 °C
Glu(NADH)[c]	−5.5	−30 °C
Lys (6-OH tetrahydro NMeN)[b,d]	−3.0	15 °C
Glu($FADH_2$)	−2.5	25 °C
Glu(AMP)	+1.0	70 °C
Ser(-O-SO_3H)	+1.5	80 °C
Thr(-O-SO_3H)	+2.0	100 °C
Glu(NAD)[c]	+2.0	120 °C
Lys(NMeN, oxidized)[b,d]	+2.0	120 °C
Glu(FAD)	+2.0	120 °C
Tyr(-O-SO_3H)[e]	+2.5	140 °C
Tyr(-O-NO_2^-)[f]	+3.5	220 °C
Ser($PO_4^{=}$)	+8.0	860 °C

[a] Usual conditions are 40 mg/ml polymer, 0.15N NaCl and 0.01M phosphate at pH 7.4.

[b] NMeN is for N-methyl nicotinamide at a lysyl side chain, i.e., N-methyl-nicotinate attached by amide linkage to the ε-NH_2 of Lys and the most hydrophobic reduced state is N-methyl-1,6-dihydronicotinamide (dihydro NMeN), and the second reduced state is N-methyl-6-OH 1,4,5,6-tetrahydronicotinamide or (6-OH tetrahydro NMeN).

[c] For the oxidized and reduced nicotinamide adenine dinucleotides, the conditions were 2.5 mg/ml polymer, 0.2M sodium bicarbonate buffer at pH 9.2.

[d] For the oxidized and reduced N-methyl nicotinamide, the conditions were 5.0 mg/ml polymer, 0.1M potassium bicarbonate buffer at pH 9.5, 0.1M potassium chloride.

[e] The pK_a of polymer bound -O-SO_3H is 8.2.

[f] The pK_a of Tyr(-O-NO_2) is 7.2.

[g] Gross estimates of $\mathbf{\Delta G^\circ_{HA}}$ using the $\mathbf{T_t}$-values in the right column in combination with the $\mathbf{T_b}$ versus $\mathbf{\Delta G^\circ_{HA}}$ values from Urry (2006c).

2.3 Calculation of the Change in Gibbs Free Energy for Hydrophobic Association

The basis set for the model proteins can be described as poly[f_V(GVGVP), f_X(GXGVP)] where f_V and f_X are the mole fractions of the pentamers with $f_V + f_X = 1$. Accordingly, experimental values of T_t and ΔH_t are plotted for a set of values of f_X, and the line is linearly extrapolated to $f_X = 1$. At the intercept for $f_X = 1$, at the values for the heat and temperature of the transitions are given as bold-faced quantities, i.e., $\mathbf{T_t}$ and $\mathbf{\Delta H_t}$, to signify per mole of (GXGVP). As has been derived elsewhere at the values of $\mathbf{T_t}$ (Urry DW, 2004), the change in Gibbs free energy for hydrophobic association, $\mathbf{\Delta G^\circ_{HA}}$, due to replacement of a G residue with an X residue (G → X) can be obtained from the heats of the inverse temperature transitions as follows,

$$(1) \qquad \mathbf{\Delta G^\circ_{HA}}(G \rightarrow X) \equiv [\mathbf{\Delta H_t}(GGGVP) - \mathbf{\Delta H_t}(GXGVP)]$$

2.4 $\Delta G°_{HA}$-Hydrophobicity Scale for Amino Acid Residues and Prosthetic Groups Including in Different Functional States

The $\Delta G°_{HA}$-Hydrophobicity Scale for the amino acid residues (and where relevant for amino acid residues in different functional states, for example, uncharged and ionized) is given in Table 1 (Urry, 2004; 2006c) and for chemical modifications is listed in Table 2, which includes oxidized and reduced states of selected redox groups. In general, the reference solvent conditions are 0.15 N NaCl and 0.01 M phosphate. Now it becomes possible to examine interesting hydrophobic domains of protein machines of biology and to obtain a sense of the relative change in Gibbs free energy that could occur on hydrophobic association and also the relative capacity of surfaces to repulse charged species.

2.5 ΔG_{ap} is the Operative Element within ΔG_{HA}

One means of observing ΔG_{ap} is to see the effect of charge formation on ΔG_{HA}, for example, to see the change in $\Delta G°_{HA}$ on ionization of the side chain of the glutamic acid residue. As seen in Table 1, $\Delta G°_{HA}$(glutamic acid) $= -1.5$ kcal/mole-(GEGVP) and $\Delta G°_{HA}$(glutamate) $= +3.7$ kcal/mole-(GE$^-$ GVP), where E^- stands for the glutamate residue having the charged side chain, -CH_2-CH_2-COO^-. Thus, the effect of ionization of the carboxyl side chain of glutamic acid, $\Delta G_{ap}(E \rightarrow E^-) = \Delta G°_{HA}(\text{glutamate}) - \Delta G°_{HA}(\text{glutamicacid}) = +5.2\,\text{kcal/mole-}(E \rightarrow E^-)$.

The carboxylate containing side chain, -CH_2-CH_2-COO^-, of glutamate (E^-) competes for hydration with the V and P side chains of poly[f_V(GVGVP), f_E(GEGVP)] and in doing so destroys hydrophobic hydration (Urry, Peng, Xu, McPherson, 1997). In the noted model protein, the competition resulting from ionization of the E residue favors hydrophobic dissociation by 5.2 kcal/mole-Glu. Now consider the following scenario. During transient hydrophobic dissociations hydrophobic hydration forms. As too much hydrophobic hydration forms, hydrophobic re-association occurs. When a carboxylate forms proximal (within a few nm) to the transient dissociation, it recruits the nascent hydrophobic hydration for its own charge hydration, and the hydrophobic dissociation stands. *Protein function often derives from those energy inputs that change the values of ΔG_{ap}.*

2.6 Direct Quantification of Apolar-Polar Repulsion, ΔG_{ap}, from pKa Shifts

Our approach determines the effect on the pKa when replacing the less hydrophobic valyl (Val, V) residue with a more hydrophobic phenylalanyl (Phe, F) residue. From Table 1 one calculates that $\Delta G°_{HA}(V \rightarrow F) = \Delta G°_{HA}(\text{phenylalanine}) - \Delta G°_{HA}(\text{valine}) = [-6.15\text{ kcal/mol.-F} - (-2.5\text{ kcal/mol.-V}] = -3.65$ kcal/mole. As seen in the ECMP of Table 3, there is one E per 30 mer. The reference ECMP has no F residues such that it is labeled E/0F. Then two, three, four, and five V residues are replaced as indicated by F residues to give the series, E/0F, E/2F, E/3F, E/4F, and E/5F as short-hand representations of **Model Proteins, I, II, III, IV**, and **V**, respectively.

Table 3. Hydrophobic-induced pKa shifts in Elastic-contractile Model Proteins (ECMP) by Systematic Replacement of V by F

			n	pKa	ΔpKa	ΔG_{ap}
ECMP I:	$(GVGVP\ GVGVP\ G\underline{E}GVP\ GVGVP\ GVGVP\ GVGVP)_{36}$	:E/0F;	1.5	4.5	0.5	0.7
ECMP II:	$(GVGVP\ GVGFP\ G\underline{E}GFP\ GVGVP\ GVGVP\ GVGVP)_{40}$	:E/2F;	1.6	4.8	0.8	1.1
ECMP III:	$(GVGVP\ GVGVP\ G\underline{E}GVP\ GVGVP\ GVGFP\ GFGFP)_{39}$	:E/3F;	1.9	5.2	1.2	1.6
ECMP VI:	$(GVGVP\ GVGFP\ G\underline{E}GFP\ GVGVP\ GVGFP\ GVGFP)_{15}$	:E/4F;	2.7	5.6	1.6	2.2
ECMP V:	$(GVGVP\ GVGFP\ G\underline{E}GFP\ GVGVP\ GVGFP\ GFGFP)_{42}$	:E/5F;	8.0	6.4	2.4	3.3

Also seen in the Table 3 are the values for the Hill coefficient, **n**, the pKa, the ΔpKa, and the $\mathbf{\Delta G_{ap}}$, where $\mathbf{\Delta G_{ap}}$ = 2.3RT ΔpKa. As the number of V residues are replaced by F residues, there occurs a supra-linear shift in pKa as V residues are progressively replaced by more hydrophobic F residues. The result is called a hydrophobic-induced pKa shift, and from the ΔpKa, the apolar-polar repulsive free energy of hydration, $\mathbf{\Delta G_{ap}}$, can be calculated.

The $\mathbf{\Delta G_{ap}}$ so calculated from the acid-base titration curve that utilizes the pKa shift can be compared with the $\mathbf{\Delta G_{ap}}$ calculated from the heats of the transition as determined experimentally from differential scanning calorimetry data. In **Model Protein I**, the pKa is 4.5, whereas the pKa for an unperturbed glutamic acid residue is between 3.8 and 4.0. Thus, the hydrophobic-induced pKa shift due to the presence of the V and P residues in **Model Protein I** is 0.5 to 0.7 pH units. As the value obtained from Table 1 is per (GXGVP), the value from the acid-base titration data should be multiplied by six. This gives a pKa shift of 3.0 to 4.2. On conversion to $\mathbf{\Delta G_{ap}}$ this gives +4.3 to +6.0 kcal/mol-E, which correlates well with the value of +5.2 kcal/mol-E as obtained from the $\mathbf{\Delta G^{\circ}_{HA}}$-Hydrophobicity Scale, based on Equation (1) and evaluation using differential scanning calorimetry data.

2.7 Additivity of $\mathbf{\Delta G_{ap}}$ due to Multiple sources, i.e., $\mathbf{\Sigma_i(\Delta G_{ap})_i}$

Another way to induce pKa shifts is to stretch a cross-linked hydrophobically associated matrix of ECMP. As stretching of the hydrophobically associated matrix exposes hydrophobic groups to the entering water of hydration, the stretch-induced pKa shift is another way to achieve a hydrophobic-induced pKa shift. The two types of induced pKa shifts should be additive, as found in Table 4 for the poly[f_V(GVGIP),f_E(GEGIP)] model protein system (Urry, 2005b).

In fact, any perturbation that introduces a $\mathbf{\Delta G_{ap}}$ simply adds or subtracts for the final $\mathbf{\Delta G_{ap}}$ acting on the hydrophobic domain of interest. This brings in the added changes noted in Table 2, as well as mechanical force, chemical energy (of all forms including changes in salt concentrations), pressure, electromagnetic radiation, and so on and allows the general statement,

$$(2) \qquad \mathbf{\Delta G_{ap}}\text{(hydrophobic domain)} = \mathbf{\Sigma_i(\Delta G_{ap})_i}$$

where the summation over i covers all of the energy inputs that alter $\mathbf{T_t}$ and $\mathbf{\Delta G_{ap}}$.

Table 4. Additivity of Hydrophobic-induced and Stretch-induced pKa shifts

		f	pKa	ΔpKa	ΔG_{ap}
I : (GVGVP GVGVP G$\underline{E}$GVP GVGVP GVGVP GVGVP)$_{36}$:	E/0I,	0	4.5	0.5	0.7
II : (GVGVP GVGIP G$\underline{E}$GIP GVGVP GVGVP GVGVP)$_n$:	E/2I;	0	4.7*	0.7	1.6
III : (GVGVP GVGIP G$\underline{E}$GIP GVGVP GVGIP GVGIP)$_n$:	E/4I;	0	4.9*	0.9	2.2
IV : (GVGIP GVGIP G$\underline{E}$GIP GVGIP GVGIP GVGIP)$_{23}$:	E/6I;	0	5.4	1.4	1.9
V : X^{20}-poly[0.83(GVGIP),0.17(GEGIP)] :	E/6I;	0	6.3	2.3	3.2
V' : X^{20}-poly[0.83(GVGIP),0.17(GEGIP)] :	E/6I;	3.6	6.6	2.6	3.6
V' : X^{20}-poly[0.83(GVGIP),0.17(GEGIP)] :	E/6I;	5.4	6.9	2.9	4.1
V' : X^{20}-poly[0.83(GVGIP),0.17(GEGIP)] :	E/6I;	6.4	7.4	3.4	4.9
V' : X^{20}-poly[0.83(GVGIP),0.17(GEGIP)] :	E/6I;	7.3	8.2	4.2	5.9
V' : X^{20}-poly[0.83(GVGIP),0.17(GEGIP)] :	E/6I;	8.0	9.0	5.0	7.1

$\mathbf{\Delta G_{ap}}$ (= 2.3 RT ΔpKa) is in kcal/mol-carboxylate. **f** is force in units of 10^5 dynes/cm^2.
X^{20} is a 20 Mrad γ-irradiation dose for forming cross-linked elastic matrix. *Calculated value.
Reproduced from Urry (2005b).

2.8 Obtaining Crystal Structure Data and Approach to Visual Analysis

Preferably, two crystal structures are obtained for each protein machine for which mechanism is to be derived. These are obtained from the Protein Data Bank at http://www.rcsb.org/pdb (Berman et al., 2000) and are referred to as the Structure Files, such as 1BMF and 1H8E for the F_1-motor of ATP synthase (Urry, 2006b; 2006c). The FrontDoor to Protein Explorer, due to Eric Martz (2002), which is available at no cost from http://www.proteinexplorer.org, is used to examine the crystal structures. By this means structures are accessed and analyzed to develop insightful perspectives. Important in developing a sense of structure and mechanism is the capacity to visualize the structures in three dimensions. In our case this is achieved by utilizing stereo views set for cross-eye viewing.

In order to achieve easy visual delineation of hydrophobic regions from polar regions of charged amino acid residues, a gray code is used, whereby the most hydrophobic (aromatic) residues are black; the remaining hydrophobic residues are given in gray; the neutral residues are shown in light gray, and the charged residues (both positive and negative) are white. In this way predominantly hydrophobic (apolar) regions are immediately and visually distinguishable from charged (polar) regions. Cropping of the illustrations and setting conditions for the structure utilized Adobe ® Photoshop® 5.5 and labeling and otherwise achieving informative structure representations utilized Microsoft Power Point.

2.9 Evaluation of Relative Hydrophobicity of Hydrophobic Domains by Means of $\Sigma_i[\Delta G^\circ_{HA}(X)_i]$

When evaluating hydrophobic domains of proteins by utilizing the data in Table 1, it becomes possible to sum over i, where i stands for all of the amino acid residues that comprise the domain. Hydrophobic domains of interest in the set of

Table 5. Values for $\Sigma\Delta\mathbf{G}_{\mathbf{HA}}$ (γ-rotor faces)

γ-rotor at catalytic site	β-empty face Res. No. : ΔG_{HA}		β-ATP face Res. No. : ΔG_{HA}		β-ADP face Res. No. : ΔG_{HA}	
Residue number and ΔG_{HA} values in kcal/mol	Thr 2/3 :	−0.20	Ala 1 :	−0.75	Ala 1 :	−0.75
	Leu 3 :	−4.05	Thr 2/3 :	−0.20	Thr 2 :	−0.60
	Lys 4 :	+2.94	Asp 5 :	+3.40	Leu 3/3 :	−1.30
	Thr 7 :	−0.60	Ile 6/3 :	−1.20	Lys 4 :	+2.94
	Leu 10/2 :	−2.00	Glu 264 :	+3.72	Asp 5 :	+3.40
	Ile 263/3 :	−1.20	Ile 263/3 :	−1.20	Thr 7/2 :	−0.30
	Leu 262/2 :	−2.00	Lys 260 :	+2.94	Glu 264 :	+3.72
	Glu 261/2 :	+1.85	Thr 259 :	−0.60	Glu 261 :	+3.72
	Thr 259 :	−0.60	Ile 258/3 :	−1.20	Lys 260 :	+2.94
	Ile 258 :	−3.65	Ala 256 :	−0.75	Ile 258 /2 :	−1.80
	Val 257/2 :	−1.25	Gln 255 :	+0.75	Val 257 :	−2.50
	Gln 255 :	+0.75	Val 257/3:	−0.80	Ala 256/3:	−0.25
	Arg 254 :	+0.70	Thr 253 :	−0.60	Arg 254/2 :	+0.35
	Arg 252/2 :	+0.35	Arg 252 :	+0.70	Thr 253/3:	−0.20
	Asn 251 :	−0.05	Asn 251 :	−0.05	Thr 249/3:	−0.20
	Phe 250 :	−6.15	Leu 248/2:	−2.00		
	Leu 248 :	−4.05	Thr 249 :	−0.60		
	Thr 247 :	−0.60				
	Sum :	−19.8	Sum :	+0.4	Sum :	+9.2

$\Sigma\Delta G_{HA}$ (β-empty face) ≈ −20 kcal/mol; $\Sigma\Delta G_{HA}$ (β-ATP face) ≈ +0 kcal/mol;
$\Sigma\Delta G_{HA}$ (β-ADP face) ≈ +9 kcal/mol
Reproduced from Urry (2006b).

protein-based machines of this article include: 1) those within Complex III, the FeS center/hydrophobic tip of the globular component of the Rieske Iron Protein and the Q_o-site and heme $\mathbf{c}_1$-site to which the tip hydrophobically associates and dissociates, 2) the three faces of the γ-rotor of the F_1-motor of ATP synthase, which is detailed in Table 5 below, 3) the myosin II motor of muscle contraction, which include the hydrophobic domains that hydrophobically associate and dissociate during function such as the hydrophobic association of the myosin cross-bridge to the actin filament and the hydrophobic association of the underside of the N-terminal with the head of the lever arm, 4) in kinesin the hydrophobic association of the foot with the microtubular binding site, and 5) in the calcium-ion activated potassium channel the change in hydrophobic association between the four large extra-membranous globular protein subunits that accompany calcium ion binding.

3. RESULTS AND DISCUSSION (LESSONS FROM BIOLOGY)

From the study of diverse energy conversions by *de novo*-designed, experimentally characterized and data analyzed ECMP, two distinct but interlinked physical processes arose; these describe the nature of a comprehensive hydrophobic effect and the mechanisms of near ideal elasticity. For each of the physical processes we speak of a consilient mechanism, that is, of a 'common groundwork of explanation' (Wilson, 1998). So there are hydrophobic and elastic consilient mechanisms. The hydrophobic consilient mechanism applies to all amphiphilic polymers in water. Protein is the most extraordinary amphiphilic polymer because of twenty different natural side chains, of strict control of sequence, and of retention of a single optical isomer. The elastic consilient mechanism applies to all polymers of whatever composition as long as there is sufficient freedom of motion in the backbone of even a single chain that becomes damped in its amplitude on deformation.

Common to protein function are hydrophobic associations that stretch interconnecting chain segments to store deformation energy that is then used to achieve movement. This is particularly apparent in the function of the Rieske Iron Protein within Complex III of the electron transport chain that participates in electron transfer and results in proton translocation across the inner mitochondrial membrane. It is also apparent in the myosin II motor of muscle contraction, in the function of kinesin, and in the function of calcium-ion activated potassium channel. In the function of ATP synthase the apolar-polar repulsion becomes the operative component of the comprehensive hydrophobic effect that causes synthesis of ATP from ADP and P_i (inorganic phosphate).

3.1 Complex III of the Electron Transport Chain of the Inner Mitochondrial Membrane

Figure 1A contains a stereo overview of the dimer of Complex III (ubiquinone: cytochrome **c** oxidoreductase, also called the cytochrome $\mathbf{bc_1}$ complex) with the protein subunits in ribbon representation and with one molecule of cytochrome **c** attached to one of the dimers (Lange and Hunte, 2002). As depicted in Figure 1B, it is the protein machine that receives electrons from ubiquinone and transfers them to cytochrome **c**, and in the process pumps protons across the inner mitochondrial membrane to the space between the inner and outer membranes of the mitochondrion.

The redox components of Complex III are cytochrome **b** with two hemes $\mathbf{b_L}$ and $\mathbf{b_H}$, cytochrome $\mathbf{c_1}$ with its heme $\mathbf{c_1}$ and the very important Rieske Iron Protein (RIP) with its redox FeS center. RIP is anchored by a single chain in the membrane portion of one monomer and angles across to function as a movable globular part containing the FeS redox center that achieves electron transfer from the ubiquinol in the Q_O-site to the heme $\mathbf{c_1}$, both being in the second monomer.

As depicted in Figure 1B, at the Q_O-site ubiquinol gives up one electron to the FeS center and the second electron to heme $\mathbf{b_L}$ leaving the ubiquinol with two positive charges, which for reasons of developing an apolar-polar repulsion allows the FeS center to be moved by an elastic retraction to the heme $\mathbf{c_1}$-site where it releases its electron, and the reduced heme $\mathbf{c_1}$ passes its electron to the heme of cytochrome **c**. Accordingly, the former is the basis for the name of ubiquinone:cytochrome c oxidoreductase for this is a protein machine that performs electro-chemical transduction.

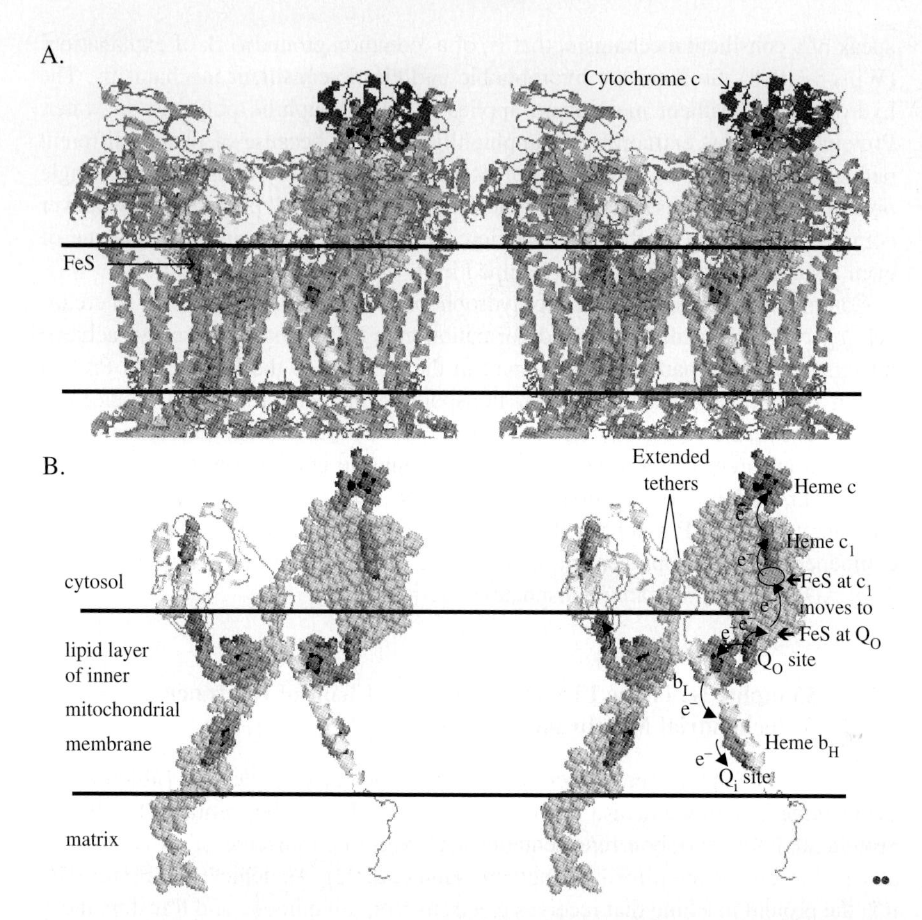

Figure 1. (A) Cross-eye stereo view of the dimer of Complex III (cytochrome $\mathbf{bc}_1$ complex) with a single cytochrome **c** attached at the upper right. The protein subunits are given in gray ribbon representation with the exception of the Rieske Iron Protein, which is in white, and cytochrome **c**, which is in dark gray. Also buried within the structure are the redox ligands. (B) Cross-eye stereo view of the Rieske Iron Protein (RIP) and the redox ligands for the purpose of demonstrating the electron transfers. The RIP subunits anchor in the membrane with one monomer but cross over and function in the second monomer. The RIP rising from left to right is in space filling representation and the RIP rising from right to left is given in ribbon representation in order to see the FeS center within the tip at the Q_O-site. Protein Data Bank, Structure File 1KYO due to Lange and Hunte, 2002. Adapted from (Urry 2006c)

Importantly, for the conversion of electrical energy derived from electron transfer to the chemical energy of proton pumping, the ubiquinol develops two positive charges that by means of the resulting $\mathbf{\Delta G_{ap}}$ repulses the globular component of RIP out of the way, and releases its two protons to regenerate ubiquinone and to have pumped two protons to the inner membrane space.

Figures 2, 3, and 4 provide a more complete description of these key processes centered at the Q_O-site where electron transfer couples to proton pumping by means of the coupling of the hydrophobic and elastic consilient mechanisms.

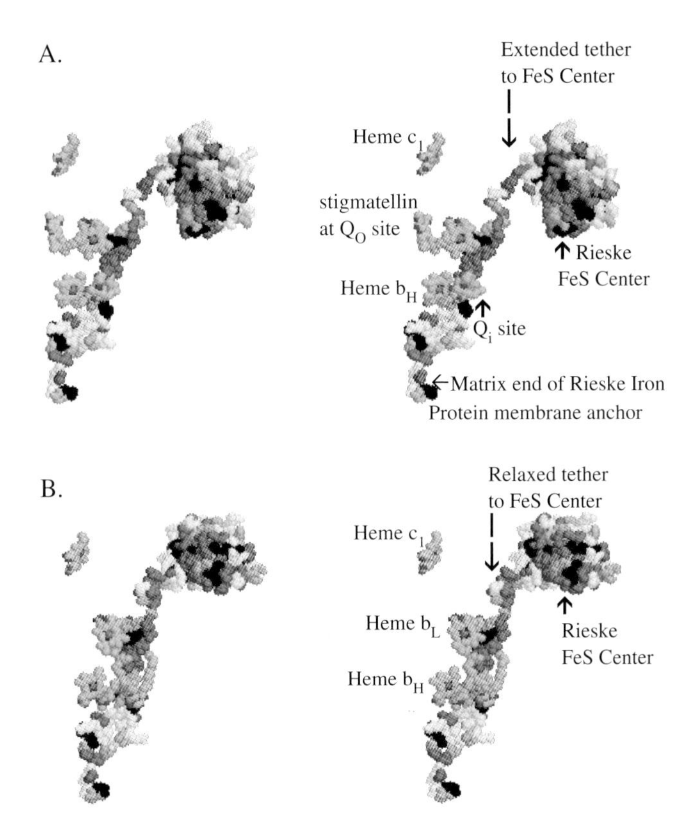

Figure 2. Cross-eye stereo view of a monomer of the Rieske Iron Protein (RIP) in space filling representation with the redox centers (ligands) included. Since the RIP anchors with one monomer but reaches over to function in the second monomer, it will be necessary to look across at the ligands to the left of the RIP to relate the position of the tip of the globular component to the positions of the site at which FeS center is interacting. The amino acids are gray coded so the relative hydrophobicities of the globular portion of the RIP are apparent. In particular, the most hydrophobic aromatic residues are black, the other hydrophobics are gray, the neutral residues are light gray and the charged residues are white. The dark tip of the RIP shows that it is very hydrophobic. Protein Data Bank, Structure File 1KYO due to Lange and Hunte, 2002. (A) The FeS center is positioned at the Q_o-site and the tether is extended (stretched). (B) The FeS center is positioned at the heme $\mathbf{c_1}$-site and the tether relaxed (contracted). From Urry, 2006c

Figure 2A depicts the Rieske Iron Protein, with its FeS center just inside the very hydrophobic tip of the globular component, as positioned for hydrophobic association at the Q_O-site (Zhang et al., 1998). The second part of the hydrophobic association involving the Q_O-site is shown in Figure 3A. When the FeS center is at the Q_O-site, the single chain tether is extended. On the other hand in Figure 2B, when the FeS center is at the heme $\mathbf{c_1}$-site, the single chain tether is contracted (Zhang et al., 1998). Figure 3B shows the heme $\mathbf{c_1}$-site as when hydrophobically paired with the very hydrophobic tip of the globular component of RIP.

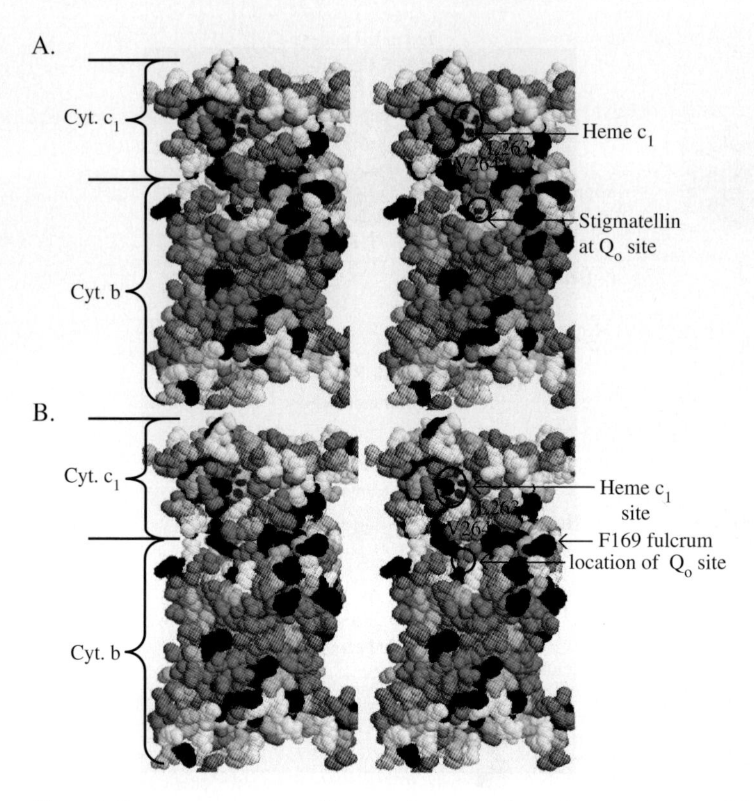

Figure 3. Cross-eye stereo view of the cytochrome **b** and cytochrome $\mathbf{c_1}$ in space filling representation oriented to allow view of both the Q_o and heme $\mathbf{c_1}$ binding sites for RIP. The amino acids are gray coded so that the relative hydrophobicities of the domains are apparent. The darker regions are more hydrophobic whereas the white residues are charged. Protein Data Bank, Structure File 1BCC and 3BCC due to Zhang et al., 1998. (A) Q_o-site containing stigmatellin as it occurs when hydrophobically associated with the FeS center of RIP. (B) State of heme $\mathbf{c_1}$ binding site as it occurs when hydrophobically associated with the FeS center of RIP. From Urry, 2006b

Figure 4 combines the ribbon representation of RIP and space filling representation of redox centers of the opposite monomer with a cartoon in order to depict mechanism in four steps. On the one hand there is the coupling of hydrophobic association to the stretching of the single interconnecting chain tether, and on the other hand there is the electron transfer to give two positive charges on the ubiquinol that (by means of an increase in apolar-polar repulsion, $\mathbf{\Delta G_{ap}}$) disrupt the hydrophobic association of RIP hydrophobic tip with Q_O-site and allow the release of protons for completing the coupling of electron flow to proton pumping.

Elaboration of the Four Steps of Figure 4:
STEP 1: The hydrophobic tip of the globular component of the Rieske Iron Protein (RIP) is situated with its FeS center at the heme $\mathbf{c_1}$-site and with the most distal part of the relaxed tether making hydrophobic contact with the hydrophobic residue,

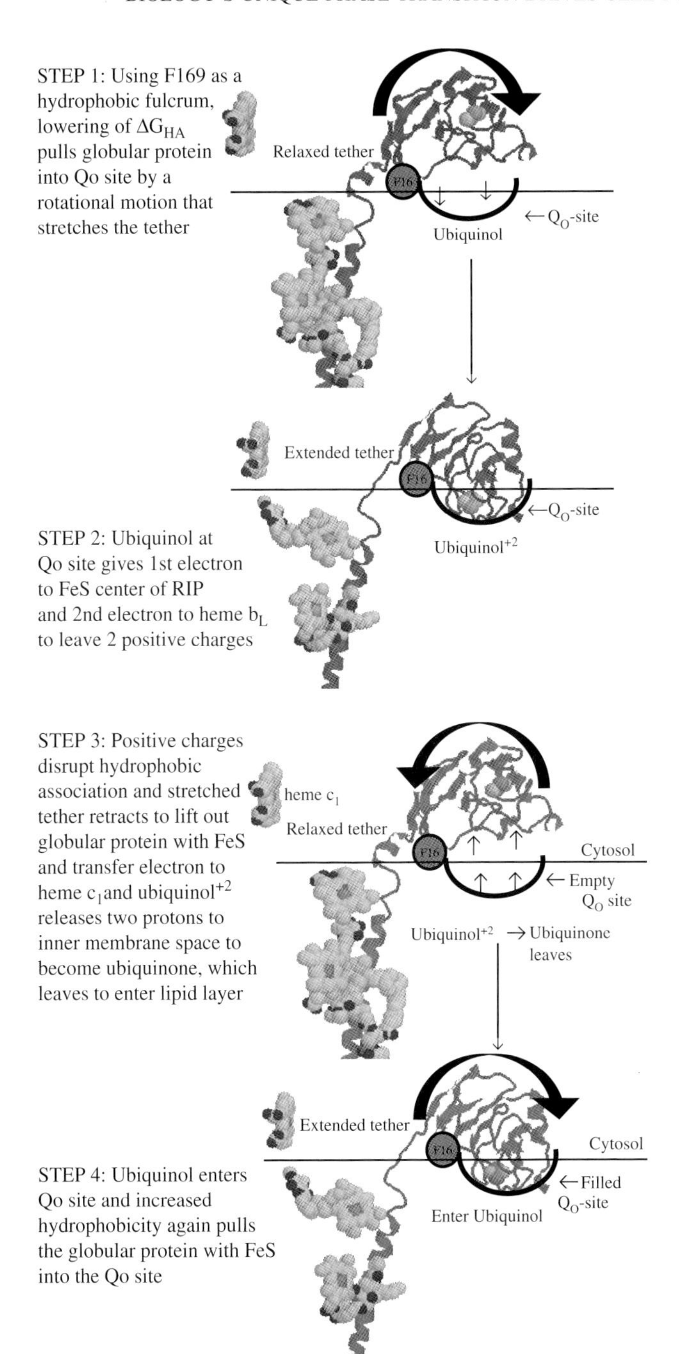

Figure 4. Proposed cycle in four steps at Q_o-site of electron transfer coupling to proton translocation due to hydrophobic association/dissociation coupling to stretching/relaxation of single chain tether. Utilizing the Protein Data Bank, Structure Files 1BCC and 3BCC due to Zhang et al., 1998, for the RIP structure. See text for a more complete description of the cycle. STEPS 1 and 2 adapted from Urry, 2005a

F169, such that of the hydrophobic tip of RIP can rotate into gradually increasing hydrophobic association with the Q_o-site. The improved ΔG°_{HA}(hydrophobic tip→Q_o -site) occurs at the cost of stretching the relaxed tether.

STEP 2: At full hydrophobic association of RIP hydrophobic tip with Q_o-site, the single chain tether has become extended with energy storage as an elastic deformation; the FeS center is in position to accept a single electron from the underlying ubiquinol, which passes another electron to the heme $\mathbf{b_L}$ of the opposite monomer from which it is anchored. The result is a ubiquinol with two positive charges.

STEP 3: The two positive charges of the ubiquinol at the Q_o-site, by means of the apolar-polar repulsion ($\mathbf{\Delta G_{ap}}$), disrupt the hydrophobic association of hydrophobic tip with Q_o-site, and the extended tether contracts, lifts the hydrophobic tip of RIP with its FeS center out of the Q_o-site, places the FeS center at the heme $\mathbf{c_1}$-site, reduces the heme $\mathbf{c_1}$ which passes its electron to heme **c** and thereby reduces cytochrome **c**. With the Q_o-site open, the two protons of ubiquinol^{+2} are released to the cytosolic side of the membrane, and ubiquinone leaves the Q_o-site to diffuse in the lipid bilayer. Thus, by the coupling of the hydrophobic and elastic consilient mechanisms, electron transfer couples to proton pumping.

STEP 4: Ubiquinol enters the Q_o-site, and the increased hydrophobicity of the Q_o-site draws the hydrophobic tip containing the oxidized FeS site of RIP back into hydrophobic association again stretching the single chain tether and bringing the cycle again into **STEPS 1–2**.

3.2 ATP Synthase of the Inner Mitochondrial Membrane

Figure 5A presents in space filling representation a cross-eye stereo view of the crystal structure of the F_1-motor of ATP synthase (Menz, Walker and Leslie, 2001). The perspective is from the top axis; it shows part of the γ-rotor that couples to the base of the F_0-motor, which drives clockwise rotation of the γ-rotor. The subunit composition is $\gamma(\alpha\beta)_3$. The α subunits, with each normally containing ATP, are non-catalytic sites, and the β subunits contain the catalytic sites, which may contain ADP, ADP plus Pi (inorganic phosphate), or ATP, or may be empty. Under physiological conditions, due to the perturbation of the γ-rotor and different occupancies of the three β catalytic subunits, the six globular subunits, $(\alpha\beta)_3$, exhibit a pseudo three-fold symmetry.

In Figure 5A the structure (1H8E) due to Menz et al., (2001) has all sites occupied by nucleotides and nucleotide analogues in such a way that the most hydrophobic side of the γ-rotor (to be defined below) faces a β catalytic subunit containing ADP plus SO_4^{-2} (an analogue of inorganic phosphate). This becomes key to the analysis

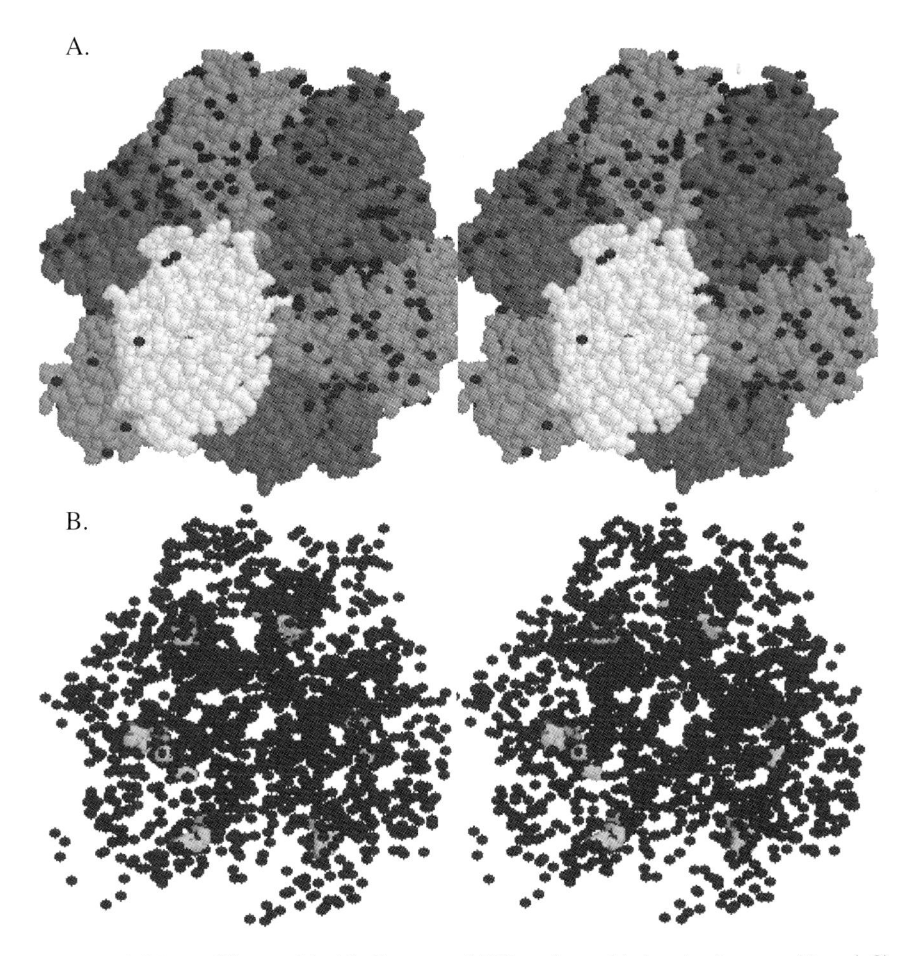

Figure 5. (A) Space filling model of the F_1-motor of ATP synthase with the subunit composition $\gamma(\alpha\beta)_3$. The structure shows the detectable water molecules on its surface. (B) After removal of the protein subunits, the space filling view of all water molecules detected by X-ray diffraction demonstrates the abundance of internal "waters of Thales" available to function by the hydrophobic consilient mechanism of apolar-polar repulsion for protein-based machines. See text for further discussion. Protein Data Bank, Structure File 1H8E due to Menz et al., 2001. Adapted from Urry, 2006b

that follows using the concept of an apolar-polar repulsion, $\mathbf{\Delta G_{ap}}$. Obviously for $\mathbf{\Delta G_{ap}}$ to be operative, there must be adequate interfacial water within the F_1-motor. In Figure 5B the protein subunits of 1H8E are removed leaving only the ligands and the detectable internal interfacial water. These we refer to as the 'Waters of Thales.' By comparison Figure 5A also contains but a few detectable water molecules that remain sufficiently in place on the surface of $\gamma(\alpha\beta)_3$ during data collection.

Three faces of the γ-rotor may be defined by structure 1BMF (Abrahams, Leslie, Lutter and Walker, 1994), which has an empty β catalytic site that would be expected to define the most hydrophobic face of the γ-rotor, and more polar ADP and ATP-containing β catalytic sites that would define more polar faces. The three faces of the γ-rotor defined in structure 1BMF are calculated by the summation, $\Sigma_i[\Delta G°_{HA}(X)_i]$, using the values for $\Delta G°_{HA}$ obtained from Table 1 with consideration that a residue at the edge of a particular face would have its effect split between two faces. The interest will be to see whether the most hydrophobic face does indeed correspond to that expected from the location of the empty β catalytic site.

The individual residue values, as from calculated differential calorimetry data using the derivation for $\Delta G°_{HA}$ (Urry, 2004), are summed as listed in Table 5 to obtain hydrophobicity values for the three faces. The results are truly striking. As listed in Table 5 and shown in Figure 6, there is a very hydrophobic face of −20 kcal/mol-face that opposed the empty β catalytic site, a neutral face with a hydrophobicity of 0 kcal/mol-face, and a somewhat polar face with a +9 kcal/mol-face. In fact, by visual observation with the gray-coding scheme where the more hydrophobic domains are darker, and the more polar domains are lighter, a similar qualitative result becomes apparent (Urry, 2006b; 2006c).

In structure 1H8E (Menz, Walker and Leslie, 2001), when the most hydrophobic face is juxtaposed to the very polar analogue, $ADP + SO_4^{-2}$, in the β catalytic site, there are two issues that can be considered. One issue is whether or not a ΔG_{ap} repulsion between the most hydrophobic face and the β catalytic site results in an increase in distance between the two. The answer to this issue is that there is indeed an increase in distance as reported by Menz et al., (2001) and also noted in a different way in our previous work (Urry, 2006b). The second issue is whether or not a prediction can be made as to the direction of rotation of the γ-rotor when the F_1-motor is functioning as an ATPase and when the rotation of the γ-rotor is driven by the F_0-motor in the inner mitochondrial membrane, which uses the energy

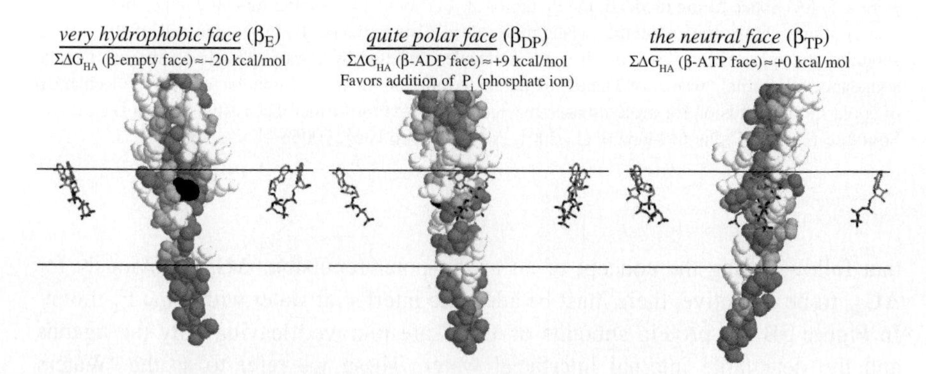

Figure 6. Hydrophobicities of the three faces of the γ-rotor of the F_1-motor of ATP synthase as defined using the Protein Data Bank, Structure File 1BMF due to Abrahams et al., 1994, and as calculated using the amino acid hydrophobicity values of ΔG_{HA} in Table. From Urry, 2006b

derived from the proton gradient developed by Complexes I, II, III, and IV of the electron transport chain to produce ATP.

Figure 7 provides a cross-eye stereo view looking through the γ-rotor in ribbon representation at the inner wall of the housing of the F_1-motor with the β catalytic site containing ADP plus SO_4^{-2} at near center and with one α subunit on each side (Urry, 2006b; 2006c). Several striking features require noting. The sulfate analogue for phosphate is fully exposed at the base of an aqueous cleft that opens out into the aqueous chamber and is directly opposed through the aqueous chamber to the most hydrophobic side of the γ-rotor. On the other hand when ATP is in the β catalytic site there is only a very small peephole to the β phosphate of ATP and no other phosphate is seen from the γ-rotor. Remarkably, on ATP hydrolysis the water-starved analogue of phosphate bursts to the surface of the housing in order

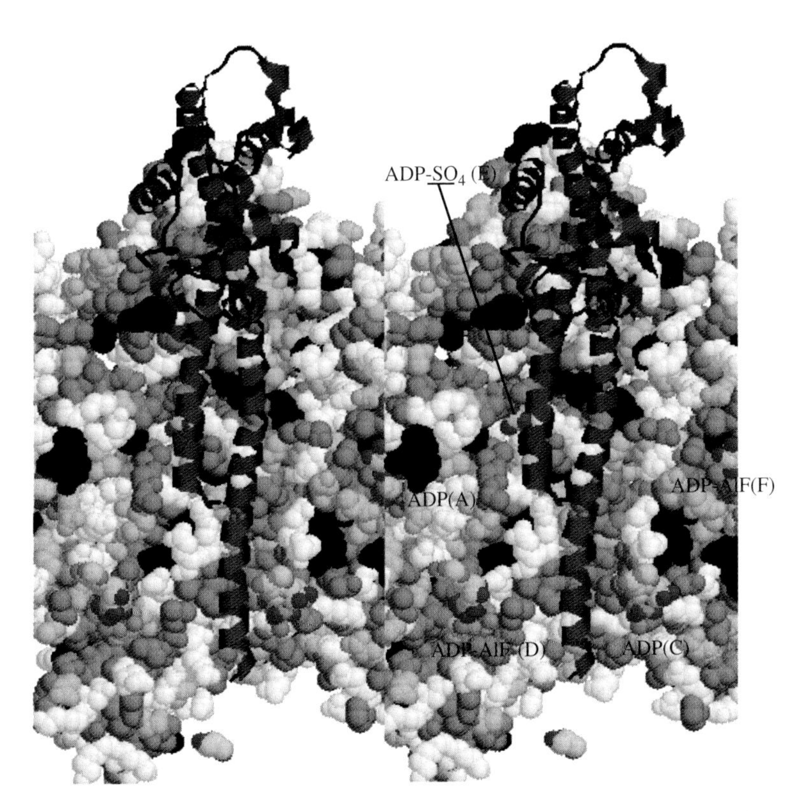

Figure 7. Stereo view of F_1-motor of ATP synthase in space-filling representation with neutral residues light gray; aromatic residues black, other hydrophobic residues gray and charged residues white. Shows view of SO_4^{-2} analogue of P_i from the perspective of the γ-rotor that occurs in chain the β-catalytic site containing ADP plus SO_4^{-2}. The overlying γ-rotor in dark gray ribbon representation has its most hydrophobic face opposed to the highly charged site. The ΔGap due to SO_4 applies its torque to N-terminal sequence of γ-rotor that would give a counter-clockwise rotation as ATPase. Protein Data Bank, Structure File 1H8E due to Menz et al., 2001. Adapted from Urry, 2006b

to access hydration of the internal aqueous chamber. In accessing water, however, it must compete with the most hydrophobic face of the γ-rotor. The result is a dramatic apolar-polar repulsion that allows an understanding of ATP synthesis and allows a prediction of the direction of rotation of the γ-rotor when the F1-motor functions as the F_1-ATPase.

In Figure 7, the competition for hydration between the most charged state, ADP plus SO_4^{-2}, of the β catalytic site and the most hydrophobic surface of the γ-rotor drives the two sites to move away from each other as each attempts to obtain hydration less perturbed by the other (Urry, 2006b; 2006c). Looking at Figures 7 and 8, the $\mathbf{\Delta G_{ap}}$ repulsion is directed almost entirely at the shorter α-helix and would drive the γ-rotor in a counterclockwise direction, as found experimentally in the remarkable studies of Noji et al., (1997). As the driving force works through an aqueous chamber with no essential friction between subunits, the conversion of chemical energy into motion can be expected to occur at high efficiency (Kinosita, Yasuda and Noji, 2000). For more extensive treatment and analyses of the mechanism of the F_1-motor of ATP synthase (See Urry, 2006b; 2006c).

3.3 Myosin II Motor of Muscle Contraction

The objective here, in a brief discussion of the myosin II motor, is to note physical processes of the hydrophobic and elastic consilient mechanisms that bear analogy to those of the other protein-based machines. This motor is viewed as one that converts chemical energy derived from the hydrolysis of ATP to ADP and P_i to produce motion by a sliding filament mechanism. Myosin filaments, by means of cross-bridges to actin filaments, increase overlap of the filaments during contraction and decrease the extent of filament overlap during relaxation. In our view developed from the crystal structures of Himmel et al., (2002), the operative structure is the cross-bridge that reaches out along the actin filament by hydrophobic dissociations, binds to the actin filament by hydrophobic association and contracts by primarily an intra-cross-bridge twisting hydrophobic association that stretches single chains between the two hydrophobic associations and shortens the cross-bridge by the contraction of the extended single chains to draw the myosin filament into greater overlap with the actin filament.

More specifically in our view (Urry, 2006b; 2006c), the function of the myosin II motor derives from the action of two hydrophobic associations, one at the cross-bridge-actin binding site and a second within the cross-bridge (involving the head of the lever arm and the underside of the N-terminal domain). These hydrophobic associations form on decreasing charge (primarily due to release of P_i resulting in a decrease in $\mathbf{\Delta G_{ap}}$) with stretching of interconnecting chain segments, and contraction of the stretched interconnecting chain segments drives the actin and myosin filaments into greater overlap. The most polar state of ADP plus Pi disrupts the hydrophobic associations due to cleft-directed apolar-polar repulsions, $(\mathbf{\Delta G_{ap}})_i$. that allow the cross-bridge to reach forward along the actin filament to its next attachment site.

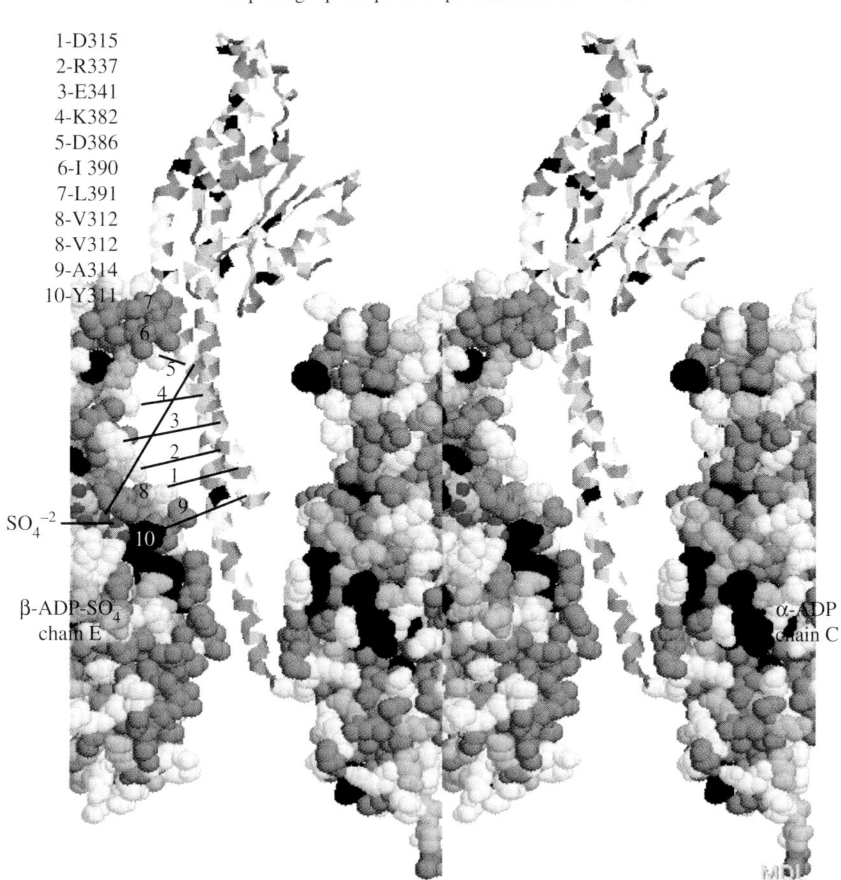

Figure 8. Cross-eye stereo view of the F_1-motor of ATP synthase with γ-rotor in ribbon and E and C chains in space-filling representations with residues indicated as neutrals light gray, hydrophobics in black and gray and charged residues in white. Lines to indicate a repulsive interaction between hydrophobic γ-rotor and SO_4^{-2}, in place of hydrolyzed γ-phosphate, and augmented by newly emerged charged groups that occur in the β-catalytic chain. The ΔGap due to SO_4 and emerged charged groups apply a torque to the double-stranded portion of γ-rotor that would give a counter-clockwise rotation when functioning as an ATPase. Protein Data Bank, Structure File 1H8E due to Menz et al., 2001. Adapted from Urry, 2006b

3.4 Kinesin Bipedal Walking Motor

Utilizing the structure of Kozielski et al., (1997) for kinesin, an analogy is drawn between the behavior of the Rieske Iron Protein (RIP) and the kinesin bipedal motor that walks along microtubules transporting its cargo on its back. In Figure 9A, viewing only the two RIP subunits of the Complex III structure of Lange and Hunte (2002), the two subunits of RIP can be viewed almost as though the two α-helix membrane

A.

B.

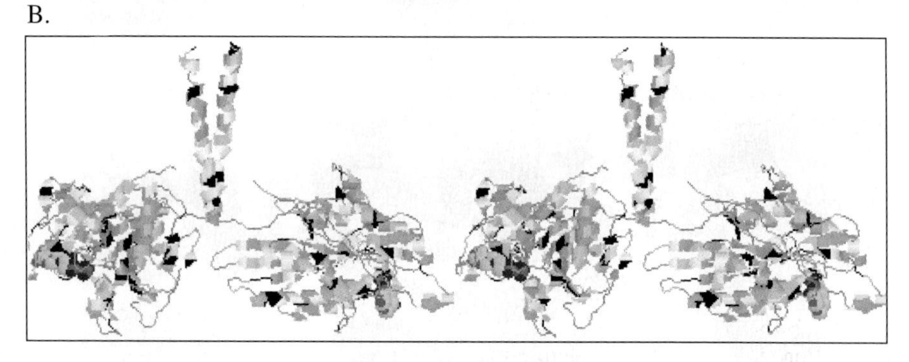

Figure 9. (A) Rieske Iron Protein subunit of Complex III to show transitional argument to walking kinesins. The tether becomes stretched on optimization of hydrophobic association by rolling the hydrophobic globular tip of the Rieske Iron Protein component into an optimized hydrophobic association at the Q_O-binding site. Protein Data Bank, Structure File 1KYO due to Lange and Hunte, 2002 (B) Kinesin dimer with right ATP-containing subunit at binding site but with the left subunit that has yet to rotate into and optimize hydrophobic association with the microtubular binding site thereby stretching connecting single chain tethers. See text for further discussion. Protein Data Bank, Structure File 1LNQ due to Jiang et al., 2002. From Urry, 2005a

anchors were associated as a double stranded α-helical coiled coil as occurs for kinesin in Figure 9B. With this orientation the analogy of enhanced hydrophobic association coupling with stretching of a single chain tether can be developed.

Recall in **STEP 1** of Figure 4, we have proposed (Urry, 2006a; 2006c), that the initial association of the globular component of RIP with the Q_O-site occurs by a peripheral hydrophobic contact with residue F169, that the hydrophobic association increases as the hydrophobic tip of RIP rolls into the hydrophobic depression which is the Q_O-site, and that in doing so the single chain tether becomes stretched. In Figure 9B, the hydrophobic surface is on the underside, i.e., on the bottoms of the feet of the bipedal motor, whereas in the orientation of Figure 9A, the hydrophobic tip of RIP is on the top side. Nonetheless, the physical operation would be the same for kinesin. In Figure 9B, the foot on the right side would be planted in hydrophobic association with a site on the microtubule and connected to the coiled coil, which attaches the two feet, by a single somewhat extended single chain.

Here we postulate that the foot on the left side of Figure 9B would have a poor hydrophobic contact with the microtubule, but one that could be enhanced as it rolls forward and into the binding site for the leading foot of kinesin on the microtubule. Again in rolling into a more favorable hydrophobic association the single chain 'linkers' (equivalent to the tether of RIP, but in this case attached at the coiled coil instead of anchored in the membrane) become stretched. Postulating further, the forward stretching of the linkers triggers hydrolysis in the trailing foot, and the formation of the most polar state of ADP plus P_i disrupts the hydrophobic association of trailing foot with microtubule by the resulting apolar-polar repulsion, $\mathbf{\Delta G_{ap}}$. The serially aligned stretched linkers propel the trailing foot forward to take one step along the microtubule to a limited hydrophobic association as the forward foot, and the newly positioned forward foot rolls into optimized hydrophobic association at the binding site for a leading foot. This again stretches the linkers and so on. As in general, hydrophobic association stretches an interconnecting chain segment and formation of the most polar state, ADP plus P_i, disrupts hydrophobic association.

3.5 Calcium-Gated Potassium Channel

Without having analyzed the structure of the calcium-gated potassium channel, but with a general sense of structure and its change on calcium ion binding, the channel gating process is proposed based on the common perspective whereby hydrophobic association stretches interconnecting chain segments to achieve function, as argued above in particular for the RIP of Complex III, the myosin II motor, and kinesin.

The calcium-gated potassium channel is a four-fold symmetric structure with a membrane component containing the selectivity filter formed from a highly conserved sequence called the "K^+ channel signature sequence" and with an extra-membranous component to which calcium ion binds to initiate conductance (Jiang, et al., 2002).

The potassium ion selectivity filter spans less than half way across the lipid bilayer and does not change shape during the opening and closing of the K^+ channel. The transmembrane channel beyond the selectivity filter is closed by the association of one α-helix from each of the four subunits of the fourfold symmetric structure to form a four-helix bundle. The four chains leave the four-helix bundle within the membrane to form four tethers that connect to a fourfold symmetric arrangement of hydrophobically associated globular extra-membranous components.

Our perspective, which has yet to be examined, is that calcium ion binding to the extra-membranous globular components changes their hydrophobic associations in a way that pulls on and stretches the tethers connecting to the intra-membrane four-helix bundle and causes the bundle to separate with the result of opening the channel. In our elastic-contractile model proteins, we have designed sequences wherein calcium binding at carboxylates drives hydrophobic association (Urry, 2006c). Thus, we propose that calcium ion binding changes hydrophobic association of the extra-membranous globular components in a manner that extends the four interconnecting chain segments and thereby opens the channel. Again, an energy input, in this case of calcium ion changes hydrophobic association of the extramembanous assembly of globular units in such a way as to stretch interconnecting chain segments. This concept of hydrophobic association stretching interconnecting chain segments to achieve function was first explicitly stated in reference (Urry and Parker, 2002).

ACKNOWLEDGEMENT

The author gratefully acknowledges support of the Office of Naval Research (ONR) by means of Grant No. N00014-98-1-0656 and Contract No. N00014-00-C-0404.

REFERENCES

Abrahams JP, Leslie AGW, Lutter R, Walker JE (1994) Structure at 2.8 Å of F_1–ATPase from bovine heart mitochondria. Nature (London) 370:621–628. Protein Data Bank, Structure File 1BMF

Berman HM, Westbrook J, Feng Z, Gilliland G, Bhat TN, Weissig H, Shindyalov IN, Bourne PE (2000) The Protein Data Bank. Nucleic Acids Research 28:235–242

Himmel DM, Gourinath S, Reshetnikova L, Shen Y, Szent-Gyorgyi AG, Cohen C (2002) Crystallographic findings on the internally uncoupled and near-rigor states of myosin: Further insights into the mechanics of the motor. Proc Natl Acad Sci USA 99:12645–12650. Protein Data Bank, Structure Files 1KK7 and 1KK8

Jiang Y, Lee A, Chen J, Cadene M, Chait BT, MacKinnon R(2002) Crystal structure and mechanism of a calcium-gated potassium channel. Nature 417:515–522 Protein Data Bank, Structure File 1LNQ

Kinosita K, Yasuda R, Noji H (2000) F_1-ATPase: A highly efficient rotary ATP machine. In: Banting G, Higgins SJ (eds), Essays in Biochemistry, Molecular Motors. Portland Press, 35: 3–18

Kozielski F, Sack S, Marx A, Thormählen M, Schönbrunn E, Biou V, Mandelkow E-M, Mandlekow M (1997) The crystal structure of dimeric kinesin and implications for microtubule-dependent motility. Cell 91:985–994. Protein Data Bank, Structure File 3KIN

Lange C, Hunte C (2002) Crystal structure of the yeast cytochrome bc_1 complex with its bound substrate cytochrome c. Proc Natl Acad Sci USA 99:2800–2805. Protein Data Bank, Structure File 1KYO

Martz E (2002) "FrontDoor to Protein Explorer 1.982 Beta" Copyright © 2002, proteinexplorer.org

Menz RI, Walker JE, Leslie AGW (2001) Structure of bovine mitochondrial F_1-ATPase with nucleotide bound to all three catalytic sites: Implications for mechanism of rotary catalysis. Cell 106:331–341. Protein Data Bank, Structure File 1H8E

Mota F, Teixeira M (2005) Crystal structure and mechanism of a calcium-gated potassium channel: MthK. Report for the Post-graduate Training Course: Biology's Engineering Principles for Design of Protein-based Machines and Materials. University of Minho, Braga, Portugal, Spring

Noji H, Yasuda R, Yoshida M, Kinosita K (1997) Direct observation of the rotation of F_1–ATPase. Nature (London) 386:299–302

Pollack GH (2001) Cells, Gels and the Engines of Life: A New Unifying Approach to Cell Function Ebner and Sons, Seattle

Stackelberg Mv, Müller HR (1951) Zur Struktur der Gashydrate. Naturwissenschaften 38:456

Stackelberg Mv, Müller HR (1954). "Feste Gashydrate II: Struktur und Raumchemie." Zeitschrift für Elektochemie 54:25–39. (now Berichte der Bunsengesellschaft für physicalische Chemie)

Teeter MM (1984) Hydrophobic protein at atomic resolution: Pentagonal rings of water molecules in crystals of Crambin. Proc Natl Acad Sci USA 81:6014–6018

Urry DW (1992) Free energy transduction in polypeptides and proteins based on inverse temperature transitions. Prog Biophys Mol Biol 57:23–57

Urry DW (1997) Physical chemistry of biological free energy transduction as demonstrated by elastic protein-based polymers. J Phys Chem B 101:11007–11028

Urry DW (2004) The change in Gibbs free energy for hydrophobic association: Derivation and evaluation by means of inverse temperature transitions. Chem Phys Lett 399:177–183

Urry DW (2006a) Deciphering engineering principles for the design of protein-based nanomachines. In: Renugopalakrishnan V, Lewis R, Dhar PK (eds), Protein-Based Nanotechnology Springer-Verlag (Kluwer Academic Publishers) (in press)

Urry DW (2006b) Function of the F_1-motor (F_1-ATPase) of ATP synthase by apolar-polar repulsion through internal interfacial water. Cell Biol Int 30:44–55

Urry DW (2005a) Hydrophobic and elastic mechanisms in Complex III/Rieske Iron Protein (RIP), walking protein motors and protein-based materials. In: Shimohigashi Y(ed.), The Japanese Peptide Society, Proceedings of Asian Pacific International Peptide Symposium, APIPS-JPS 2004, pp 115–118 (ISSN 1344 7661)

Urry DW (2005b) Protein-based polymers: Mechanistic foundations for design and processing. Proceedings of the 21st Annual Meeting of the Polymer Processing Society, Leipzig, Germany, June 19–23

Urry DW (2006c) What Sustains Life? Consilient mechanisms for protein-based machines and materials. Springer-Verlag, LLC, New York, ISBN: 081764346X

Urry DW, Parker TM (2002). Mechanics of elastin: Molecular mechanism of biological elasticity and its relevance to contraction. J Muscle Res Cell Mobility 23:541–547; Special Issue: Mechanics of Elastic Biomolecules, Henk Granzier, Miklos Kellermayer Jr., Wolfgang Linke, Eds

Urry DW, Peng S-Q, Xu J, McPherson DT (1997) Characterization of waters of hydrophobic hydration by microwave dielectric relaxation. J. Am. Chem. Soc 119:1161–1162

Wilson EO (1998) Consilience, The Unity of Knowledge Alfred E. Knopf, New York, p 8

Zhang Z, Huang L, Shulmeister VM, Chi YI, Kim KK, Hung LW, Crofts AR, Berry EA, Kim SH (1998) Electron transfer by domain movement in cytochrome bc_1. Nature 392:677–684. Protein Data Bank, Structure Files 1BCC and 3BBC

CHAPTER 7

THE EFFECTS OF STATIC MAGNETIC FIELDS, LOW FREQUENCY ELECTROMAGNETIC FIELDS AND MECHANICAL VIBRATION ON SOME PHYSICOCHEMICAL PROPERTIES OF WATER

SINERIK N. AYRAPETYAN*, ARMINE M. AMYAN AND GAYANE S. AYRAPETYAN

UNESCO Chair-Life Sciences International Postgraduate Educational Center, 31 Acharyan St., Yerevan, 375040, Armenia

Abstract: At present the biological effect of SMF and LF EMF can be considered as a proven fact; however, the question how such a low-energy of EMF radiation could modulate the functional activity of cell and organism still remains unanswered. Numerous hypotheses on molecular mechanisms of the specific biological effect of EMF have been proposed, but none have provided a reliable and exhaustive explanation of the experimental findings. The oldest hypothesis is that EMF-induced structural changes of the cell bathing solution could serve as a primary target for the biological effect of EMF. As water is the main medium where the major part of biochemical reactions are taking place, it is predicted that a slight changes of physico-chemical properties of both intracellular and extracellular water could dramatically change the metabolic activity of cells and organisms

Therefore, extension of the knowledge on the mechanisms of SMF and EMF effects on physicochemical properties of water seems extremely important for understanding the biological effect of these factors, which are realized through water structural changes

Keywords: water structure; valence angle; distilled water; thermal capacity; melting point; specific electrical conductivity

* Corresponding author. UNESCO Chair-Life Sciences International Postgraduate Educational Center; 31 Acharyan St., Yerevan, 375040, Armenia. Tel.: +374 10 624170/612461; fax: +374 10 624170; E-mail address: life@arminco.com (S.N. Ayrapetyan)

G. Pollack et al. (eds.), Water and the Cell, 151–164.

1. GENERAL NOTES ON WATER STRUCTURE

The structure of a single water molecule is well described in literature. From 5 pairs of electrons one pair is localized near the oxygen nucleus and the rest 4 pairs are socialized between protons and oxygen nucleus. The oxygen nucleus partly attracts to the electrons moving them away from the hydrogen nuclei. The latter acquires a weak positive charge. The other two corners of the imaginary tetrahedron acquire a weak negative charge near the oxygen atom. Moreover, 2 pairs are polarized and directed to the peaks of the tetrad opposite the protons. These unshared pairs of electrons have a crucial role in generation of intermolecular hydrogen bounds (Figure 1). Hydrogen bounds continuously form and disrupt giving the "water polymer" a high surface tension, high specific heat, high vaporization heat and high dielectric constant ($\varepsilon = 80$ at 20 °C). According to the quant-mechanical calculations the valence angle in water molecules between O-H bounds must be 90°, however,

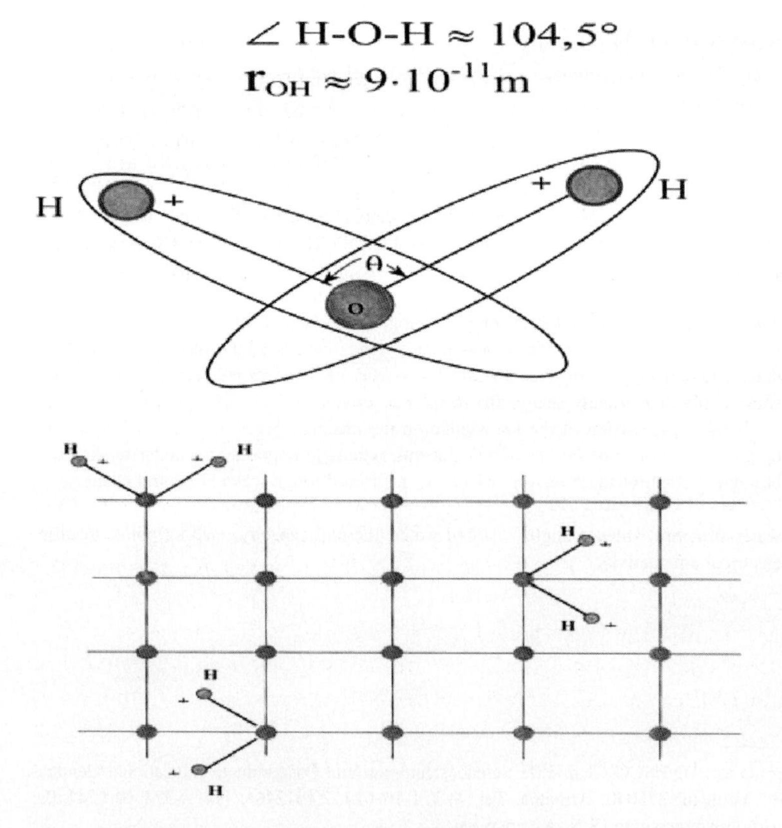

Figure 1. The theoretical conception of water structure. Each H_2O is labile linked to other four molecules with hydrogen bonds: the result is a polymeric structure of water

in reality this valence angle is near 105°, because in water, due to the strong polarity of the H-O bounds, the minimal repulsion of the positively charged hydrogen atoms increases the angle (Pullman and Pullman, 1963).

Because of the long hydrogen bound (0.28 nm) in water having an electrostatic nature and a comparatively weak energy (14.2–20.9 k joule) the water structure is very labile and sensitive to different environmental factors. The structure of liquid water is being continuously changed from the moment of its forming. The character of such changes depends on the physical and chemical characteristics of the environmental medium. Even by keeping the distilled water in constant medium its structure is being changed depending on its 'aging' (Stepanyan et al., 1999). Therefore the structure of the water could be considered as a currier of a big 'memory' on the previous effects of various environmental factors.

2. THE EFFECT LF EMF, LF MV AND SMF ON THERMAL PROPERTIES OF WATER

From the point of present knowledge on water structure, the LF EMF could modify the water structure by two pathways: a) by changing the valence angle in water molecules and b) by mechanical vibration (MV) of dipole molecules of water. To estimate the role of each of these pathways in EMF-induced water structure changes the effects of SMF and MV on water physicochemical properties were studied. It is suggested that SMF effect would imitate the valence angle changes, while the effect of MV – the mechanical vibration of dipole molecules of water. It is predicted that EMF- and MV-induced water structure changes would accompanied by the thermal release in the result of broken hydrogen bounds between water molecules.

A special setup allowing the ***treatment of distilled water*** by SMF, ELF EMF and LF MV was assembled (Institute of Radiophysics and Electronics (IRPhE) of Armenian NAS, Yerevan, Armenia). The block scheme of the setup is presented in Figure 2.

Glass test tube (1) with diameter 10 mm and volume 10 ml, was used. The vibrator was controlled by the sine-wave generator (6) (GZ-118, Made in Russian Federation), the signal went to the double pole switch (8): in position I the generator functions as EMF and LF MV sources, while in position II – as LF MV sources. To obtain MV waves the vibrating device (3) was used generating vertical vibrations by set frequency and intensity. The vibrator was constructed in the department of engineering at LSIPEC on the basis of the IVCh-01 device (Russian production) To keep vibration intensity constant (30 dB) at different frequencies, a coil (4) with a feedback amplifier system (IRPhE, Yerevan, Armenia) was used. Thus, MV was transmitted to the test tube containing DW with insignificant power dissipation. For concordance of high impedance output of generator to low impedance input of vibrator, a special power amplifier (IRPhE, Yerevan, Armenia) was used. MV

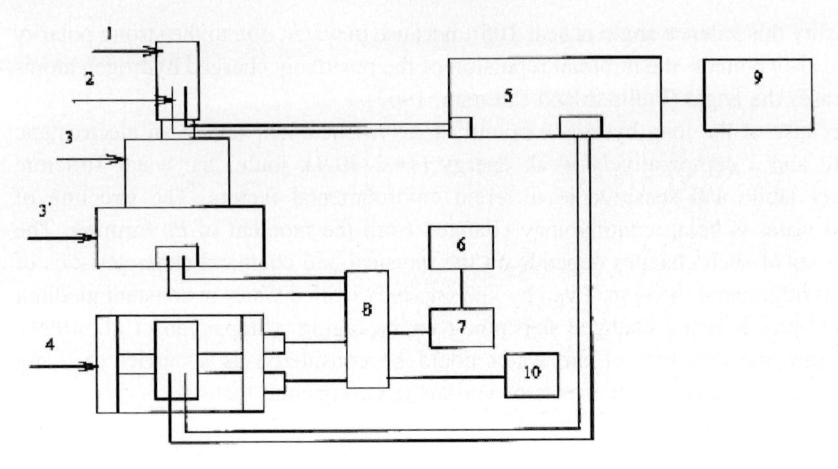

Figure 2. The setup for treatment of DW by LF EMF, SMF and LF MV. 1. Glass test tube with diameter 10 mm and volume 10 ml. 2. Platinum electrodes. 3. Mobile part of the vibrator. 3′. Motionless part of the vibrator. 4. The coil. 5. The device for the measurement of DW SEC (conductometer). 6. Generator of sinusoid vibration. 7. The low-noise amplifier. 8. The switch (has 2 positions: I and II, where I- EMF and MV and II- EMF). 9. Personal Computer. 10. The generator of a constant field

frequency was controlled by a cymometer (CZ-47D, production of Russian Federation), while the intensity was measured by a measuring device (IRPhE, Yerevan, Armenia) having a sensor on the vibration table. It was possible to keep the intensity of MV on stable level at all frequencies, including resonance frequency (more than 200 Hz for the given setup).

EMF was generated by the controlled generator (6) and low-noise amplifier (7) on the coil (4) (IRPhE, Yerevan, Armenia). The coil had a cylindrical form with 154 mm in diameter and 106 mm in height. The coil consisted of Helmholtz rings generating the homogeneous magnetic field. Rings of Helmholtz were formed by two equal ring coils located coaxially and parallel. The distance between ring coils was equal to their radius (77 mm). The magnetic field created by these rings had high homogeneity, for example, at a distance of 0.25sm from the center of an axis strength differs from computed by formula only on $0.5\% H = 71.6 \cdot \omega \cdot \frac{I}{R}$. SMF was generated by the generator of a static field (10) and transferred to the coil.

For ***determination of the thermal characteristics*** of DW during EMF exposure the following works were performed: new created DW (10 ml.) was placed into the Helmholtz rings for EMF exposure. A needle thermo-sensor of the measuring device Biophys-TT (LSIPEC, Armenia) was placed in the test-tube. During the EMF exposure the following frequencies were used: 4,10,15,20 and 50 Hz. The device Biophys-TT was connected to the personal computer through Digidata 1322A data acquisition system (Axon Instruments, USA). The data recording was carried out with the help of computer program Axsoscope 8.1.

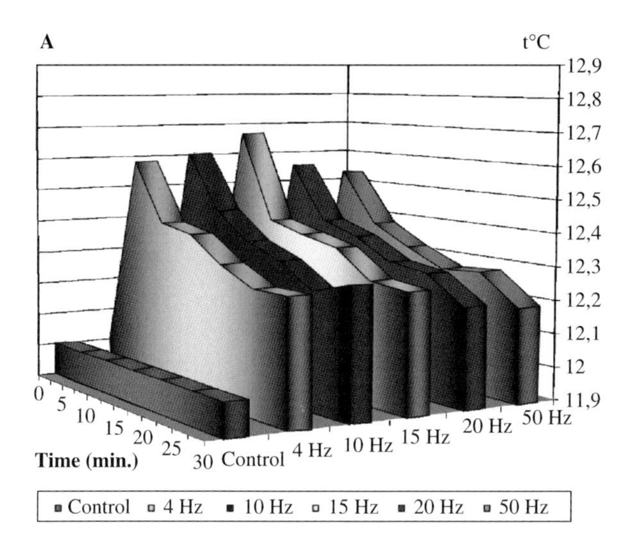

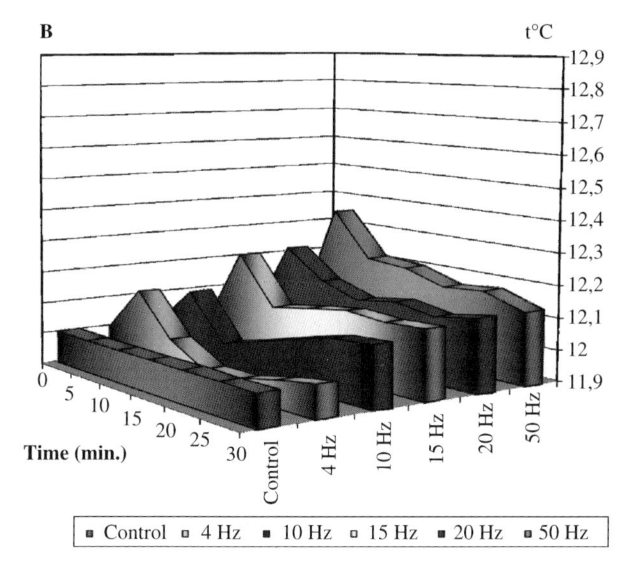

Figure 3. The time- and frequency-dependent heat release from the water samples treated by EMF (2.5 mT) (A), MV (B) and MV (30 dB) after 30 min pre-treated by SMF (12 mT) (C). Initial temperature – 11.9 °C

As it can be seen from the presented data in Figure 3, the character of frequency-dependency of heat release is changed during EMF (A) and MV (B) exposure, as well as it could be modulated by preliminary exposure to SMF (C).

These data strongly suggest that the sensitivity of water structure to these factors depends on the preliminary state of the water.

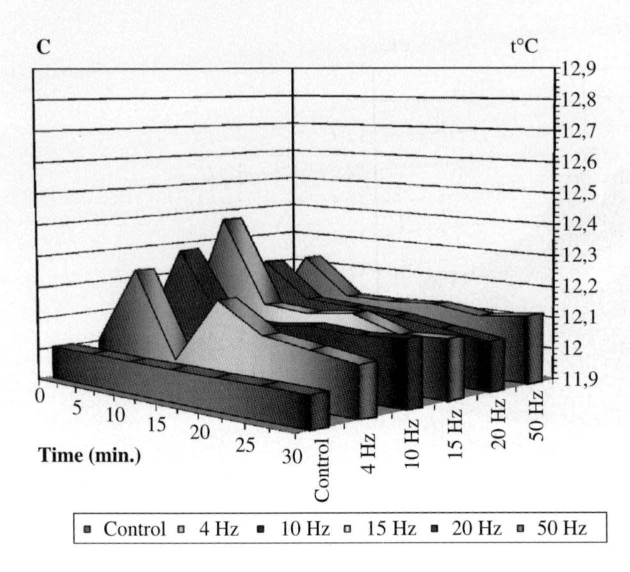

Figure 3. (Continued)

The results of studying the melting processes of water pretreated by EMF, MV and SMF after freezing in liquid N_2 brought us to the same conclusion.

For studying the ***time-dependence changes of thermal capacity*** of EMF-, MV- and SMF-pretreated DW after freezing in liquid N_2 the following method was used: the plastic tube (Vol. 1 ml) with a hermetic cup having a thermo-sensor at the bottom was fixed in another plastic tube (vol. 100 ml) and was inserted into the well containing liquid N_2. After withdrawing the tube from the liquid N_2 the hermetic cup of the tube was opened and left for melting at room temperature. The temperate recording was performed by extra sensitive thermometer Biophys-TT (production of LSIPEC, Armenia), connected to the PC through Digidata 1322A data acquisition system (Axon Instruments, USA).

The family of curves of time-dependent temperature raising at room temperature (18 °C) of EMF- (A and A*), MV- (B and B*) and SMF- (C and C*) pretreated 1 ml water after its freezing in liquid N_2 are demonstrated in Figure 4.

As it can be seen from the presented data the melting point (when the temperature keeps constant) and the time of reaching to 0 °C (marker for the thermal capacity of frozen crystals), as well as thermal capacity and thermal anomaly properties of liquid water are frequency (A,B) and intensity (C)-dependant. Comparing the family of curves of the left and right columns, the "aging" effect on frequency and intensity-dependence of the water thermal properties can be seen. From these data we can conclude that water "memory" on the effect of various factors could be modified by water 'aging'.

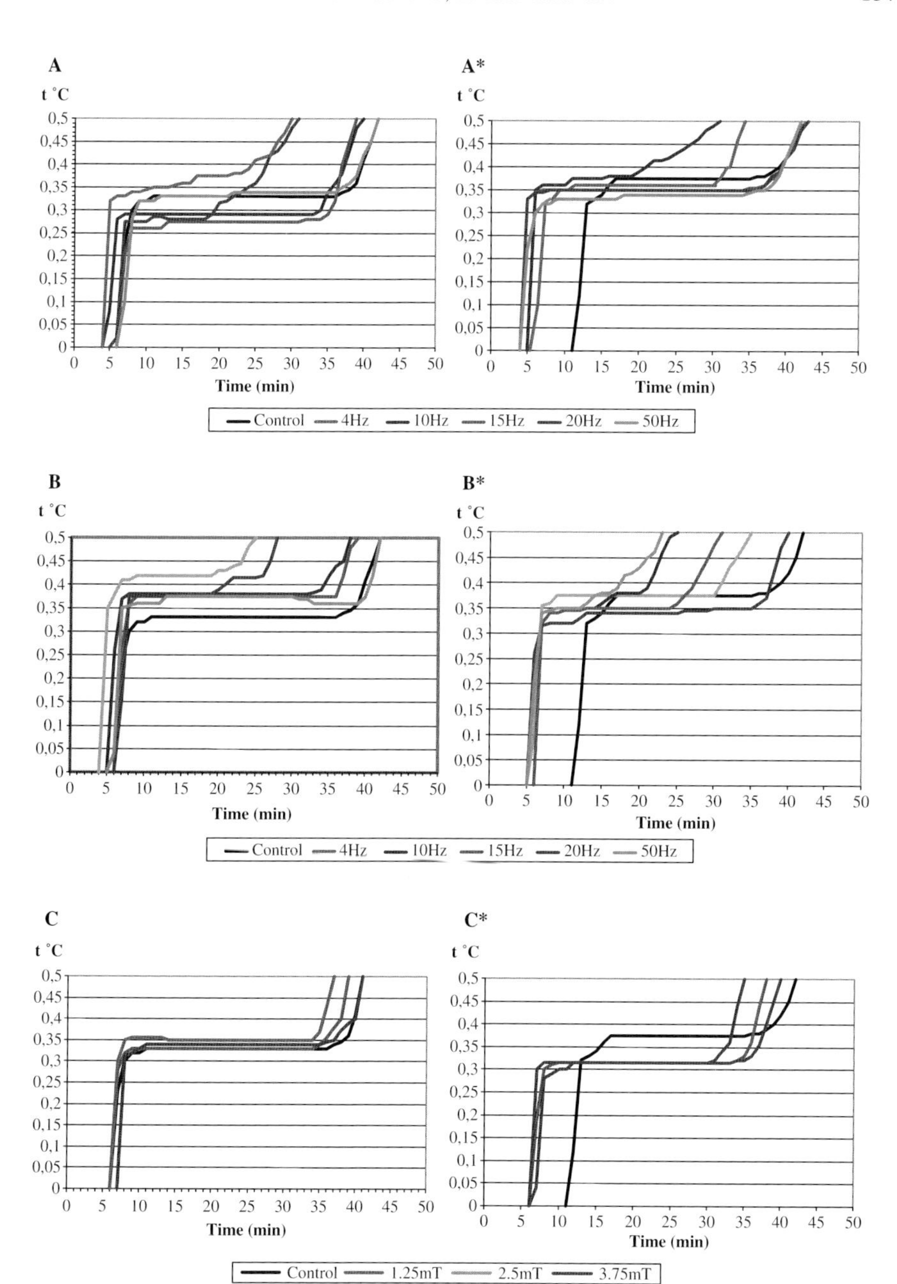

Figure 4. Time-dependent temperature rising of EMF- (A and A*), MV- (B and B*) and SMF- (C and C*) pretreated 1 ml DW at room temperature (18 °C) after freezing in liquid N_2. In the left column (A,B,C) – the one-hour water was 30 minutes treated by EMF, MV, SMF and immediately frozen in liquid N_2. In the right column (A*, B* and C*) – 30 minutes EMF, MV and SMF-treated water was frozen after 72 hours remaining at room temperature. EMF and MV have intensity 2.5 mT and 30 dB, correspondingly

3. THE EFFECT LF EMF, LF MV AND SMF ON SPECIFIC ELECTRICAL CONDUCTIVITY OF WATER

As SEC of water depends on the degree of its dissociation, SEC can be considered as a marker for studying the effect of different factors on water structure (Klassen, 1982; Ayrapetyan, 1994a). To estimate the contribution of valence angle changes and mechanical vibration of dipole moments of water molecules in LF EMF-induced water structure changes, the SMF, LF MV and LF EMF effects on SEC of DW were studied (Ayrapetyan, 1994a; Stepanyan et al., 1999; Hakobyan and Ayrapetyan, 2001).

The block scheme of the setup for these studies is presented in Figure 2. Three glass test tubes (1) with diameter of 10 mm and volume of 10 ml, with two platinum electrodes inside were used. Platinum electrodes-plates with the area 100 mm^2, located on 5 mm distance from each other, were connected with the conductivity-measuring device (5) capable to determine SEC of water at currents less than the 10^{-9} A. As the conductivity of water was measured in micro power modes, the application of low-noisy voltage amplifier of alternating current in the device raises the accuracy of measurement due to exception of self-heating influence. For the continuous recording of SEC the output of a measuring device was connected to the PC (9) through Digidata 1322A data acquisition system.

The presented data in Figure 5 show that EMF at 4, 10, 20 and 50 Hz has depressing effect on SEC of one-day DW, while in case of six-day DW only 4 and 20 Hz EMF has depressing effect on it. It is extremely interesting that the 20 Hz frequency 'window' was less pronounced at higher intensity of EMF (>10 mT) (Figure 6) (Stepanyan et al., 2000).

The similar frequency 'windows' were observed by studying the LF MV effect on SEC of DW (Figure 7). However, in case of MV effect on SEC of one-day DW, comparing to EMF, 15 Hz also has depressing effect on water SEC (Figure 7A).

As in case of EMF effect, MV at 20 Hz has less expressed depressing effect on water SEC at higher intensity (75 dB) (Figure 8) than at a weak intensity (30 dB) (Figure 7).

SMF also had a depressing effect on SEC of DW however, this effect was less sensitive to water 'aging', than in case of EMF and MV (Figure 9).

In order to find out whether these factors have specific effect on water SEC, the combined effect of 4 Hz EMF (2,5 mT), 4 Hz MV (30 dB) and SMF (2,5 mT) in different orders was investigated on one-day DW. These results are shown in Figure 10.

As it can be seen on the presented data there are no significant differences between various combinations of factors-induced depressing effect on SEC of DW, which shows that all these three factors lead to the packing of the water molecules that brings to the decrease of SEC of DW. However, whether the LF EMF-, LF MV- and SMF-induced decrease of SEC of DW has the same biological mining, could serve as a subject for future investigations.

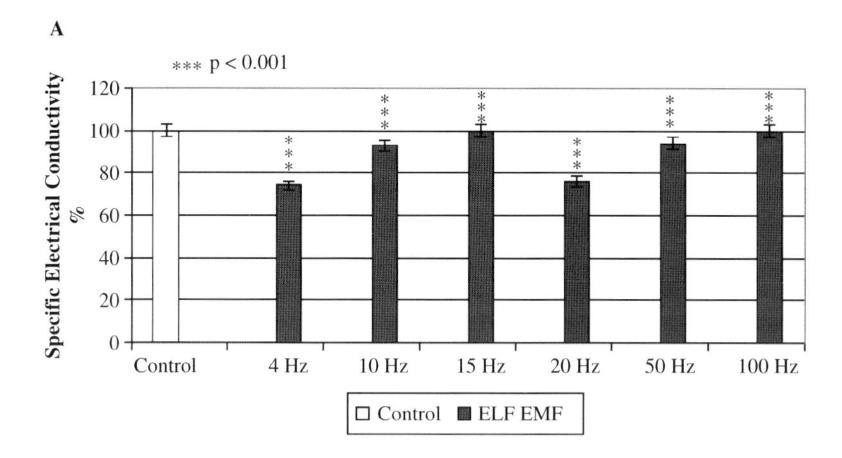

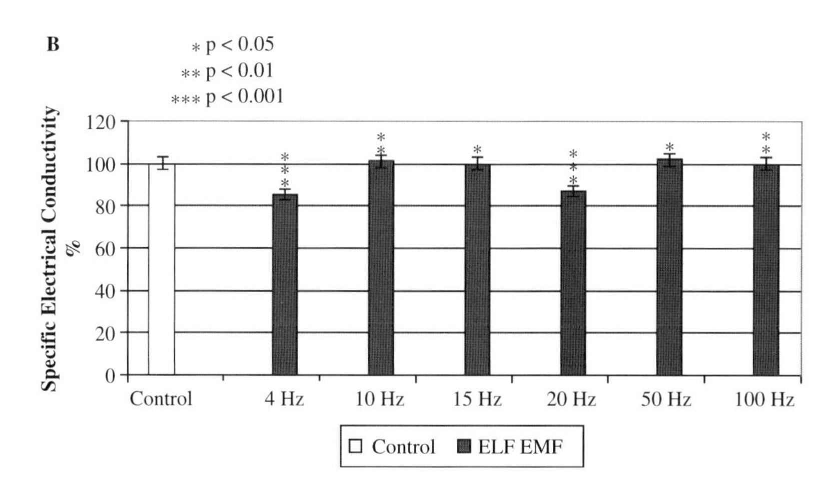

Figure 5. The effect of EMF (2,5mT) exposure at different frequencies on specific electrical conductivity of one-day (A) and six-day (B) distilled water at 18 °C

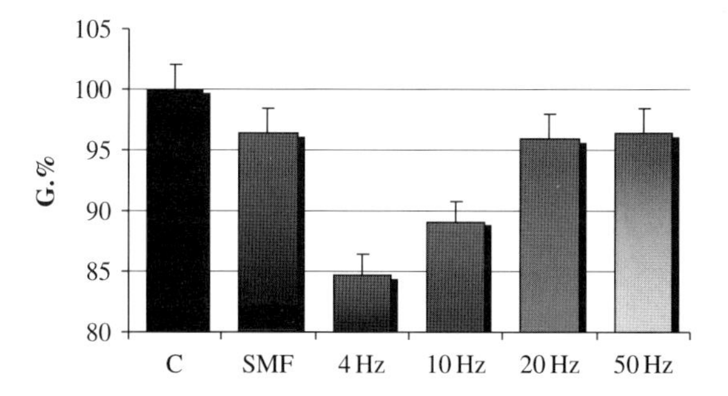

Figure 6. The effect of EMF (12 mT) exposure at different frequencies on specific electrical conductivity of one-day distilled water at 18 °C

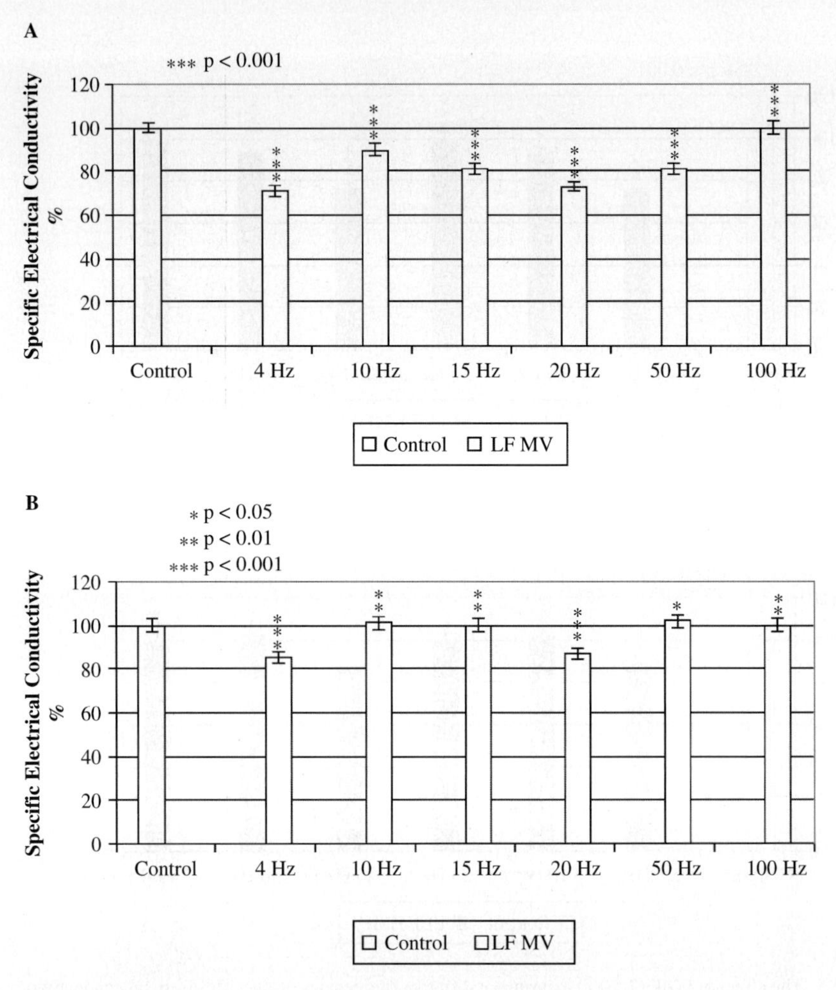

Figure 7. The effect of EL MV exposure at different frequencies on specific electrical conductivity of one-day (A) and six-day (B) distilled water at 18 °C

The preliminary studies of our laboratory have shown that LF EMF-, LF MV- and SMF-induced water structure changes have different biological effects on growth and development of Escherichia Coli (Stepanyan et al., 2000; Ayrapetyan et al., 2001) and plant seed germination potentials (Amyan and Ayrapetyan, 2004). It was shown that pretreatment of wort by EMF and SMF has depressing effect on growth and development of microbes (Stepanyan et al., 2000), while MV has activation effect on it (Ayrapetyan et al., 2001). Different effects of the mentioned factors on plant seed germination potential have also been observed. The metabolic-depended seed hydration was elevated in EMF-treated DW, while in MV-treated DW seed hydration was decreased (Amyan and Ayrapetyan, 2004). The comparative

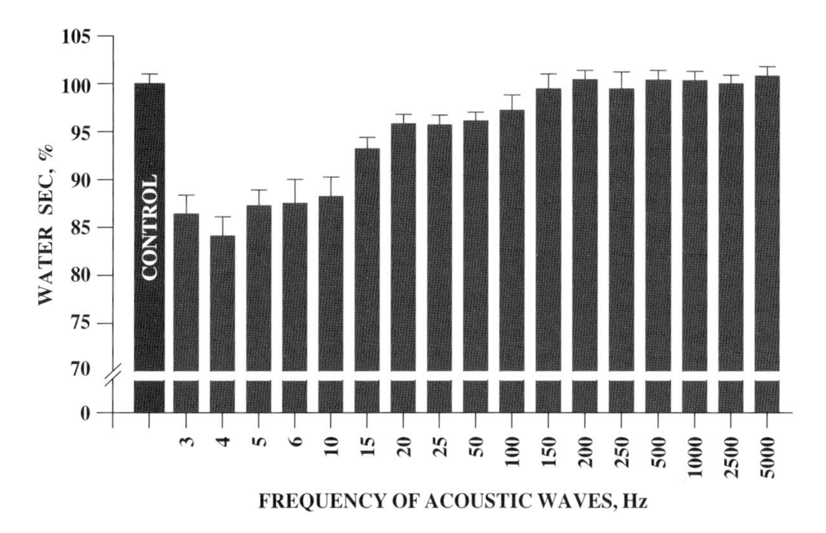

Figure 8. The effect of mechanical vibrations at different frequencies (at the intensity of 75 dB) on the specific electrical conductivity (SEC) of distilled water of the intermediate age

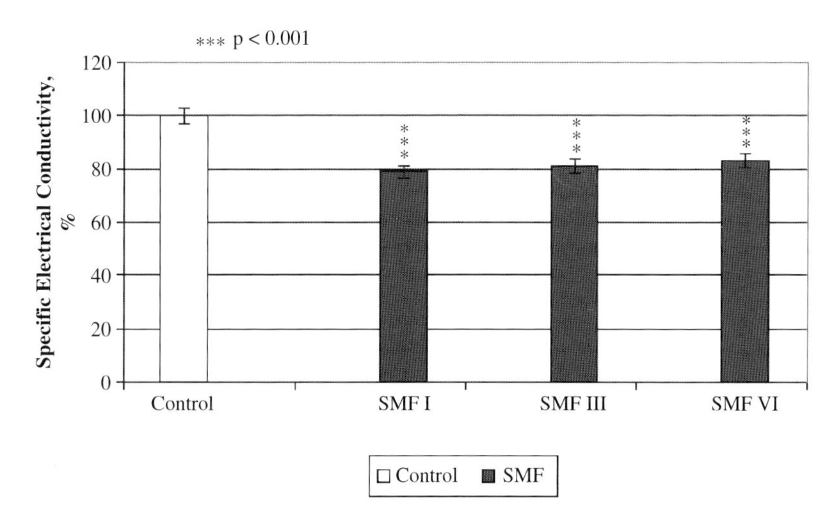

Figure 9. The effect of SMF exposure on specific electrical conductivity of one-day (I), three-day (III) and six-day (VI) distilled water at 18 °C

study of the biological effect of EMF and MV on high-level organized organisms could be the subject for future investigation.

As in reality water contains its dissociation products and soluble gasses, it is predicted that its structure could be extremely sensitive to the effect of any environmental factors. Water can be considered as an open thermo-dynamical system with

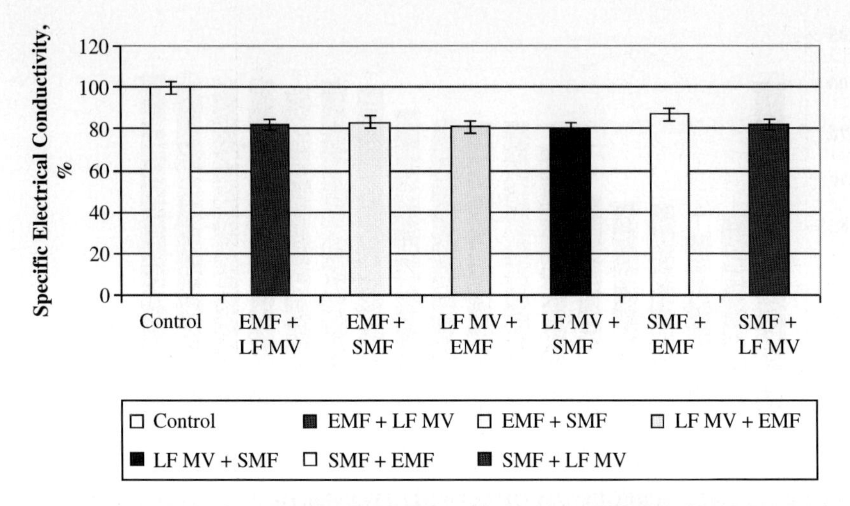

Figure 10. The combined effect of 4 Hz EMF (2.5 mT), 4 Hz MV (30 dB) and SMF (2.5 mT) on one-day DW at 18 °C exposed in different order. The exposure time for each factor was 30 min. The interval between exposures was less than 1 min

energy and substances exchange with the medium leading to continuous structural changes. The latter could appear even without breaking the hydrogen bounds, just by their deformation (Klassen, 1982). Therefore it is extremely difficult to suggest the exact value of energy which is necessary to change the water structure, however it should be less than the energy of hydrogen bounds (16, 7 - 25, 1 kDj). Such variability of water properties is the main barrier for precise reproduction of the experimental results in water studies. This picture becomes more complicated in case of water solutions containing electrolytes, non-electrolytes, solid particles and air-bladders. The increase of a number of ions in water leads to the increase of its entropy, instead of its predicted decrease, because of the hydration-induced disorders of the water structure. Two groups of ions could be distinguished depending on their effect on water structure: ordering and disordering the water structure (Kireev, 1968). As velocity and chemical activity of ions are determined by the degree of their hydration, the knowledge on the effect of magnetic fields on ions hydrations is important to understand the mechanism of its effect. It was shown that the hydration of ions is highly sensitive to the effect of EMF. The hydration of diamagnetic ions is decreased, while in case of paramagnetic ions it is increased. In this aspect Ca ions play a crucial role in realization of biological effect of EMF, because of forming the aqua-complexes $[Ca(H_2O_6)]^{2+}$ in water making it very sensitive to EMF. Therefore the character of magnetic field effect on water structure depends on the concentration of Ca ions. Early our works have shown that the direction of SMF-induced changes of water SEC could be changed depending on $CaCl_2$ concentration in water (Ayrapetyan, 1994a).

The sensitivity of water structure to EMF and MV significantly depends on the effect of solute gases in it. The solubility of even neutral gazes in water leads

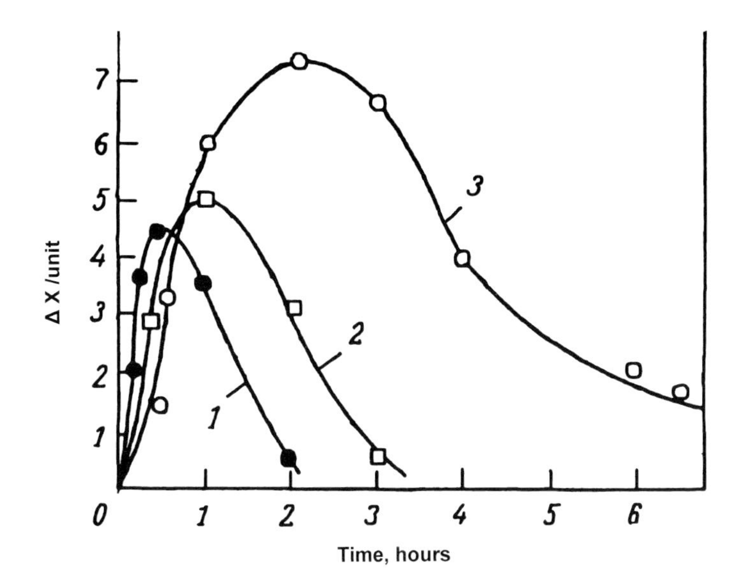

Figure 11. The kinetics of IR-specter of magnetized bi-distilled (1), distilled (2) and natural water (3) after SMF exposure

to the deformation of hydrogen bounds in result of which the formation of new hydrogen bounds is taking place. The degree of solubility of CO_2 and O_2 in water is very sensitive to EMF and MV (Stepanyan et al., 1999). It was shown that SMF has depressing effect on CO_2 and elevation effect on O_2 solubility in water (Klassen, 1982).

The 'memory' of EMF-induced water structure changes from the point of its biological effects is extremely important. From the point of equilibrium thermodynamic system, it is predicted that after EMF expose, its effect on water should disappear immediately, however, the experimental results show that the 'trace' effect of EMF stays incomparably longer than the exposure time. This memory is much longer in water solutions than in pure water (Klassen, 1982).

As can be seen in the presented Figure 11 the rate of SMF effect on be-distillated water in IR-specters (magnetic susceptibility) is higher than in case of distillate and natural water, and the spontaneous relaxing period after SMF-exposure for natural water is much longer than for bi-distilled and distilled waters.

REFERENCES

Amyan AM, Ayrapetyan SN (2004) On the modulation effect of pulsing and static magnetic fields and mechanical vibrations on barley seed hydration. Physiol Chem Phys Med NMR 36:69–84

Ayrapetyan SN, Avanesian AS, Avetisian T, Majinian S (1994a) Physiological effects of magnetic fields may be mediated through actions on the state of calcium ions in solution. In: Carpenter D, Ayrapetyan SN (eds) Biological effects of electrical and magnetic fields, vol 1, pp 181–192, Academic Press

Ayrapetyan SN, Stepanyan RS, Oganesyan HG, Barseghyan AA, Alaverdyan ZhR, Arakelyan AG, Markosyan LS (2001) Effect of mechanical vibration on the lon mutant of *Escherichia coli* K-12. Microbiology 70:248–252 (in Russian)

Hakobyan SN, Ayrapetyan SN (2001) The effect of EMF on water specific electrical conductivity and wheat sprouting. WHO Meeting on EMF Biological Effects and Standards Harmonization in Asia and Oceania, 123

Kireev V (1968) Physical chemistry. Higher School Publishing House, Moscow

Klassen VI (1982) Magnetizing of water systems. Chemistry Press, Moscow, 296 (in Russian)

Pullman B, Pullman A (1963) Quantum biochemistry. Interscience Publisher, New York

Stepanyan RS, Ayrapetyan GS, Arakelyan AG, Ayrapetyan SN (1999) The effect of mechanical vibration on the water conductivity. Biophysics 2(44):197–202 (in Russian)

Stepanyan RS, Alaverdyan ZhR, Oganesyan HG, Markosyan LS, Ayrapetyan SN (2000) The effect of magnetic fields on lon mutant of *Escherichia coli* K-12 growth and division. Radiational Biology, Radioekology 3(40):319–322 (in Russian)

CHAPTER 8

SOLUTE EXCLUSION AND POTENTIAL DISTRIBUTION NEAR HYDROPHILIC SURFACES

JIANMING ZHENG AND GERALD H. POLLACK*

Department of Bioengineering, Box 355061, University of Washington, Seattle WA 98195

Abstract: Long-range interaction between polymeric surfaces and charged solutes in aqueous solution were observed microscopically. At low ionic strength, solutes were excluded from zones on the order of several hundred microns from the surface. Solutes ranged in size from single molecules up to colloidal polystyrene particles 2 μm in diameter. The unexpectedly large exclusion zones regularly observed seem to contradict classical DLVO theory, which predicts only nanometer-scale effects arising from the presence of the surface. Using tapered glass microelectrodes similar to those employed for cell-biological investigations, we also measured electrical potentials as a function of distance from the polymeric surface. Large negative potentials were observed – on the order of 100 mV or more – and these potentials diminished with distance from the surface with a space constant on the order of hundreds of microns. The relation between potential distribution and solute exclusion is discussed

Keywords: exclusion zone; solute exclusion; surface potential; hydrophilic surface

1. INTRODUCTION

Interaction between charged surfaces in aqueous solution has been a subject of intensive investigation not only because of its biological relevance, but also for potential industrial relevance (Israelachvili, 1991). Interaction between like-charged colloidal particles is a fundamental force underlying colloid science, lubrication, and friction (Klein et al., 1993; Raviv et al., 2003), and is centrally relevant to the question of how proteins and other like-charged particles maintain their independence inside the cell.

---

* FAX: 206 685-3300. Email address: ghp@u.washington.edu.

G. Pollack et al. (eds.), Water and the Cell, 165–174.

Among the phenomena revealed using colloidal particles, some of the most interesting are the formation of colloid crystallites (Ise et al., 1999; Larsen and Grier, 1996) and voids (Ito et al., 1994; Yoshida et al., 1995). In the latter, certain regions are found to be particle free – much like vacuoles in cells. In the former, colloidal particles coalesce into crystallites with fixed, regular inter-particle spacing, the crystallite growing with time through a mechanism similar to an Ostwald ripening process. Self-organization of this kind occurs regularly in cells, where proteins self-assemble into regular supramolecular arrays. The observation of voids and crystallites have led to mechanistic debates, between those on the one hand who present evidence that crystallite formation conforms to standard DLVO theory, and others who propose the Ise-Sogami potential as an explanation (Grier and Crocker, 2000; Sogami and Ise, 1985; Tata and Ise, 1998; Tata and Ise, 2000). Other factors as well have been proposed to play a role in the interaction between surfaces, including hydration, structural factors, hydrophobic depletion, and protrusion forces (Henderson, 1992; Huang and Ruckenstein, 2004; Malomuzh and Morozov, 1999; Ruckenstein and Manciu, 2003; Yaminsky, 1999; Ye et al., 1996).

The role of water in such interactions has not been seriously explored. In classical theories, water is treated as a continuous, homogeneous medium, commonly referred to as bulk water (Israelachvili, 1991). Although layers of tightly bound water organize around hydrophilic polymers, the number of layers has been thought to lie in the single digits. The possibility of more substantial layering is left open (Fisher et al., 1981; Xu and Yeung, 1998), and some investigators are actively considering long-range water-structure layers (Bohme et al., 2001; Siroma et al., 2004; Zheng and Pollack, 2003). The results presented here support the possibility of long-range ordering. They also support the possibility that such ordering might be involved not only in crystallite formation, but also in the formation of particle-free voids, both of which have counterparts in the cell.

2. MATERIALS AND METHODS

Nafion-117 sheets in protonated form were obtained from Sigma-Aldrich, and were further treated by bathing with ion-exchange resins in deionized water at room temperature for one week. The Nafion molecule has a Teflon backbone with perfluorine side chains containing sulfonic acid groups. Nafion film is partially swollen in water, and the sulfonic acid groups will be dissociated, leaving the surface negatively charged. Hydrated sheets were approximately 200 μm thick. Nafion is reported to be extremely insoluble in water (Siroma et al., 2004).

Polyacrylic acid (PAAc) gels were synthesized in this laboratory. A solution was prepared by diluting 30 ml of 99% acrylic acid (Sigma-Aldrich) with 10 ml deionized water. Then, 20 mg *N*, *N*′-methylenebisacrylamide (Sigma-Aldrich) was added as a cross-linking agent, and 90 mg potassium persulfate (Sigma-Aldrich) as an initiator. The solution was vigorously stirred at room temperature until all solutes were completely dissolved, and then introduced into capillary tubes and sealed. Gelation took place as the temperature was slowly raised to about 70 °C.

The temperature was then maintained at 80 °C for one hour to ensure complete gelation. Synthesized gels were carefully removed from the capillary tubes, rinsed with deionized water, and stored in a large volume of deionized water, refreshed daily, for one week.

Negatively charged sulfated microsphere suspensions were obtained from Interfacial Dynamics (Portland OR). Microsphere charge density was 11.4 $\mu C/cm^2$, and remains stable over a wide range of pH. The concentrated suspension was diluted with deionized water, and deionized by bathing with ion-exchange resins for at least two weeks.

Amidine microsphere suspensions (Interfacial Dynamics) were also used. They were treated in the same way as the sulfated colloidal suspensions. Amidine surface charge varies with pH of the bathing solution: positive below pH 7 as used in the experiments, and reported to be uncertain at higher pH (Bohme et al., 2001).

Colloidal particle behavior in the vicinity of the Nafion or gel surface was observed on the stage of an inverted microscope (Zeiss Axiovert 35). The Nafion sheets were oriented normal to the optical axis. The sample was viewed in bright field with a 20X objective, and/or in dark field generally with a 5X objective. A small specimen was put on a cleaned cover slip mounted on the microscope stage, and covered with another cover slip. A few drops of colloidal suspension with a colloid volume fraction about 0.01% were injected into the intervening space. Image recording began immediately. Then the focus was adjusted to the middle plane of the specimen, and recording continued.

3. RESULTS

3.1 Solute Exclusion

Figure 1 shows the time course of particle disposition near the PAAc surface (left) and the Nafion surface (right). Initially, the particles were disposed relatively uniformly. Immediately, they began translating away from the surface, leaving a particle-free zone that was detectable within seconds. The width of the zone increased with time for several minutes, whereupon zone growth stopped. Typically, the zone grew to about 300 μm in the vicinity of PAAc gels, and to 400 μm in the vicinity of Nafion. Following that, it remained stable.

Oppositely charged amidine microsphere suspensions also showed exclusion zones. However, the boundaries were less clear than with the negatively charged particles. The sizes of the exclusion zones in the vicinity of Nafion and PAAc were approximately 400 μm and 300 μm, respectively; similar to the typical values using sulfated microspheres. An example is shown in Figure 2.

Exclusion zones were also seen with small fluorescent molecules, including sodium fluorescein (mw. 376.3), and 6-methoxy-*N*-(3-sulfopropyl) quinolinium (Molecular Probes), mw. 281.33. A clear, non-fluorescent zone adjacent to the Nafion developed progressively with time, as viewed under a UV lamp. An example is shown in Figure 3. The size was similar in magnitude to typical values observed using colloidal particles.

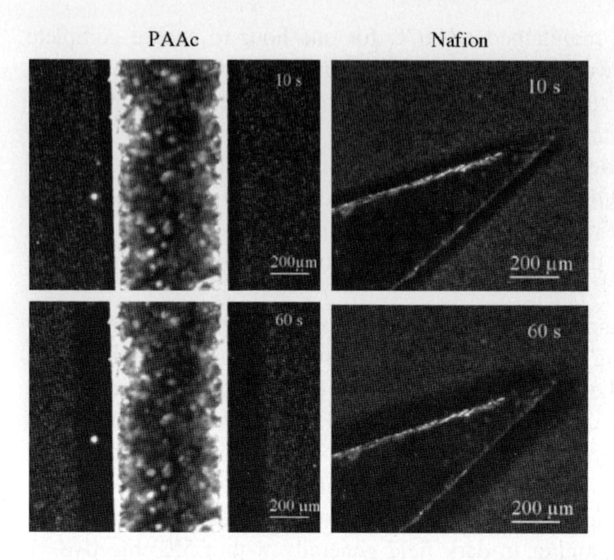

Figure 1. Colloidal particle exclusion formed near by the charged surfaces of polyacrylic acid gel (left) and Nafion sheet (right). Negatively charged sulfated microspheres with diameter of 1 μm were used. Darker regions adjacent to samples are microsphere free

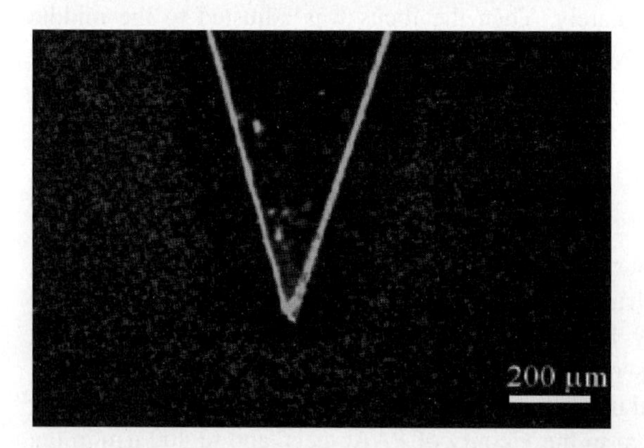

Figure 2. Exclusion zone formed near the Nafion surface with suspension of amidine microspheres, 1 μm in diameter

Particle exclusion was observed not only in ordinary distilled water, but also in water with extremely low concentration of ions. To remove ions from the suspension, ion-exchange resins were introduced. A sample of Nafion film 3 × 3 mm was introduced into a polymethyl methacrylate UV cuvette 10 × 10 × 40 mm that contained a high density of AG501-X8(D) (Bio-Rad) ion-exchange beads. The cuvette was sealed using acetone solvent and strongly shaken for 30 minutes before observation. The ion concentration in the container was estimated to be as low as tens of micromolar from measurements of ion conductivity relative

Figure 3. Exclusion zone near the Nafion surface observed after 5 minutes using 6-methoxy-*N*-(3-sulfopropyl) quinolinium

to a reference solution. Following the lowering of ion concentration in this manner, the exclusion zone became even larger than in ordinary distilled water, $425 + / - 40$ vs. $380 + / - 35\,\mu\text{m}(n = 5)$.

Figure 4 shows the effects of added salts on the size of the exclusion zone. The chloride salts of K^+, Na^+, Li^+, and Ca^{2+} were examined, and the results were obtained in the same way as those depicted in Figure 1.

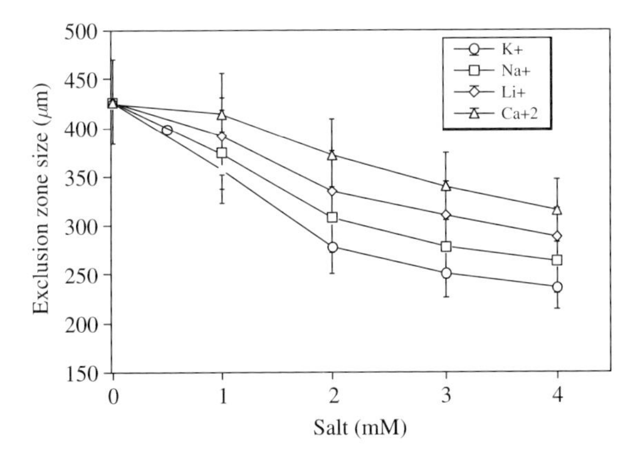

Figure 4. Size dependence of the exclusion zone on salt concentration

The results show that the various salts have a negative impact on the size of the exclusion zone. Among the salts used, K^+ was the most effective in reducing exclusion-zone size, followed by Na^+, Li^+ and Ca^{2+}.

3.2 Electrical Potential Measurements

In addition to measuring particle/solute exclusion, we measured the electrical potential in the vicinity of gel and Nafion surfaces. Standard, 3M-KCl glass micro-electrodes with Ag/AgCl wires, tip diameter $\sim 0.1\,\mu m$, were used, as schematized in Figure 5. One electrode was placed near the sample, the other at a remote position in the chamber, at least 20 mm from the first. The signals were input into a high-impedance amplifier (Electro 705, VWR), low-pass filtered, and stored on a computer.

In typical runs, the two electrodes were first manually positioned. Then, the near-sample probe was driven by an electric motor to move along the vertical axis toward the sample at a constant rate of 36 $\mu m/s$. The specimen sat on the chamber floor, and the chamber was filled to ~ 4 mm above the sample. Zero potential difference was set as the probe electrode just pierced the solution surface, after which the probe was driven downward. In chambers with no sample, the potential difference between any two points in the chamber was zero, except with the probe near the air/water interface, where fluctuations were detected.

Figure 6 shows the potential distributions in the vicinity of Nafion and the PAAc gel in distilled water. Negative potentials were observed in both cases. The magnitude of the negative potential increased as the probe approached the specimen surface, gradually at first, and then more steeply within several hundred micrometers of the surface. The (negative) potential at the surface was ~ 200 mV for Nafion, and ~ 125 mV for PAAc. In the latter case, with a soft gel into which the electrode

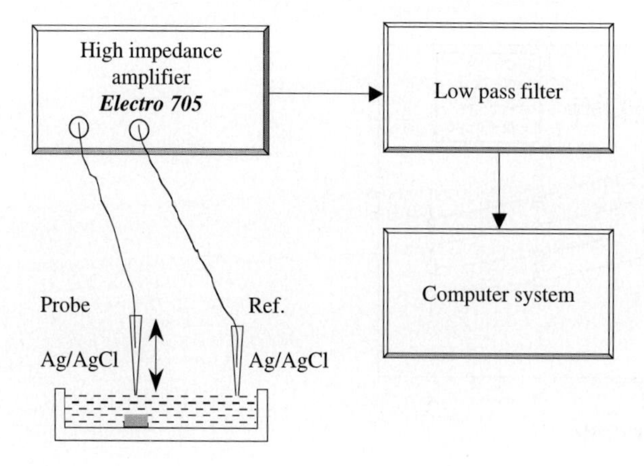

Figure 5. Schematic of instrumentation used for measuring electrical potential of charged surfaces

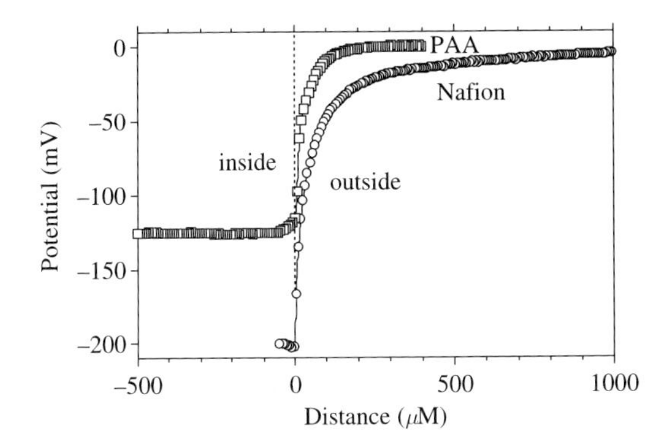

Figure 6. Electrical potentials measured in the vicinity of Nafion and PAAc gel surfaces

could easily penetrate, we found that the ~125 mV potential remained constant throughout the gel interior. Potentials inside the Nafion specimen could not be measured because attempts at penetration caused consistent electrode-tip breakage.

The measured electrical potential depended on the ion concentration in the surrounding solution. Figure 7 shows the potential distribution in the vicinity of the Nafion surface as a function of KCl concentration in the bathing medium. All potentials were measured after a fresh sample had been exposed to a fresh salt solution for five minutes. With an increase of salt concentration, the potential evidently diminished, as did the space constant. The zero-potential point thus moved toward the surface.

Similar experiments were carried out using other salts. Figure 8 shows the results of measurements carried out in chloride salts of different cations. While potential values at the surface are too close to be clearly distinguishable, the magnitude of the potential at various distances from the surface decreased in the order $Ca^{2+} > Li^{+} > Na^{+} > K^{+}$.

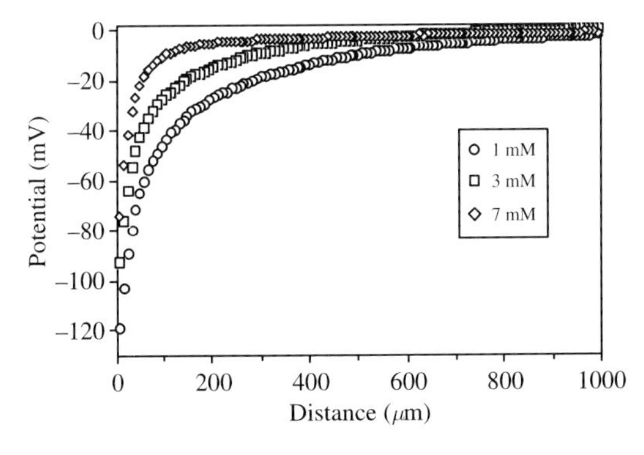

Figure 7. Electrical potentials measured near the Nafion surface in solutions of various KCl concentration

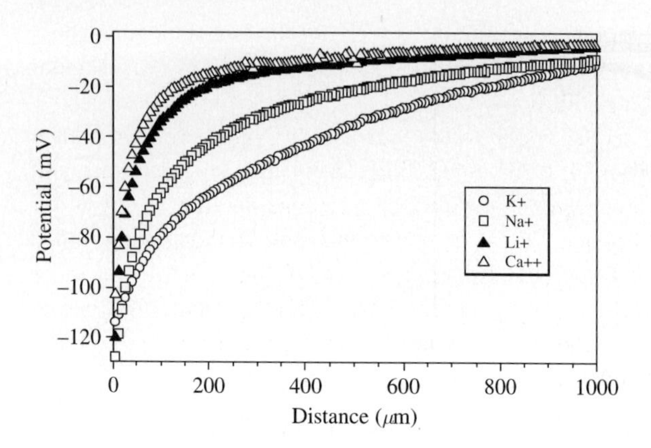

Figure 8. Electrical potential measured near the Nafion surface in the presence of various salts. Concentration 1 mM throughout

4. DISCUSSION

The finding of a solute-exclusion zone extending several hundreds of micrometers from the surface is unanticipated. Various potential artifacts have been considered and ruled out (Zheng and Pollack, 2003). Ordinary DLVO theory suggests that surface-induced effects should exist, but are anticipated to extend no more than nanometers from the surface (Israelachvili, 1991). The observations here show surface effects extending many orders of magnitude farther.

Also unanticipated is observation that adjacent to the gel or Nafion surface, negative potentials exist. These potentials extend hundreds of microns from the surface. While exact quantitative relations between the extent of the negative potential zone and the extent of the exclusion zone are not yet established, the two are roughly the same order of magnitude, implying that the two may well be related to one other.

In explaining the presence of the exclusion zone, the hypothesis that comes immediately to mind is that the negative surface potential may give rise to the exclusion zone. That is, the negative surface potential creates an electric field that exerts a force on the solutes, and drives them away from the surface. The field's long span might be created by a non-uniform distribution of ions extending over an appreciable distance. One can envision a higher concentration of anions nearer to the surface than farther from the surface, or cation distribution directed oppositely, yielding the observed potential gradient. Even in the case of scrupulously de-ionized water, some ions may still be present and could play a role.

However, the results of Figure 4 show that ions are not likely to be responsible for the exclusion zone and the potential gradient: As the ion concentration increases, the exclusion-zone width decreases, as does the potential magnitude (Figure 7). If ions gave rise to these phenomena, the opposite would be expected.

Another possibility is the polymer from the gel or from the Nafion diffuses into solution. Commercially available Nafion is considered insoluble in water under ambient conditions, and our preliminary experiments using a mass spectrograph show no traces of polymer in the water solution in which Nafion film had been stored for one week. The possibility of contamination by diffusing polymer and also by polymer brushes extending from the surface was considered in detail in our previous paper (Zheng and Pollack, 2003), as were other potential artifacts.

Another possibility is that these phenomena arise out of some change of water structure induced by the Nafion or gel surface. The liquid crystalline structure of that region would force solutes into the less ordered region, away from the solid-liquid interface. This could account for the solute-free zone. Ordered water has been proposed to exist in many structural variants, and it is of interest that one of these variants long discounted because of its association with polywater (Lippincott et al., 1969), carries a net negative charge. Potentially, such structure could account for the observed negative potential, and could explain why the negative potential extends for roughly the distance of the exclusion zone. The diminution of potential with distance from the surface would be explained as cations from the bulk penetrate the lattice and bridge the existing negative charges.

Such an explanation might also be relevant for the potential measured inside gels (see paper by Safronov et al., this volume), as well as the potential measured inside cells. Indeed, the fact that the potential is continuous across the gel-water interface (Figure 6) implies that the potential arises from some component that is also continuous across the interface – which reduces to water alone. In other words, it is possible that the negative potential inside the gel (or the cell) arises at least in part from the negative charge carried by the water.

This explanation for the exclusion zone and negative potential is at present speculative, and we are searching for alternative hypotheses to consider. Whatever the mechanism, the two long-range phenomena are novel and striking, and may have important implications not only for surface chemistry and physics, but also for cell biology, where solute distributions and potential gradients are fundamental to virtually all processes.

REFERENCES

Bohme F, Klinger C, Bellmann C (2001) Surface properties of polyamidines. Colloids and Surfaces A: Physicochemical and Engineering Aspects, 189:21–27

Fisher IR, Gamble RA, Middlehurst J (1981) The Kelvin equation and the capillary condensation of water. Nature 290:575

Grier DG, Crocker JC (2000) Comment on "Monte Carlo study of structural ordering in charged colloids using a long-range attractive interaction". Phys Rev E61:980

Henderson D (1992) Interaction between macrospheres in solution. Paper presented at: Proceedings of the 11th International Conference on Thermophysical Properties, Jun 23–27, 1991, Boulder, CO, USA

Huang H, Ruckenstein E (2004) Double-layer interaction between two plates with hairy surfaces. Colloid Interface Sci 273:181–190

Ise N, Konishi T, Tata BVR (1999) How homogeneous are 'homogeneous dispersions'? Counterion-mediated attraction between like-charged species. Proc. 1998 2nd International Symposium on Polyelectrolytes (ISP), May 31–Jun 3, 1998, Inuyama, Jpn, ACS, Washington, DC, USA

Israelachvili J (1991) Intermolecular and Surface Forces, 2nd edn. Elsevier

Ito K, Yoshida H, Ise N (1994) Void structure in colloidal dispersions. Science 263:66–68

Klein J, Kamiyama Y, Yoshizawa H, Israelachvili JN, Fredrickson GH, Pincus P, Fetters LJ (1993) Lubrication forces between surfaces bearing polymer brushes. Macromolecules 26:5552–5560

Larsen AE, Grier DG (1996) Melting of metastable crystallites in charge-stabilized colloidal suspensions. Phys Rev Lett 76:3862

Lippincott E, Stromberg R, Grant W, Cessac G (1969) Polywater. Science 164:1482–1487

Malomuzh NP, Morozov AN (1999) Fluctuation multipole mechanism of long-range interaction in solutions of colloidal particles. Colloid Journal of the Russian Academy of Sciences: Kolloidnyi Zhurnal 61:332–341

Raviv U, Giasson S, Kampf N, Gohy J-F, Jerome R, Klein J (2003) Lubrication by charged polymers. Nature 425:163–165

Ruckenstein E, Manciu M (2003) Specific ion effects via ion hydration: II. Double layer interaction. Adv Colloid and Interface Sci 105:177–200

Siroma Z, Fujiwara N, Ioroi T, Yamazaki S, Yasuda K, Miyazaki Y (2004) Dissolution of Nafion [registered trademark] membrane and recast Nafion [registered trademark] film in mixtures of methanol and water. Journal of Power Sources 126:41–45

Sogami I, Ise N (1985) On the electrostatic interaction in macroionic solutions. J Chem Phys 81:6320–6332

Tata BVR, Ise N (1998) Monte Carlo study of structural ordering in charged colloids using a long-range attractive interaction. Phys Rev E Stat Phys Plasmas Fluids Relat Interdiscip Topics 58:2237

Tata BVR, Ise N (2000) Reply to "Comment on 'Monte Carlo study of structural ordering in charged colloids using a long-range attractive interaction' ". Phys Rev E Stat Phys Plasmas Fluids Relat Interdiscip Topics 61:983–985

Xu X-HN, Yeung ES (1998) Long-range electrostatic trapping of single-protein molecules at a liquid-solid interface. Science 281:1650–1653

Yaminsky VV (1999) Hydrophobic force: The constant volume capillary approximation. Colloids and Surfaces A: Physicochemical and Engineering Aspects 159:181–195

Ye X, Narayanan T, Tong P, Huang JS, Lin MY, Carvalho BL, Fetters LJ (1996) Depletion interactions in colloid-polymer mixtures. Phys Rev E Stat Phys Plasmas Fluids Relat Interdiscip Topics 54:6500–6510

Yoshida H, Ise N, Hashimoto T (1995) Void structure and vapor-liquid condensation in dilute deionized colloidal dispersions. J Chem Phys 103:10146

Zheng J-M, Pollack GH (2003) Long-range forces extending from polymer-gel surfaces. Phys Rev E 68:31408–31411

CHAPTER 9

VICINAL HYDRATION OF BIOPOLYMERS: CELL BIOLOGICAL CONSEQUENCES

W. DROST-HANSEN*

Laboratorium Drost, 5516 N.Mallard Run, Williamsburg, VA 23188, USA

Abstract: A novel type of hydration of macromolecules in aqueous solution was first suggested by Etzler and Drost-Hansen (1983). This hydration, observed for all macromolecules with a critical mass of >2000 Daltons (MW_C), seems identical with vicinal hydration of solid surfaces, possessing the same characteristics, e.g., thermal anomalies at the same temperatures [T_k] and similar shear rate dependence, as well as slow reforming after shear. Furthermore, the vicinal hydration is independent of the detailed chemistry of the macromolecules and of the presence of other solutes, electrolytes and non-electrolytes alike. Evidence for this poorly recognized and often overlooked hydration is presented

Keywords: Vicinal hydration; Thermal anomalies; Shear rate effects; MW dependence; Anomalous viscosities; Anomalous diffusion coefficients; Hydrodynamic radii; Biophysical implications; Sources of variability

1. INTRODUCTION

Water adjacent to a solid surface is structured differently from that of the bulk phase, but the burning question of the extent of this water structuring has been a highly contentious issue for ~100 years. Few disagree that the first two or three molecular layers of water will be strongly influenced by the specific chemical nature of a solid surface, particularly when ionic sites are present (as well as strongly polar groups). However, many reports indicate that this structured interfacial water can be altered over far greater distances from, say, 10 molecular layers to 10,000. Currently, more realistic estimates are in the range of 10 to 100 (ca. 5 to 50 nm). Such distances approach the (probable) maximum 'free distances' within a typical

* Correspondence to Professor Walter Drost-Hansen: wdrosthans@aol.com

G. Pollack et al. (eds.), Water and the Cell, 175–217.

biological cell (Clegg 1984a, 1984b). Much of the evidence for extensive modifications of interfacial water has been less than convincing. However, over the past 20–30 years substantial evidence now makes it more certain that extensive restructuring does occur, and such is generally referred to as 'vicinal water' [hereafter *VW*] – a summary of the unusual properties of *VW* is in Section 2. In view of the fact that the structure of bulk water continues to escape a complete and rigorous description, it is not surprising that far less is known about *VW*.

One highly characteristic feature of *VW* is the occurrence of thermal anomalies, which exist at no less than 4 distinct temperatures [T_{ks}] at ~15, 30, 45 and 60 °C. This paper largely addresses newer evidence that strongly suggests – indeed proves – that sufficiently large macromolecules in aqueous solution are also vicinally hydrated; indeed, *all* macromolecules of >2000 Da (MWc) will be vicinally hydrated. Thermal anomalies are also properties of aqueous macromolecular solutions at the T_k values just mentioned. In addition, vicinal hydration structures are highly shear rate dependent, and once sheared, the time to reform *VW* structures close to the original state may be minutes or even hours. Thus these properties exhibit notable hysteresis that can account for the apparent lack of reproducibility in much of the data.

2. WATER AT SOLID INTERFACES: PROPERTIES OF VICINAL WATER

2.1 Density of VW

Etzler, Conners and Ross (1990) have contributed greatly to the delineation of the properties of *VW*, in particular to its density and specific heat. Etzler and Fagundus (1987) explored the density of water in a number of porous silica gels of controlled pore diameters. Invariably densities were observed of less than that of bulk water, decreasing smoothly with decreasing pore diameter. For pores with an average diameter of 10nm, the density at room temperature was ~0.97 g/cm^3 (see Table 3). Until fairly recently, there was uncertainty regarding the density of interfacial water, but this data puts it now above reproach, although it is easy to see how vastly different estimates might previously have been reported. For water near an ionic solid, or a solid with numerous surfaces charges, it seems obvious that the innermost 2 or even 3 layers of water might have a notably greater density than the bulk because of electrostriction of the water molecules near the surface charges. However, even if densities as high as, say, 1.05 g/cm^3 prevailed over a distance of 3 molecular layers from the surface, this contribution to the net density of the combined innermost layers plus the *extended VW* structures is far too little to bring average value to the density of the bulk. With regard to thermal expansion coefficient, Shoufle et al. (1972) measured this for water in narrow capillaries of different diameters over a wide range of temperatures. One interesting result was that the temperature of maximum density [T_{MD}] of water decreases essentially linearly with decreasing diameter of the capillaries (and T_{MD} also decreases for

D_2O, although apparently not in a linear manner). According to Deryaguin, a decrease in T_{MD} in narrow pores of TiO_2 and silica gel. The data of Shoufle et al., were reanalyzed and it was found that the thermal expansion coefficient of the capillary-confined water was *notably* larger than the corresponding bulk value (see Drost-Hansen, 1982).

The density of *VW* is distinctly less than that of bulk water; in other words, the specific partial molar volume of *VW* is larger than that of bulk. Hence, by LeChatelier's principle, we can predict that increased pressure will tend to 'squeeze out' *VW* (presumably converting it to more bulk-like water). As discussed below, the same type of vicinal hydration at solid surfaces also occurs at water/macromolecule interfaces. Hence, the density data may be of particular interest to studies of the hydration of macromolecules in aqueous solutions by ultracentrifugation. In principle, the high pressure in the centrifuge tube may indeed diminish or completely eliminate *VW*. On the other hand, this fact may conceivably be useful in delineating the likely total contribution to the hydration of macromolecules if, for instance, low shear-rate and low-pressure means of determining the total amount of hydration are available. One such determination might be by measurements of diffusion coefficients that may be assumed to affect neither the 'classical' types of hydration nor the hydration due to the *VW*. The difference between the two values may directly relate to the amount of vicinal hydration. Interesting consequences of the volume-pressure effects on living cells, as mediated by the hydrated macromolecules, can be found in Mentre and Hui (2001).

An interesting confirmation of the low density of *VW* was provided in some high-precision dilatometric studies in our laboratory (Drost-Hansen et al., 1987). We showed that when a suspension of polystyrene spheres (of highly controlled particle diameter) settled, a distinct contraction in volume was observed, which was interpreted as the result of increasing overlap of the less dense *VW* of hydration of the particles. The compaction process thus 'squeezes' out *VW* hydration 'hulls' and this water then 'reverts' to the higher density bulk water, leading to a decrease in total volume. The same effect is seen in the sedimentation of a suspension of silica spheres ('Minusil'; radii ~5 microns); in other words, the same effect is observed for *VW* adjacent to polystyrene and silica (examples of the 'Substrate-Independence Effect'; see below).

Low et al. (1979) found densities of water near clay surfaces were lower than the bulk phase value (see also Viani et al., 1983). However, Low apparently believed that the effect was specifically due to the chemical nature of the clays employed, whereas it now appears that *VW* with low density is indeed induced by proximity to *any* solid surface (Table 1).

2.2 Specific Heat

One of the most characteristic properties of bulk water is its high specific heat of 1.0 cal/°K gram. We have made extensive measurements of the specific heat of vicinal water, measurements further perfected by Etzler and White (1987). In *all*

Table 1. Densities of vicinal water

T (°C)	ρ pore($kg^{-1}\,m^3$)	SD	mm
10	949.8	3.0	1.25
	983.55	1.8	5.85
	986.7	2.0	7.00
	993.2	2.2	12.1
	999.7	–	bulk
20	952.9	3.6	1.25
	983.3	7.9	5.85
	985.2	2.1	7.00
	991.3	3.0	12.1
	998.2	–	bulk
30	949.0	5.8	1.25
	981.3	1.0	5.85
	981.8	1.2	7.00
	990.7	1.7	12.1
	995.6	–	bulk

cases, values for the specific heat were notably larger than that of bulk water, which must surely have important implications for the structure of *VW* and merits attention by those modelling the structure of bulk water by computational means (Table 2).

Values for the specific heat of water adjacent to chemically different surfaces are all higher than for bulk water, clustering around 1.2 cal/°K gram, which is consistent with the concept of the Substrate-Independent Effect ('Paradoxical Effect'). This further endorses the conclusion that *VW is induced by proximity to any surface, regardless of the specific detailed chemical nature of the solid*. More recently, Etzler and co-workers have greatly improved on the measurements of c_p for water in dispersed systems and invariably obtained values larger than that for bulk (see under thermal anomalies; Section 3.4).

Table 2. Heat capacity of vicinal water (cal^{-1} g @ 25 °C)

Confining material	Observed value	SD	Corrected value	SD
Physical				
Bulk	1.08	0.08	1.00	–
Porous glass	1.37	0.20	1.27	0.20
Activated charcoal	1.38	0.03	1.28	0.03
Zeolite	1.31	0.08	1.21	0.03
Diamond	1.30	0.08	1.20	0.08
Biological				
Collagen			1.24	
Egg albumin			1.25	0.02
DNA			1.26	0.06
Artemia cysts			1.28	0.07

2.3 Viscosity of VW

Much data has accumulated on the viscosity of water at interfaces, usually confined in capillaries or between highly polished plates. As early as 1932, Deryaguin (1932, 1933) reported elevated values for the viscosity of water between glass plates, as determined with a torsion viscometer, using a highly polished glass plate and a slightly convex plate free to oscillate above the base plate (reviewed in Henniker (1949), Clifford (1975) and Drost-Hansen (1968). Elevated viscosity values of confined water have frequently been reported, but none has proved entirely satisfactory. For example, it is very difficult to rule out spurious effects of a speck of dust or, in some cases, the postulated existence of 'swollen' surface layers (especially where the confining solid is glass). However, Peschel and Belouschek (1976) showed in a series of elegant studies of water confined between highly polished quartz (and sapphire) plates, that *VW* had a higher viscosity than the bulk water (by a factor of 2–10). The reproducibility of Peschel's data strongly argues against both spurious dust contamination and alleged 'swollen layers'. The latter is further substantiated by results with sapphire being similar to those obtained with the highly polished quartz plates (Peschel and Aldfinger, 1969; see also Section 2.7.2).

2.4 Overview of Some Properties of VW

The properties of VW are listed in Table 3.

Table 3. Summary of some of the properties of vicinal water

Property	Bulk water	Vicinal water
Density ($g\,cm^{-3}$)	1.00	0.96–0.97
Specific heat ($cal\,kg^{-1}$)	1.00	1.25
Thermal expansion coeff. ($°C^{-1}$)	250.10^{-6}	$300–700.10^{-6}$
Adiabatic compressibility ($coeff.\,atm^{-1}$)	7.10^{-17}	35.10^{-17}*
Heat conductivity ($cal\,sec^{-1}\,°C^{-1}\,cm^{-1}$	0.0014	0.010–0.050
Viscosity (cP)	0.89	2–10
Activation energy, ionic conduction ($kcal\,mole^{-1}$)	4	5–8
Dielectric relaxation frequency (Hz)	19.10^{9}	2.10^{9}

Etzler et al. (1983, 1987, 1990, 1991).
Shoufle et al. (1976).
*Drost-Hansen (unpublished).

2.5 Historical Aspects; on the Geometric Extent of VW

The question of possible long-range structural effects in water near interfaces, as well as the question of the occurrence of thermal anomalies in the properties of aqueous systems, have had a long and contentious history. It appears that many investigators have – either by design or inadvertently – adopted an attitude of 'maximum intellectual economy', and on that basis ignored the available evidence for long-range structural effects on water at interfaces. The same holds true for thermal anomalies of aqueous systems – thermal anomalies having often been referred to as 'kinks' or 'discontinuities'. For clay-water systems, Low et al. (1979) have long argued for the existence of modified water structures at clay surfaces, and Clifford (1975) published a critical review of evidence for structural effects in water at solid surfaces. More than anyone, Deryaguin has advocated structural effects at interfaces or 'boundary layers' (Deryaguin, 1964, 1977; see also Churev and Deryaguin, 1985. But the most definitive studies of VW have probably come from Etzler's laboratory (discussed below). Very significant studies of VW have also been reported by Alpers and Hühnerfuss (1983), who eloquently argued for the existence of truly long-range structural effects at solid interfaces; they also demonstrated thermal anomalies in vicinal water. Table 4 is from a paper by Hühnerfuss (for references in this Table, see the original paper). We do not subscribe to some of the proposed estimates for the geometric extent of the vicinal hydration – especially not values above one micron. However, there seems to be sufficient evidence to accept distances of the order of *many* molecular diameters and for the purposes of this paper the suggestion is made that the vicinal hydration is likely to be several tens of molecular layers, but probably not >100–200; or in other words, the likely extent of vicinal water lies somewhere between 5 and 50 nm.

At this time, very little can be said for certain about the structure of *VW*. However, Etzler (1983) and Etzler et al. (1990) have speculated on structural aspects of *VW*., of which one feature appears clear; the degree of H-bonding in *VW* is probably greater than in the bulk, as reflected in the more open structure (the density being less than that of the bulk). As for thermal transitions, one likely explanation is that these reflect transitions (akin to higher order phase transitions) from one type of structure to another. However, in the theory of Kaivarainen (1995), the temperatures of the T_k are seen merely as those at which some structural parameter assumes an integral value.

2.6 The 'Paradoxical Effect' (Substrate Independence)

The properties of water near a solid interface do not appear to depend on either the detailed chemistry of the adjacent surface, or the presence (or absence) of most solutes. Because this is highly unexpected, the 'Substrate Independence Effect' is also sometimes referred to as the 'Paradoxical Effect'. Vicinal water is induced by mere proximity to *any* solid surface, and also by proximity to any sufficiently large macromolecule (see below). Thermal anomalies have been seen at identical temperatures in the properties of water adjacent to glass, quartz, clays, diamond, metals, polystyrene, cellulose and a large number of other organic, polymeric materials (Drost-Hansen, 1965; 1976; Kurihara and Kunitake, 1992; Lafleur et al., 1989). Thermal anomalies of *VW* are seen also in aqueous solutions (at solid surfaces) of both electrolytes and non-electrolytes, up to 1–2M. They also occur at the same

Table 4. Experimental evidence for 'long range ordering effects' within interfacial water layers at the boundary of solid surface and water

Reference	Method	Solid boundary	Penetration depth (μm)
Etzler and Lilies (1986)	Dielectric constant	Sheets of mica	2–5
Drost-Hansen (1976)	Adhesion at glass	Glass	1.5
Henniker (1949)	Disjoining pressure	Mica or steel plates	<1
Mastro and Hurley (1985)	Surface conductivity	Glass tube	0.3–0.4
Peschel and Aldfinger (1976)	Conductivity	Quartz particles	0.2–0.3
Peschel and Aldfinger (1969)	Conductivity	Quartz particles	0.2–0.3
Falk and Kell (1966)	Viscosity	Glass plates	0.25
Montejano et al. (1983)	Conductivity	Pyrex glass	0.05–0.2
Drost-Hansen (1969)	Viscosity	Pyrex glass	~0.2
Steveninck et al. (1991)	Viscosity	Quartz plates/convex	0.16
Braun and Drost-Hansen (1981)	Modulus of rigidity	Glass plates/convex	0.15
Bailey and Koleske (1976)	Disjoining pressure	Fused silica convex plates	~0.1
Nir and Stein (1971)	Disjoining pressure	Quartz plates	0.1
Deryaguin (1933)	Modulus of rigidity	Glass	0.1
Antonsen and Hoffman (1992)	Viscosity	Glass plates	~0.1
Clifford (1975)	Air-bubble flow	Glass tubes	~0.1

temperatures for acidic and alkaline solutions, at least within the range of pH from 1–2 to 12–13. *In other words, the establishment of vicinal water structures is a proximity effect and the underlying cause must therefore be sought solely in the unique properties of hydrogen bonding systems.* Surely, *VW* structures do *not* reflect epitaxial ordering due to some specific surface features. This effectively invalidates the basis, for instance, of the so-called Association-Induction hypothesis (Ling 1962; 1979), and some (but not all) of the alleged specific effects of clay surfaces (based on crystallographic similarities between the clay structure and the molecular arrangement in Ice-Ih). Vicinal hydration structures appear unaffected by the ionic strength of the aqueous phase – and thus *are* not dependent on the presence or absence of electrical double layers.

If VW was indeed the result of charge-water-charge interactions, as suggested by Ling, one would expect values for the density of VW to be greater than the value for the bulk, due to electrostriction. Furthermore, as proposed by Alpers and Hühnerfuss (1983), vicinal water exists beneath an insoluble monolayer at the air/water interface and as discussed by Drost-Hansen (1978), a distinct thermal anomaly at 15 °C is observed in the case of a monolayer of myristic acid on the surface of water (data of Phillips and Chapman, 1968). In other words, vicinal water exists beneath an insoluble monolayer and in the case of monolayers there are obviously no 'polarizing' effects from separated charges of opposite sign, as required by the Ling Association-Induction 'theory'. Vicinal water is induced merely by proximity to *any* interface; see postscript.

2.7 Thermal Properties of Vicinal Water

The idea that the properties of water exhibit thermal anomalies has been hotly debated for about a century. A distinct thermal anomaly may be seen in some very old data; for instance, the surface tension data of Brunner (1847) shows a very definite anomaly near 30 °C, although he did not comment on it. The idea that thermal anomalies might occur in the properties of (bulk) water (and aqueous systems) was mentioned by Dorsey (1940) in his monumental book on water properties. An even more extravagant claim was made by Drost-Hansen and Neill (1955), who suggested the occurrence of anomalies at no less than 4 different temperatures. At the time, it seemed that the anomalies were a manifestation of some unexpected property of bulk water and only later was it realized that the anomalies were in fact strictly associated with interfacial water (Drost-Hansen, 1968). Others have considered the occurrence of thermal anomalies, including Bernal, Magat, Forslind, Krone and several others (historical data can be found in Drost-Hansen (1971, 1978, 1981). Among some other earlier papers regarding thermal anomalies, it was proposed by the present author (Drost-Hansen, 1956, 1965, 1971) that the thermal anomalies might have been responsible for the selection of body temperatures in mammals (and birds) during the process of evolution – an idea that became firmly established in later years (Drost-Hansen, 1971, 1978, 1985). However, many authors have vigorously disputed the existence of thermal anomalies in (bulk) properties of water (and aqueous solutions), including Young (1966). Falk and Kell (1966) reviewed some of the evidence for 'kinks' and dismissed these as all falling within the limits of experimental error. Spurred on by these criticisms, we undertook a very high precision study of the viscosity of bulk water as a function of temperature (Korson et al., 1969), and observed *no* thermal anomalies whatsoever. In other words, for bulk systems the thermal anomalies are simply 'spurious effects' due to the properties of the water at the confining surfaces of the samples studied.

In the remainder of this section, some typical results demonstrating the thermal anomalies of *VW* adjacent to solid surfaces will be discussed, whereas Section 3 deals with the evidence for the thermal anomalies (and the existence of *VW*) at the macromolecular/water interface. The temperatures at which the thermal anomalies [T_k] occur are close to 15, 30, 45 and 60 °C, and are sometimes referred to as the 'Drost-Hansen Thermal Transition Temperatures'. The thermal transitions at T_k are relatively abrupt. However, they are generally not as sharp as expected of a first-order phase transition. The transitions generally occur over a narrow range (within 1–2 °C), e.g., 14–16, 29–32, 44–46, and 59–62 °C.

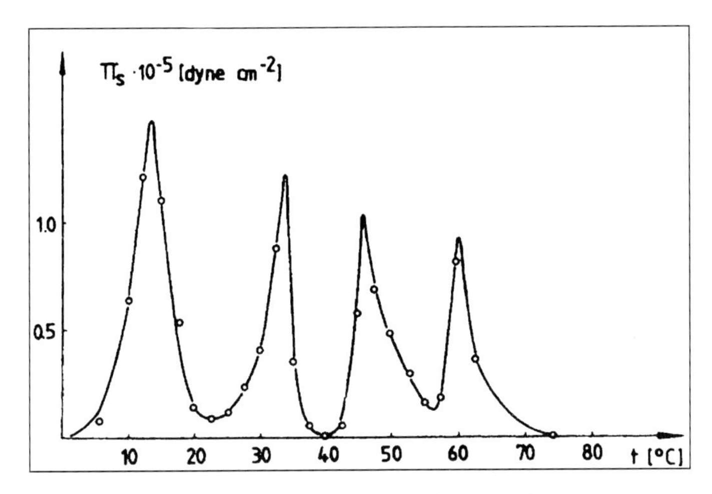

Figure 1. Disjoining pressure [π] of water (10^{-5} dynes/cm^2) between quartz plates separated by 50 nm as a function of temperature (°C). Data from Peschel and Belouschek (1979)

2.7.1 *Disjoining pressure*

Peschel and co-workers (1970, 1979) constructed an early form of a high-precision force-balance, allowing them to measure the disjoining pressure of water in thin films of water between two highly polished quartz plates, one being optically flat and the other essentially – but not quite – flat, but with a radius of curvature of ~1 meter. Typical results obtained as a function of temperature for a different plate separation of 50 nm are given in Figure 1. The results are truly remarkable, with the disjoining pressure [π] going through a series of maxima and minima. Note that the peaks in π occur at 15, 30, 45 and 60 °C – i.e., at T_k in perfect agreement with the temperatures of the thermal anomalies first postulated by Drost-Hansen and Neill (1955). Similar results were obtained using sapphire plates instead of quartz (Peschel, personal communication).

2.7.2 *Viscosity measurements*

In an alternative use of the force-balance, the dynamics of the top plate 'settling' on to the bottom plate were followed, and from these data one could calculate the viscosity of the water squeezed from between the plates. Peschel's data Peschel and Adlfinger's (1970, 1971) data are shown as Arrhenius' plots of the viscosity as a function of the reciprocal of the absolute temperature (Figure 2). Like the disjoining pressure, the viscosity goes through a number of maxima and minima.

2.7.3 *Specific heat measurements*

Etzler and Conners (1990) measured the heat capacity of water in narrow pores of silica gels (diameters of 4, 14 and 24.2 nm). In all their experiments, distinct 'heat capacity spikes' were observed in their thermograms, and the peaks of the spikes

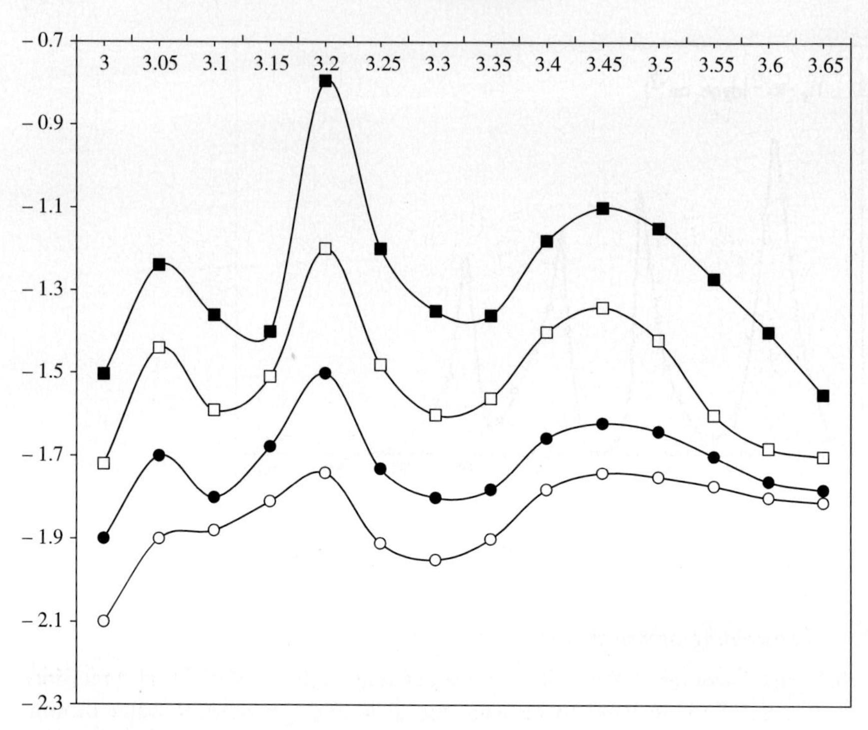

Figure 2. Arrhenius graph: log (viscosity; poise) of water between quartz plates versus 1/T (°C). Plate separations, top to bottom: 30, 50, 70 and 90 nm. Data from Peschel and Aldfinger (1969)

were at the thermal transition temperatures of *VW*. In all cases, the observed heat capacity values were above that of bulk water, consistent with the findings of Braun and Drost-Hansen (1976) and Braun (1981; Section 2.2) Figures 3 and 4 show enlargements of some of the data obtained by Etzler and Conners; the anomalies near T_k are very distinct at both the lower and the higher temperatures.

In preliminary studies, (Drost-Hansen, unpublished), distinct thermal anomalies in DSC measurements have frequently been seen on a variety of aqueous interfacial systems (similar to the characteristic peaks in the DSC curves observed by Etzler and Conners, 1991). A pronounced thermal anomaly occurs at 60 °C in thermograms of highly dilute aqueous solutions of cetyl-trimethyl ammonium salicylate ($C_{16}H_{33}(CH_3)NH_4^+ - C_6H_4(OH)COO^-$). Even at concentrations as low as 0.01%, the anomaly remains very distinct. This quaternary amine forms very large micelles that are presumably vicinally hydrated, suggesting that VW plays a role in the highly unusual rheological properties of solutions of this quaternary amine, including their viscoelastic behavior.

The distinct peaks in the specific heat curves reported by Etzler and Conners (1991) on water in porous silica particles have been seen in many other studies, including an early study of Braun and Drost-Hansen (1976), and more recently in a DSC study of 10% polystyrene sphere suspensions. In the latter study,

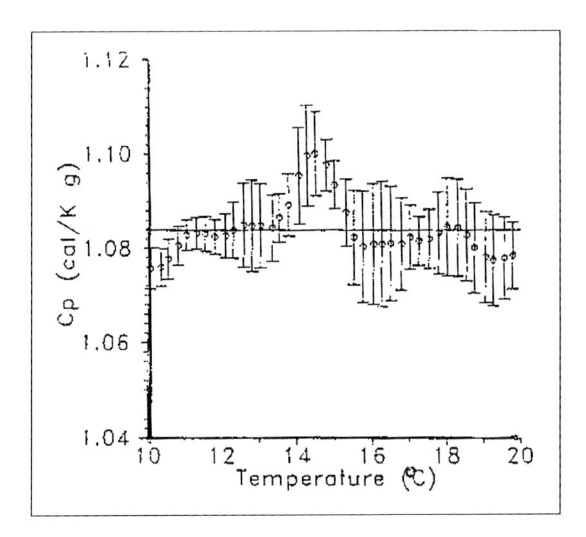

Figure 3. Specific heat of water in small pores; lower temperatures. Data from Etzler and Conners (1990)

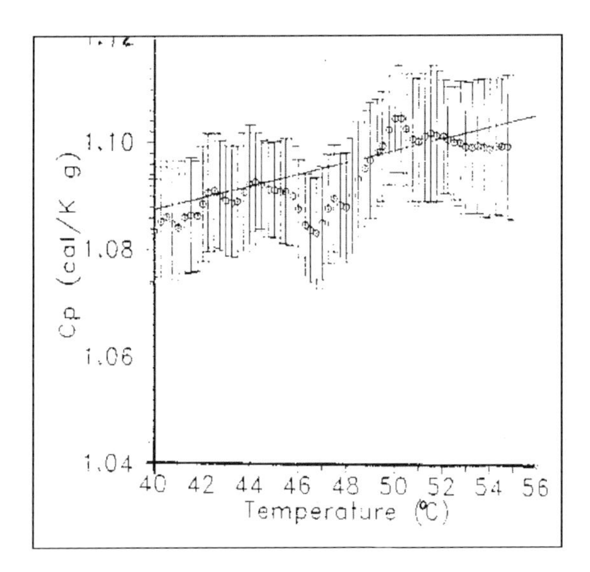

Figure 4. Specific heat of water in small pores; higher temperatures. Data from Etzler and Conners (1990)

heat flow (in a Perkin-Elmer DSC7 instrument) showed a pronounced peak near 28 °C. This is important, since it demonstrates that interfacial water does not have to be confined on all sides, as in a capillary or between plates, to exhibit *VW* characteristics. That the polystyrene spheres should be vicinally hydrated agrees with the volume contraction data for the settling of suspensions of these spheres, discussed in Section 2.1.

2.7.4 Solute distributions

Wiggins (1975) published a groundbreaking paper describing highly anomalous distributions of ions between bulk solutions of mixed electrolytes (equimolar with respect to K^+ and Na^+) and a silica gel. Three series were made with equimolar cation concentrations present as Cl^-, I^- or SO_4^{2-}. Measurements were made as a function of temperature and graphs of the selectivity coefficients between potassium and sodium ions plotted as a function of temperature. Distinct, sharp peaks were seen in the data, occurring at 15, 30 and 45 °C, i.e. exactly where thermal anomalies had been predicted by Drost-Hansen and Neill (1955), and had been observed by Peschel and Adlfinger (1971). Because of the potentially revolutionary importance of these measurements to cell physiology, we decided to repeat Wiggins' measurements (on a slightly more limited scale, using only equimolar solutions of NaCl and KCl). Our results (Hurtado and Drost-Hansen, 1979) agree *quantitatively* with Wiggins' data (Figure 5), making it exceedingly unlikely that the results reported by Wiggins are spurious or unreliable.

Further confirmation of distinct peaks in ion distribution in vicinal as a function of temperature were obtained by Etzler and Lilies (1986), using Li^+ and K^+ (as their chlorides; Figure 6A and B). This study is also important because it includes measurements of the same ion pairs in D_2O. Not surprisingly, similar and in fact dramatic and distinct peaks in the selectivity coefficients were also observed in this case, but at slightly different temperatures.

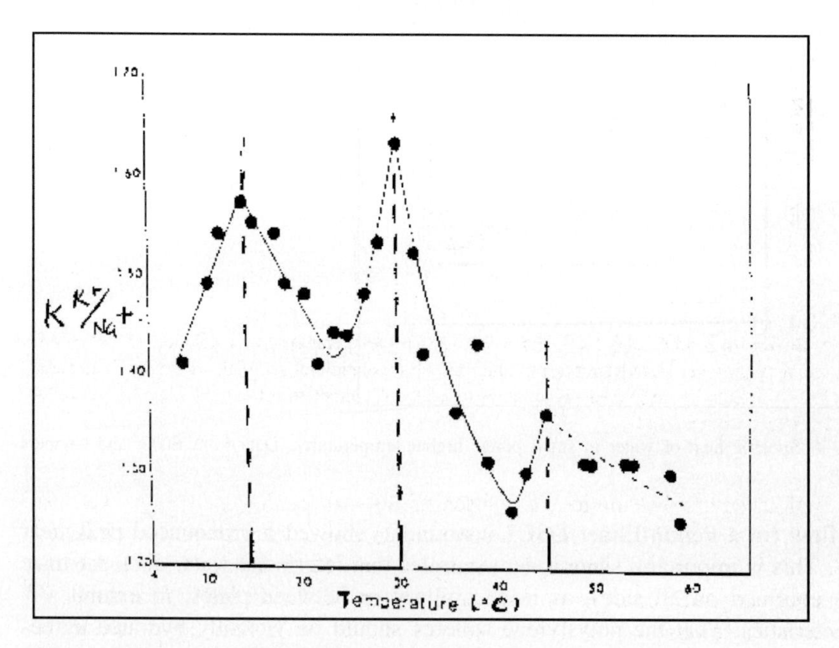

Figure 5. Selectivity coefficient, $K_{K/Na}$, as a function of temperature (°C) for potassium/sodium ions between the pores of a silica gel and the bulk solution, as a function of temperature. Data from Hurtado and Drost-Hansen (1979)

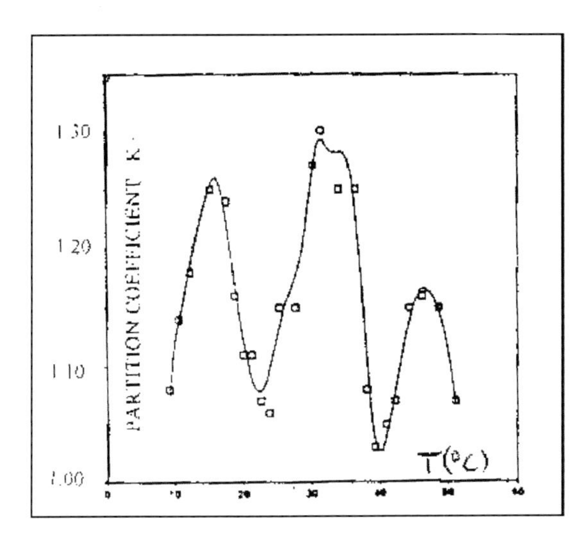

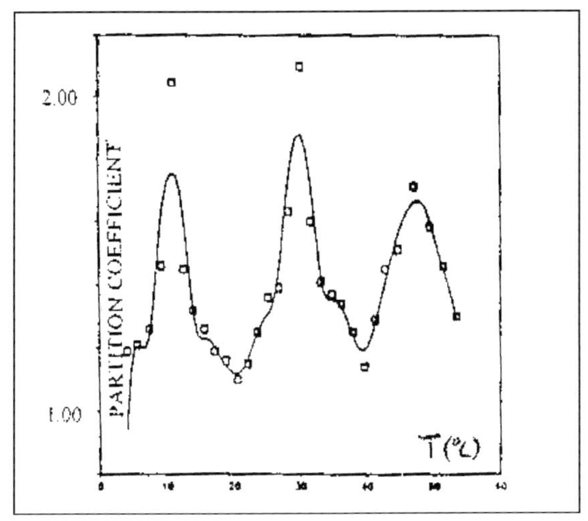

Figure 6. Selectivity coefficient, $K_{K/Li}$, as a function of temperature for a solution of pH = 2.9. Smooth curve represents a computer smoothing and interpolation of the data. Data by Etzler and Lilies. (1996). Selectivity coefficient, $K_{K/Li}$ as a function of temperature for D_2O solution, pD = 4.4 [pD = pH(meter reading) + 0.4]

Anomalous solute distributions are observed not only with ionic solutes. van van Steveninck et al. (1991) have made a series of measurements on non-electrolytes. Some of their results are shown in Figures 7 and 8. In Table 5 are listed the temperatures at which changes in slope were observed. Again the critical temperatures match those observed with ionic solutes, the T_{ks}. It is difficult to escape the conclusion that the same underlying mechanism must be operating, namely thermal transitions in the interfacial *VW*.

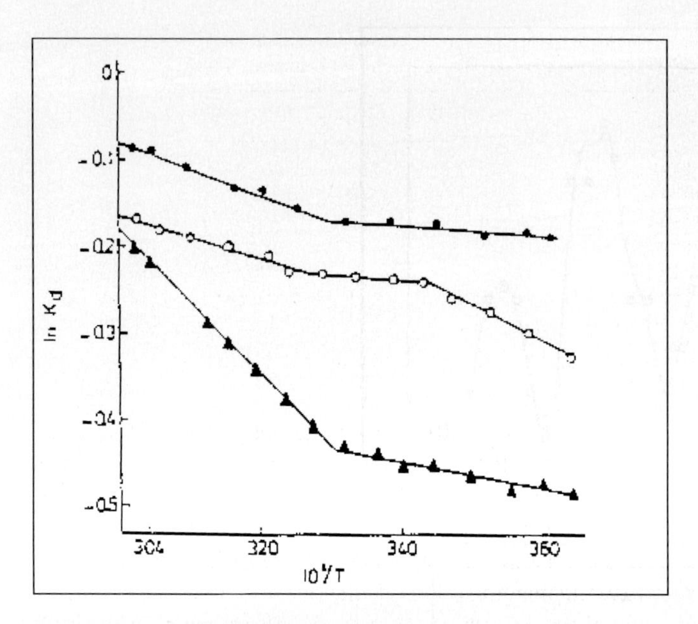

Figure 7. Log_e (selectivity coefficient) as a function of 1/T for some low-molecular weight solutes of Sephadex G10 columns. Filled circles: methanol; open circles: DMSO; triangles: galactose. Data from van Steveninck et al. (1991)

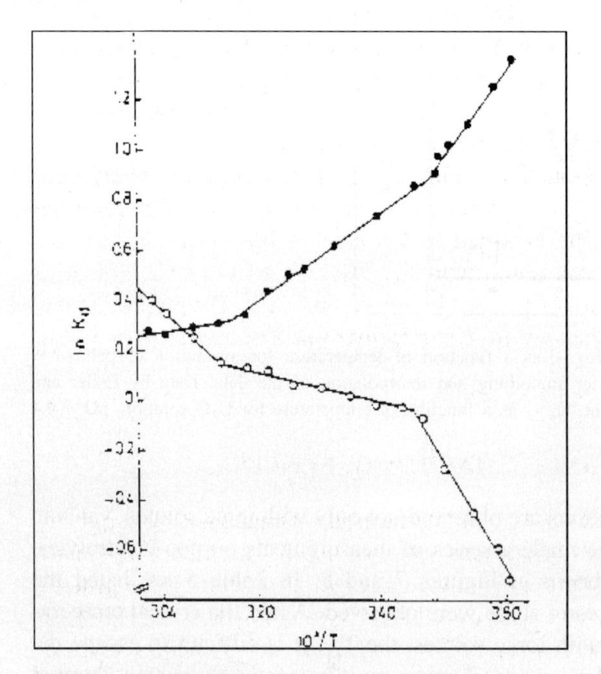

Figure 8. Log_e (selectivity coefficient) as a function of 1/T for thiourea (solid circles) and t-butanol (open circles) on Sephadex G10 columns. Data from van Steveninck et al. (1991)

Table 5. Breaking points in the curves of the partitioning coefficient (Kd) for small solutes on Sephadex columns

Solute	Breaking points (°C)		
Thiourea	13.9		43.4
Galactose		28.9	
t-Butanol	16.6		47.7
Dimethylsulfoxide	16.1	31.5	
Methanol		30.4	

2.7.5 'Missing anomalies'

In some experiments carried out over a sufficiently large temperature interval, all 4 thermal anomalies (at T_k) can be observed in any given run. However, in other experiments, only one or two thermal anomalies may be observed. The reason is not fully understood, but the explanation may well be related to the slow rate of forming of stable vicinal structure above any given transition. For instance, in experiments where measurements are made continuously as the temperature is increased, a sample initially at room temperature may exhibit an anomaly at 30 °C, but the rate of forming the new vicinal hydration structure that is stable between say 30 and 45 °C may be too low, so that when the sample temperature reaches 45 °C, the intervening stable structure may not yet have been fully formed and hence the anomaly is overlooked at that next expected T_k. (As discussed below, experiments in which measurements are made continuously while temperature is *decreasing* frequently fail to show thermal anomalies.) In addition to the anomalies reflecting the *VW* transitions, the overwhelming effects of temperature on many very large macromolecules (such as denaturation – although such processes are rarely very abrupt) must be remembered. On the other hand, great abruptness – generally not at T_k – must of course also be expected in the case of lipid phase transitions. Many of these factors may contribute with the result that nearly all biological responses to temperature inevitably become highly complicated. The purpose of this paper is to call attention to the *specific contributions one can expect from VW* in the cell.

3. VICINAL HYDRATION OF MACROMOLECULES

Since *VW* definitely exists at all solid/water interfaces, it is perhaps not too surprising that individual large molecules in aqueous solution also induce vicinal hydration structures. Obviously, small solutes are not vicinally hydrated (although, in general, both ions and non-electrolyte have their own distinct types of hydration). However, over the years we have observed thermal anomalies in the properties of most solutions of macromolecules, and indeed these thermal anomalies occur at the characteristic temperatures, T_k, Hence we have pursued the idea that *VW* does indeed occur at the macromolecule/water interface, as first proposed by Etzler and Drost-Hansen (1983) and at the same time, we began to explore the question of how large a macromolecule needs to be to become vicinally hydrated.

3.1 Critical Molecular Weight

Although a lot is known about various types of hydration/solvation of macromolecules, little attention has been paid to the possibility that this is vicinal hydration. About 20 years ago, I began investigating the properties of various macromolecules in aqueous solutions from the perspective of the existence of pronounced vicinal hydration structures at solid interfaces. Two striking observations had been reported: the first was a graph of the diffusion coefficients for 32 different solutes in water as a function of their MW (data from Stein and Nir, 1971). In Figure 9, one sees a distinct change in slope at ~1000 Daltons, and our tentative interpretation was that this MW corresponds to the onset of vicinal hydration (Etzler and Drost-Hansen, 1983; see also Drost-Hansen, 1997). The other interesting finding was the intrinsic viscosity of aqueous polyethylene oxide (PEO) solutions as a function of MW. The data in Figure 10 comes from Bailey and Koleske (1976). A distinct change in slope is seen at a MW of ~2000–3000 Daltons.

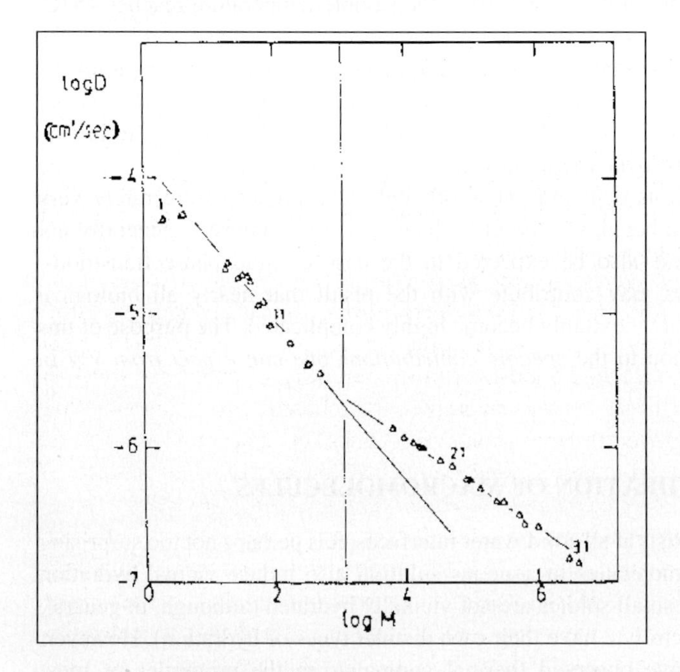

Figure 9. Log (diffusion coefficient) for solutes in water as a function of the MW of the solute at 20 °C. Solutes ranging from #1 (H_2) up to #31 (hemocyanin component, *Helix pomatia*). See original paper for detailed data, Nir and Stein (1971)

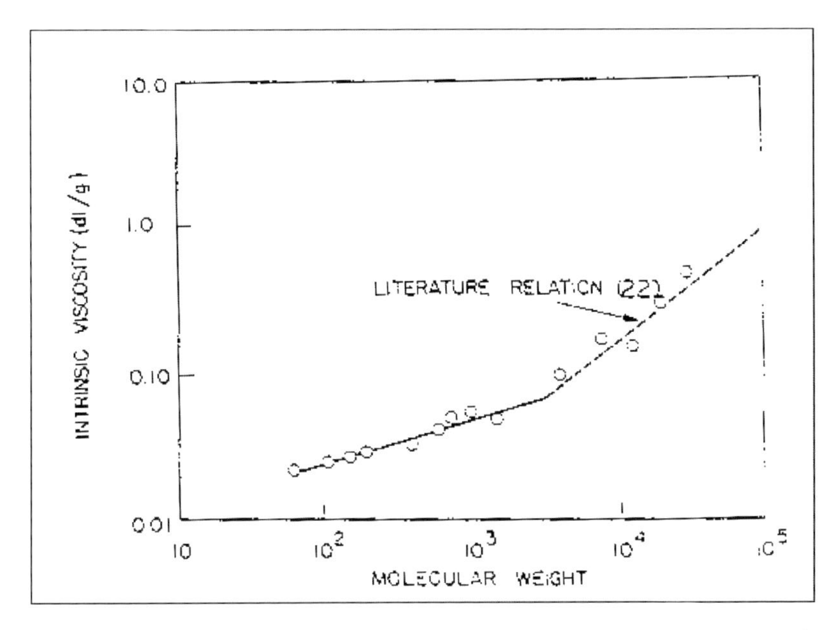

Figure 10. Intrinsic viscosity of polyethylene oxide in water as a function of MW. Data from Bailey and Koleske (1976)

3.2 Rheology

Over the years, we have studied the viscous properties of water and aqueous solutions, at first primarily of pure water and electrolytes solutions (Korson et al., 1969, Dordick et al., 1979, Dordick and Drost-Hansen, 1981), having developed a very high precision technique for such measurements. For example, we have determined the temperature dependence of the viscosity of water over a wide range with the highest precision and, as mentioned in Section 2.7, one of the important conclusions from this study was the confirmation that – contrary to earlier assertions, for instance by Magat, Bernal, Forslind, Krone, the present author and others – there were no thermal anomalies in the viscosity of pure bulk water. In part as the result of this study, it was subsequently proposed (Drost-Hansen, 1968) that the frequently reported thermal anomalies ('kinks' or 'discontinuities') in the properties of the bulk water were caused by spurious influences of the surfaces of the confining containers, as can readily be expected, for instance, from the viscosity experiments using very narrow capillaries. It was also stressed that the thermal anomalies appear to be the direct result of changes in the structural properties of the interfacial water, i.e. caused solely by its proximity to the interface. Later this was indeed expanded to mean *any* adjacent surface, regardless of the detailed, specific chemical nature of the solid (The 'Substrate Independence' or 'Paradoxical' Effect, see below; and Drost-Hansen, 1965, 1976, Kurihara and Kunitake, 1992). Lafleur et al., 1989).

More recently, it became of interest to obtain rheological data at far lower shear rates than those found in typical capillary viscometers (which are of the order of 1000–3000 sec^{-1}). For the study of the viscosities of dilute aqueous suspensions of polystyrene spheres and other suspensions, and of the rheology of aqueous macromolecular solutions, we primarily employed a Brookfield plate-and-cone variable shear viscometer (Model LVTDV-22). Because of space limitations, the results of such experiments on a wide variety of polymers will be exemplified in just one graph (Figure 11) [Note: some of these studies have not been published, although the majority have appeared in Final Grant Reports to the US Air Force, Office of Scientific Research, Bolling AF Base, Washington, DC, USA]. Our viscosity data for diverse polymers in aqueous solutions all showed distinct thermal anomalies throughout and almost invariably at the *VW* T_k values. From frequency with which these changes occur at $\sim T_k$, it is difficult to escape the conclusion that the polymers are vicinally hydrated (in particular see Drost-Hansen, 2001). In these experiments, the rate of heating in the viscometer was usually relatively low. For example, a measurement might be made at some constant temperature and after a constant viscosity value had been recorded, the temperature of the circulating bath controlling the viscometer sample cup would be increased, by $\sim 1\,°C$. The system was then left for a period of time (a few minutes or longer) and the next viscosity measurement taken. In no case were measurements made while the temperature was being lowered, and if a repeat run was to be made on any given sample, sufficient time was allowed for the vicinal water structures to reform. The time required for most systems to revert to their initial state has been from a few hours up to a day. Measurements were rarely made at shear rates above 225 sec^{-1} and many rates were as low as 11 sec^{-1}.

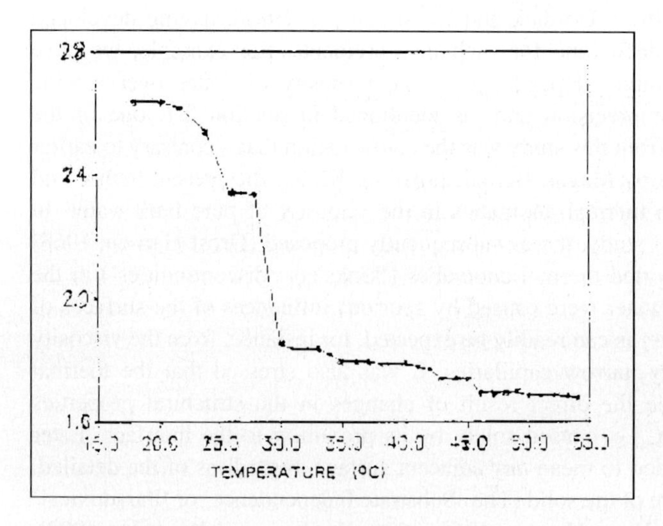

Figure 11. Viscosity versus temperature for a 5% Dextran solution. Shear rate: 90 sec^{-1}. Data from Drost-Hansen and Gamacho (unpublished)

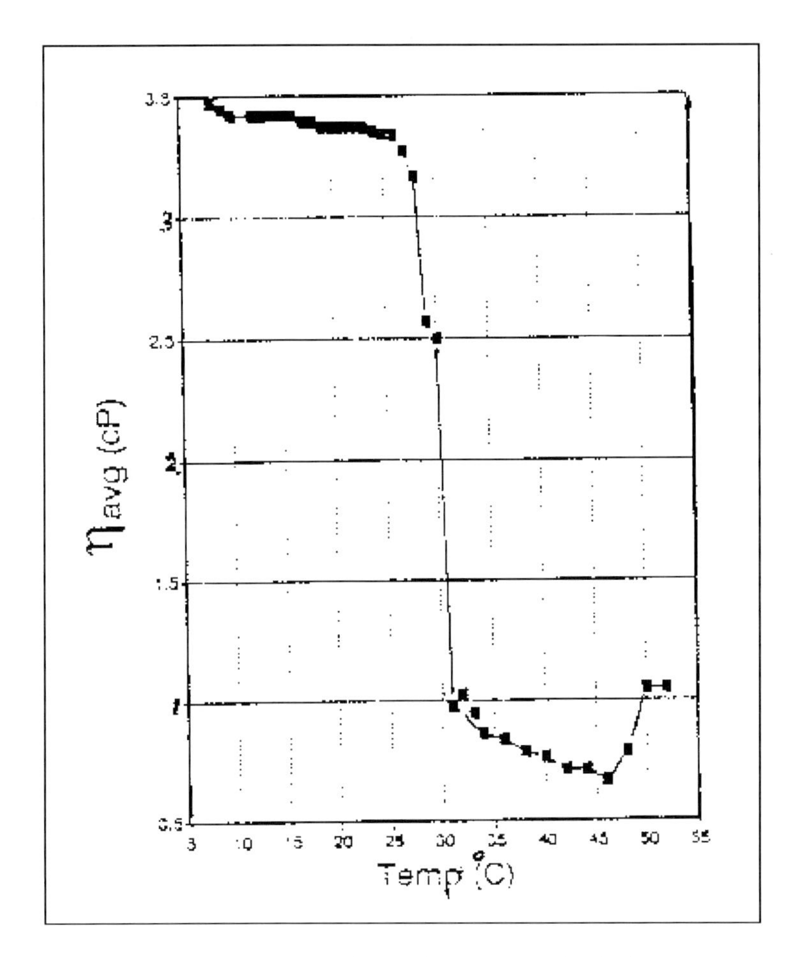

Figure 12. Viscosity versus temperature of blood plasma, Habor Seal

While *all* macromolecules in aqueous solution are vicinally hydrated (Figure 12) many biochemically important macromolecules are partly embedded in cell membranes. Again, it must be expected that those parts of such molecules which extend away from the membrane are also vicinally hydrated. Not surprisingly, thermal anomalies are also very frequently – but not invariably – seen in Arrhenius graphs of enzyme reactions (see in particular Drost-Hansen, 1971, 1978, 2001; see also Etzler and Drost-Hansen, 1979).

Anomalies are seen also in the viscosities of suspensions of solid particles., Figure 13, for example, shows the viscosity of a 0.1% polystyrene sphere suspension (particle size 0.17 microns) as a function of temperature. An anomaly is seen at 30 °C consistent with the expectations based on the evidence of vicinal hydration of such particles discussed above: the volume contraction upon settling, Section 2.1.1; the anomaly in index of refraction (see below); and the Bragg scattering results

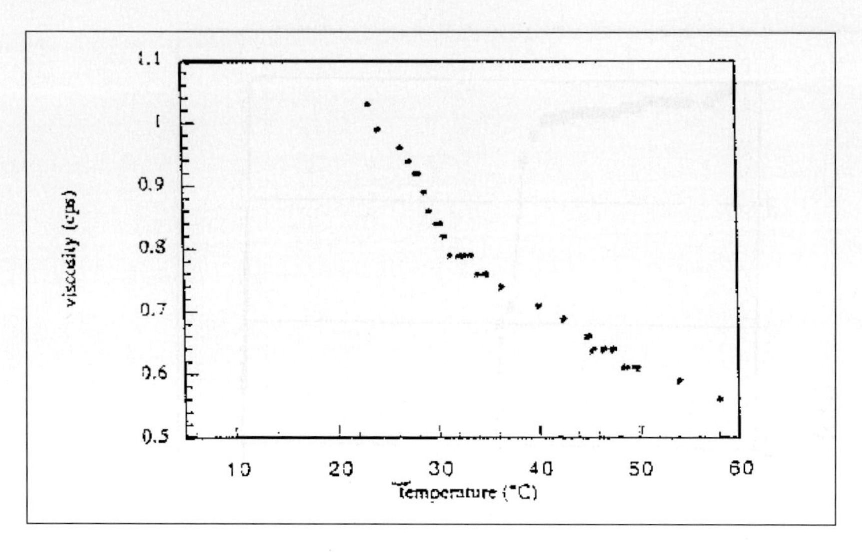

Figure 13. Viscosity versus temperature of 0.1% polystyrene sphere suspension; particle diameter 0.17 micron. Shear rate 90 sec^{-1}

(Section 3.7). The effects of vicinal hydration quite naturally become more pronounced the more concentrated the suspension, and in the case of, say, 10% kaolinite, very large anomalies are seen in the rate of sedimentation and compaction of such systems (Drost-Hansen, 1981).

3.3 Other Rheological Studies

A large number of rheological studies exist on far more complex systems than those examined so far in the present paper. Thus data ranging from cell physiology to food science and food technology have provided considerable insight into likely effects of *VW* in very complex systems. Because of space limitations, only a few examples will be discussed, e.g., Figure 14 shows the viscosity of a 1.4% actinomyosin solution.

Considering the thermal anomalies discussed above for relatively simple biopolymers, it is not surprising that complex protein systems also show distinct anomalies. In Figure 14, for instance, it is difficult to envision any obvious molecular mechanisms to explain such dramatic changes in this system over such narrow temperature intervals, except in terms of *VW* and the thermal anomalies (see also Drost-Hansen, 2001).

Finally, Figure 15 shows the viscosity of the protoplasm of *Cumingia* eggs as a function of temperature. Note the distinct (and abrupt) peak near 15 °C, and the even more dramatic change near 31°C. It seems inescapable that these effects must be due to the cell-associated *VW*.

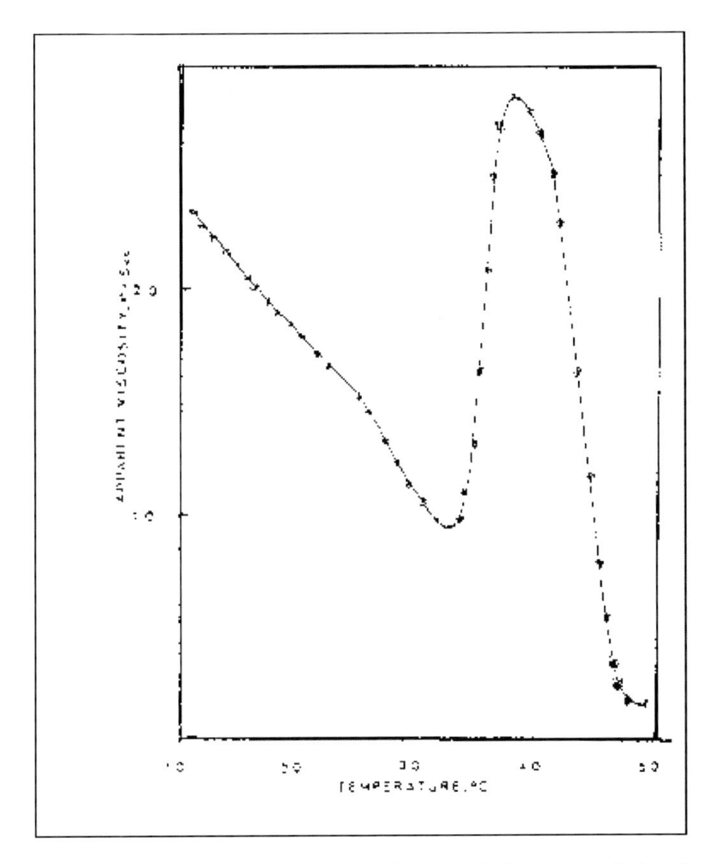

Figure 14. Viscosity versus temperature of actomyosin in aqueous solution. Continuous heating; rate = 1 °C min^{-1}. Shear rate 1.02 sec^{-1}. Data from Wu et al. (1985)

3.4 Diffusion Coefficients

Anomalous temperature dependencies of diffusion coefficients were reported as early as 1969 by Dreyer and co-workers (see, for instance, Drost-Hansen, 1972). In view of the distinct anomalies in the viscosity data discussed above, anomalies should also be expected in diffusion coefficient data, for both suspensions of particulate matter, such a polystyrene spheres, and in macromolecular solutions. The diffusion coefficient data shown here were obtained using a Photon Correlation Spectrometer (Coulter Counter Particle Analyzer Model CN4 SD). In a series of measurements, the temperature would be increased step-wise, usually in 1°C increments, allowing sufficient time (for instance, 2–5 min) between each reading to ensure that the sample had come to thermal equilibrium with the sample chamber. Figure 16 shows data obtained on rather highly diluted suspensions of polystyrene spheres. Anomalies are seen at 30 °C and in some cases at 45 °C, the spheres behaving like any other 'solid' surface (see also Sections 3.6 and 3.7).

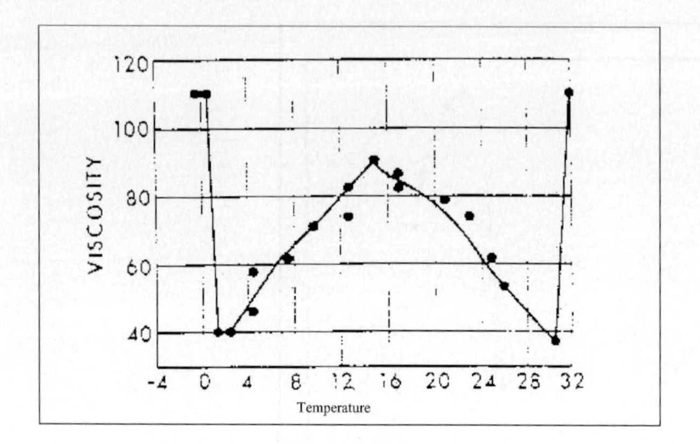

Figure 15. Viscosity versus temperature for protoplasma in *Cumingia* egg. Data of Heilbrunn, quoted in Johnson, Eyring and Polissar (1954)

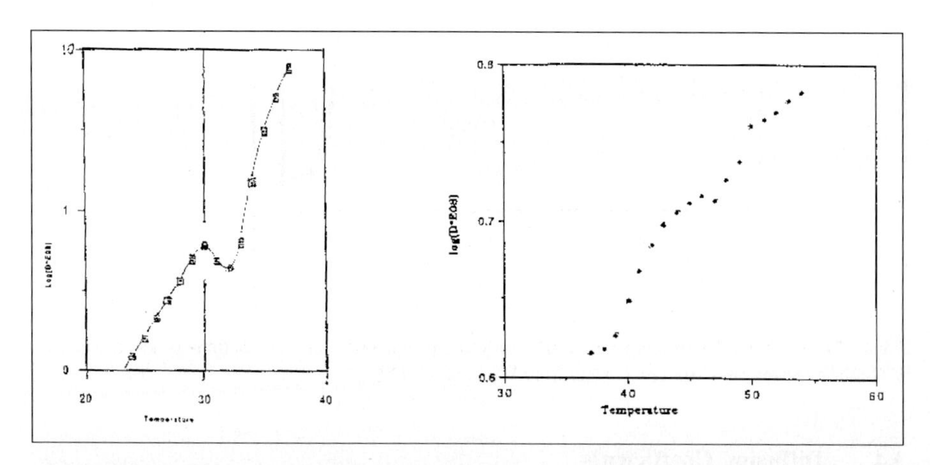

Figure 16. Log_e (diffusion coefficient) versus temperature of polystyrene spheres in water, given as a 3-point moving average. Particle diameter 0.17 micron

Recall also the volume contraction data observed upon settling of the suspensions (Section 2.1) and the DCS data (Section 3.5). Figures 17A and B give diffusion coefficients for a variety of macromolecules in water. Again distinct thermal anomalies at T_k occur. In view of the effects of viscosity on diffusion coefficients (as, for instance, implied in the Einstein relationship), the anomalies seen in diffusion coefficients should indeed be expected. Once again, the conclusion must be drawn that, as far as water structure is concerned, sufficiently large macromolecules in water behave as a 'solid surface' to promote the formation of vicinal hydration. In some cases, the anomalies are more pronounced than in other cases, this variability probably being due to slight differences in sample preparation and

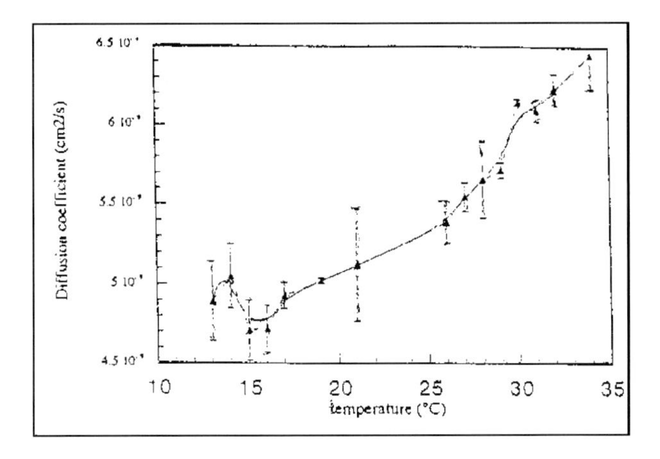

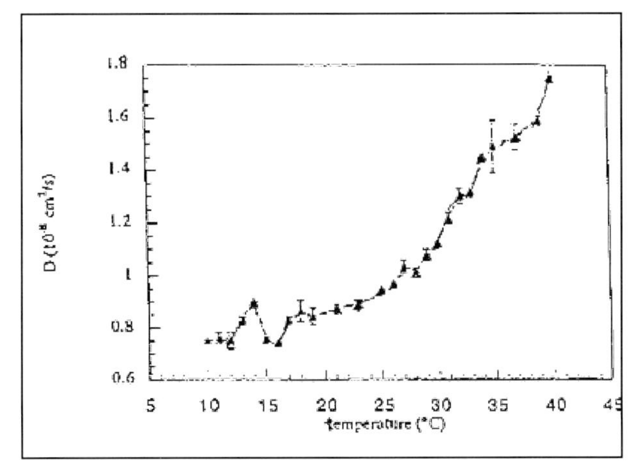

Figure 17. (A) Diffusion coefficient versus temperature of Dextran, MW = 40.8 kDa. Concentration 9.1%. (B) Diffusion coefficient versus temperature of Dextran, MW = 515 kDa. Concentration 8.18%

particularly in the time that had elapsed since the samples were last stirred and placed in the sample cuvette (see Section 3.11 on hysteresis).

3.5 Calorimetric Data

Effects similar to the distinct peaks in the specific heat curves reported by Etzler and Conners on water in porous silicate particles have been seen in many other studies, including Braun and Drost-Hansen (1976), and more recently in a DSC study of 10% suspension of polystyrene spheres (particle size 0.22 microns). The viscosity anomaly is shown in Figure 13, and the Bragg scattering results discussed in Section 3.7.

3.6 Index of Refraction

An equilibrium property has been studied in a preliminary manner, namely index of refraction, as a function of temperature. Measurements were made with an Abbe refractometer (Bausch and Lomb, Abbe 3L), with the temperature controlled by a circulating constant temperature bath. Again, data have been collected for a number of macromolecules in aqueous solution and typical results are shown

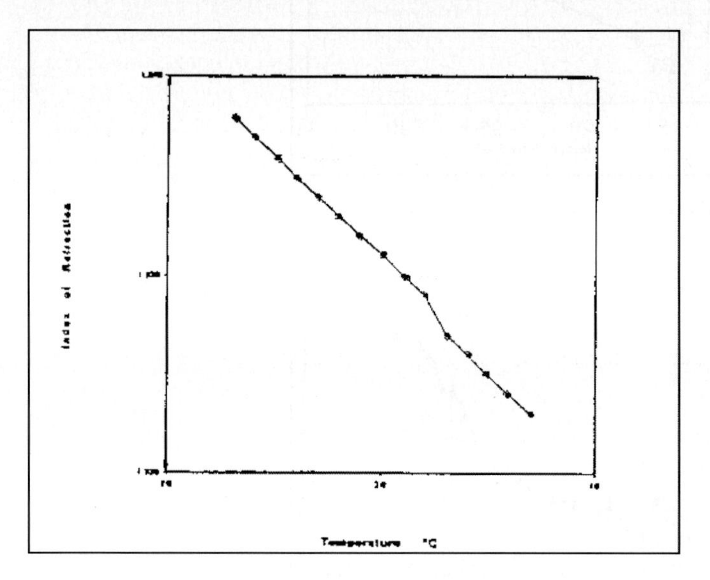

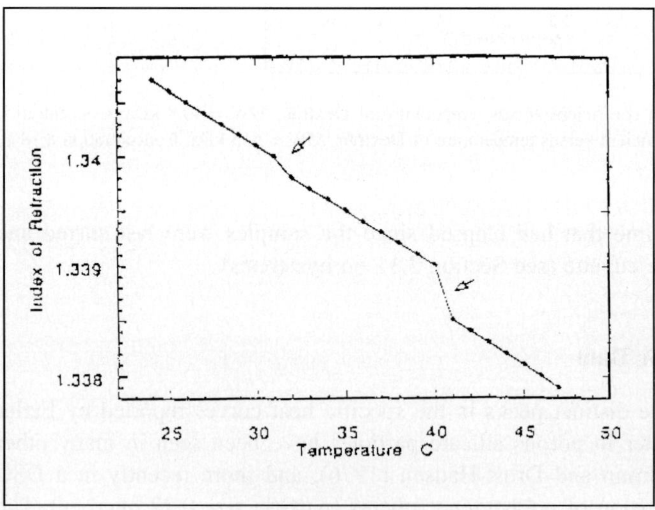

Figure 18. (A) Index of refraction of aqueous solution of Dextran in water versus temperature. Concentration 5%. (B) Index of refraction of γ-globulin in water versus temperature. Concentration 5%

in Figures 18A and B, where the expected, distinct (albeit small) anomalies are observed at T_k, regardless of the specific nature of the solutes.

3.7 Bragg Scattering from Suspensions

It has been anticipated for many years that ions in sufficiently concentrated solutions may form crystalline lattices and this has indeed been observed visually for 'giant ions' (macroions), such as negatively charged polystyrene spheres in suspension. A similar effect has also been reported for relatively concentrated solutions of turnip yellow mosaic virus. In a particularly interesting study, Daly and Hastings (1981) studied the Bragg scattering from crystallized suspensions of the readily available 'macroions', polystyrene spheres. It is important to stress that the 'crystallization process' takes place only very slowly and depends on the nearly complete absence of any mechanical shear [see also Section 3.11].

Figure 19 shows the results of Daly and Hastings (1981); in addition to the data points the calculated curves based on the theory developed in the paper are also shown. At a first approximation, the experimental points follow the trends of

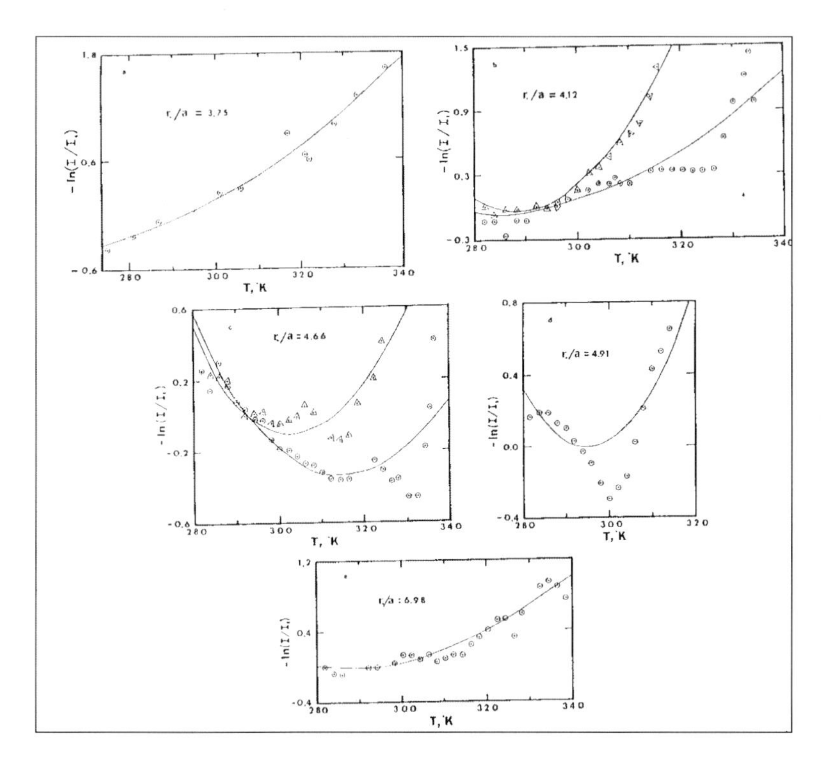

Figure 19. First-order Bragg scattering intensity versus temperature of 'crystallized' suspensions of 'macroions' (i.e., polystyrene spheres.) Data from Daly and Hastings (1981); see this reference for the experimental details and first approximation theory

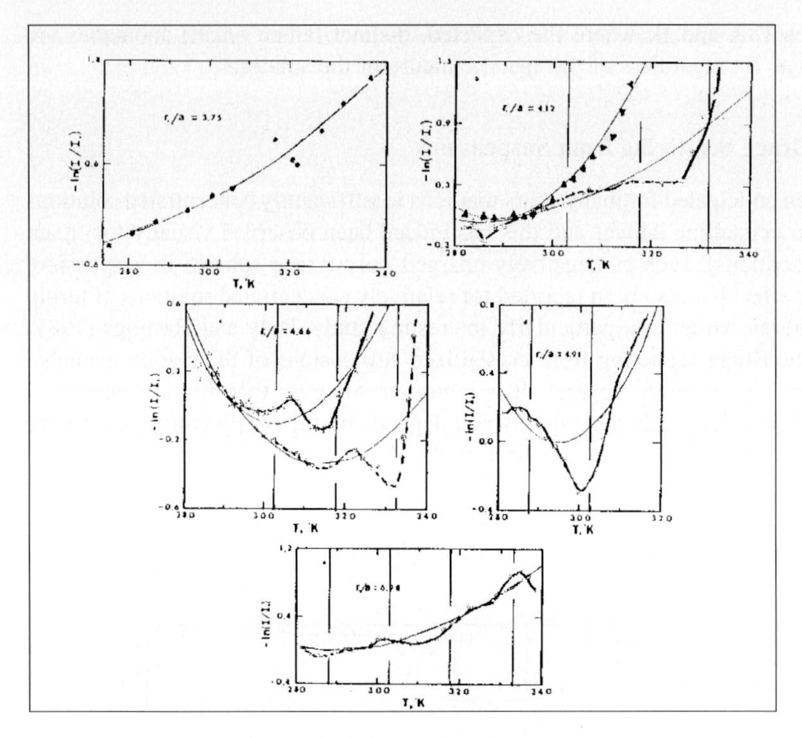

Figure 20. First-order Bragg scattering intensity versus temperature; data as in Figure 19, but with curves redrawn ('freehand') by the present author; also indicated are the temperatures of T_k (vertical lines)

the calculated curves. However, reasonably good-fit, continuous curves have been drawn in Figure 20, as well as vertical lines to indicate the temperatures of the thermal anomalies. Notable differences clearly exist between the calculated curves and the observed points. These differences demonstrate a likely role of the *VW* transitions at T_k.

3.8 Critical Molecular Weight Dependence, MWc: Detailed Considerations

Electrolytes and small non-electrolytes are obviously not vicinally hydrated, but the evidence presented in the preceding sections suggests that larger macromolecules are indeed vicinally hydrated. This poses the question: is there a critical Molecular Weight range (MWc) below which no vicinal hydration exists, but above which all molecules are vicinally hydrated? The answer seems to be that indeed there exists such a critical size, discussed in Section 3.1. As illustrated below, the MWc falls in a range 1,000 and 5,000 Daltons. Note also the *great diversity* of types of molecules with >MWc included in these graphs, consistent with the idea of the 'Substrate Independence Effect' (the 'Paradoxical Effect').

Sometimes the abruptness of the change at MWc is very pronounced. Thus Gekko and Noguchi (1971) measured the MW dependence of a number of properties of oligodextrans with different MW. Figures 21a-d show respectively:-: a Stockmayer-Fixman plot; a derivative plot of the sound velocity (with respect to concentration);

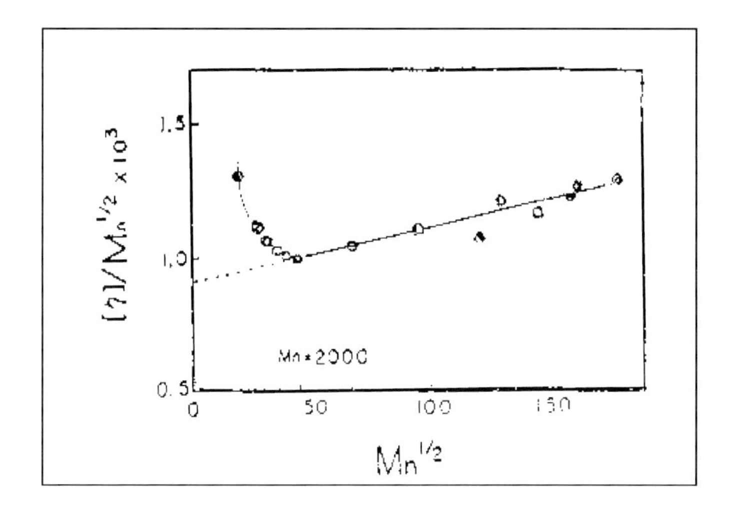

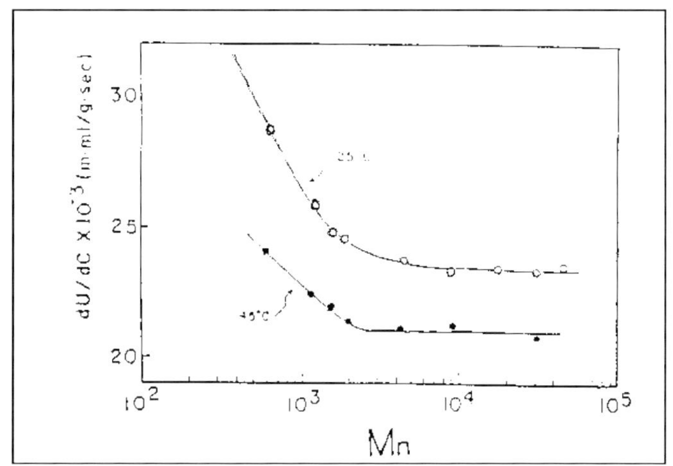

Figure 21. (A) Stockmayer-Fixman plot for aqueous solutions of Dextran versus square root of MW of the polymer. (B) Derivative of ultrasound velocity with respect to concentration for aqueous Dextran solutions versus MW. (C) Partial specific compressibility, β_t° of Dextran in water at 25 °C versus MW. (D) Fraction of 'bound water' in aqueous Dextran solutions versus MW Data from Gekko and Noguchi (1971)

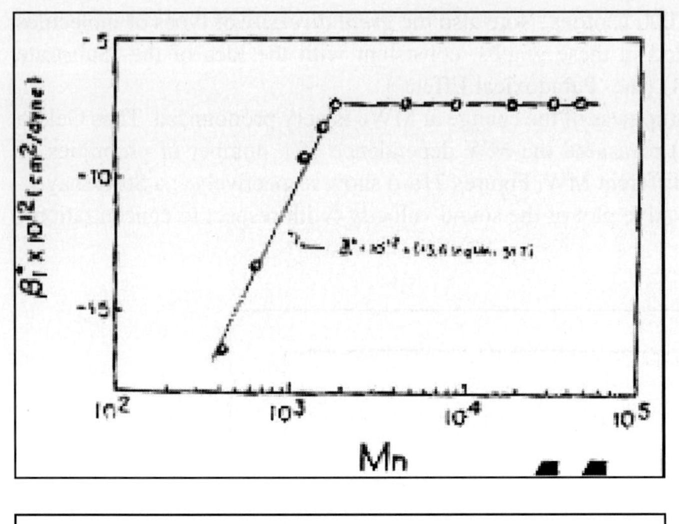

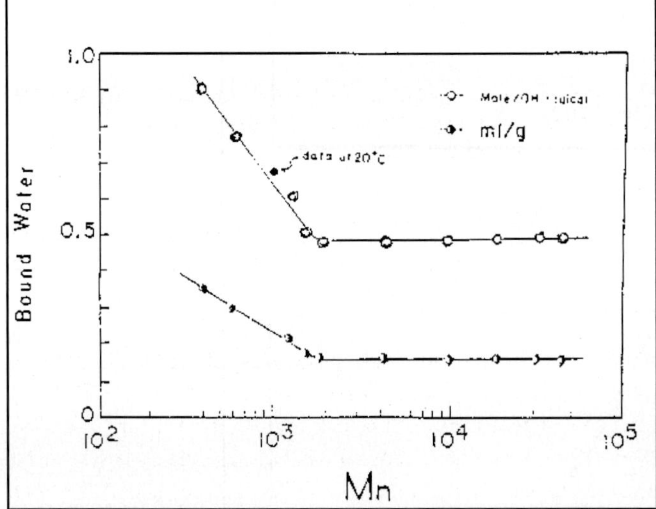

Figure 21. (Continued)

partial specific compressibilities; and the amounts of bound water – all as a function of MW. In these graphs, changes from one functional dependency to another is remarkably abrupt and always in the vicinity of a MW of $\sim 2,000$ Daltons.

As discussed in Section 3.11, vicinal hydration structures are highly shear-rate dependent. For this reason we mostly made measurements with a variable shear-rate instrument (a Brookfield cone-plate rheometer), generally using shear-rates of 100 sec^{-1} or lower. However, in some cases it was also possible to use capillary viscometers. Figure 22 shows some apparent energies of activation for viscous flow

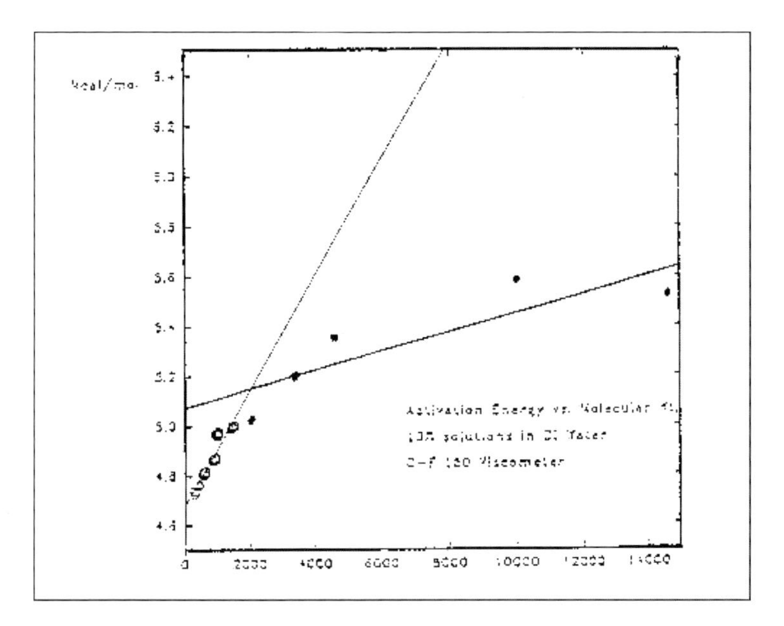

Figure 22. Apparent energies of activation for viscous flow of polyethylene oxide solutions [10%] versus MW. Data by Drost-Hansen and Vought (unpublished). See also Drost-Hansen (1992)

of 10% PEO solutions as a function of the MW of the polymer (Cannon-Fenske capillary viscometer, size 150; Drost-Hansen, 1992). A change in slope occurs at a MW of ~2, 000 to 4,000 Daltons, in agreement with the general idea that a critical MW range exists.

Antonsen and Hoffman (1992) followed the properties of PEO solutions as a function of MW. Again a small but distinct anomaly was found at ~1,000 Daltons. A far more dramatic change is seen in the transition temperature for the 30% PEO solutions as a function of MW (Figure 23). Here the abrupt change is observed for a MW of 1,200 Daltons. Antonsen and Hoffman (1992) also measured the total heat required to melt frozen 30% PEO solutions, and found a distinct change at a MW of ~1,000 Daltons.

Related to cell functioning, Mastro and Hurley (1985) have compiled data for viscosities and diffusion coefficients and viscosities of a number of 'markers' in the cytoplasm of cells. By the Walden rule, the product of viscosity and diffusion coefficients should be a constant for any given solute. Figure 24a shows the data of Mastro and Hurley, which on replotting in terms of the product of the diffusion coefficient (D) and the viscosity (η), gives the results shown in Figure 24b. In Figure 24a, a distinct inflection point is observed ~2,500 Daltons. However, Figure 24b indicates that a *very* marked change in slope occurs at or just above 1,000 Daltons – consistent with the assumption that this is the critical mass for vicinal hydration.

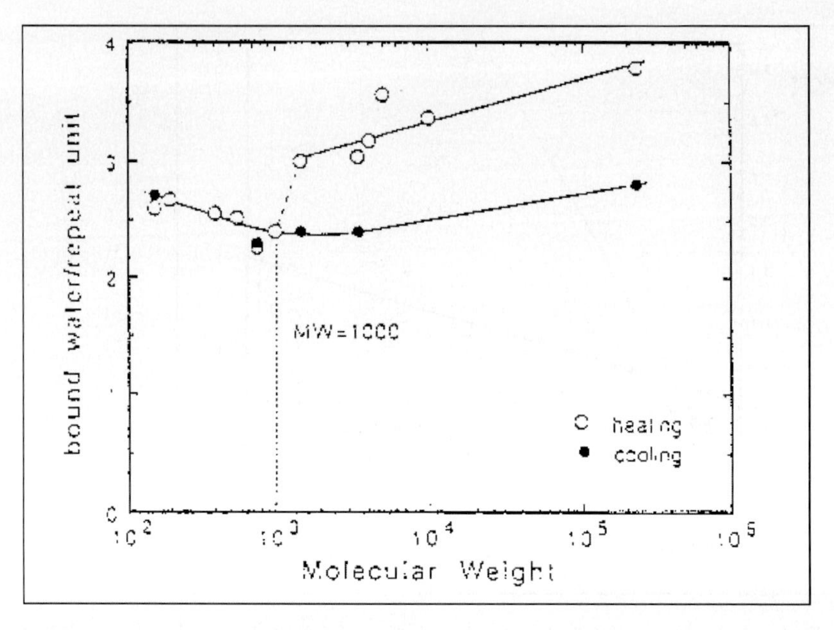

Figure 23. Amount of 'bound water' per repeat unit in 30% PEO solution as a function of MW. Data from Antonsen and Hoffman (1992)

The abrupt transition between vicinally hydrated biomacromolecules and smaller solutes finds a dramatic manifestation in cell biochemistry in terms of the normal composition of eukaryotic cells. Thus, based on data from Antonsen and Green (1975), Clegg (1979) first called attention to the remarkable coincidence of the onset of vicinal hydration and the MW of solutes in the cell. In Figure 25, the concentration of various solutes in the cell [the ordinate is the relative occurrence of solutes] are seen as a function of MW. Note the conspicuous absence of solutes in the cell in the range 1,000–10,000 Daltons. It seems as if cells, in the process of evolution, have selected for solutes that are either definitely *not* vicinally hydrated *or* distinctly vicinally hydrated. Note also that these small polypeptides (usually in very low concentrations) can have dramatic physiological effects (such as endorphins), and these are often in the range of 1,000–10,000 Daltons.

Based on the rheological data and the evidence discussed above, one must conclude that as far as water structure is concerned, a molecule with a MW of ~2,000 Daltons is 'mechanically' as 'substantial' as a 'solid surface'. Surely this information must be important to those investigators modeling the structure of water at interfaces and the idea of a 'critical MW range' must have implications for the idea of 'soft interfaces', [see De Gennes, 1997]. In this connection, see also the recent papers of Etzler et al. (2005) on particle-particle adhesion, especially in the way it is influenced by trace amounts of water at the particle interfaces.

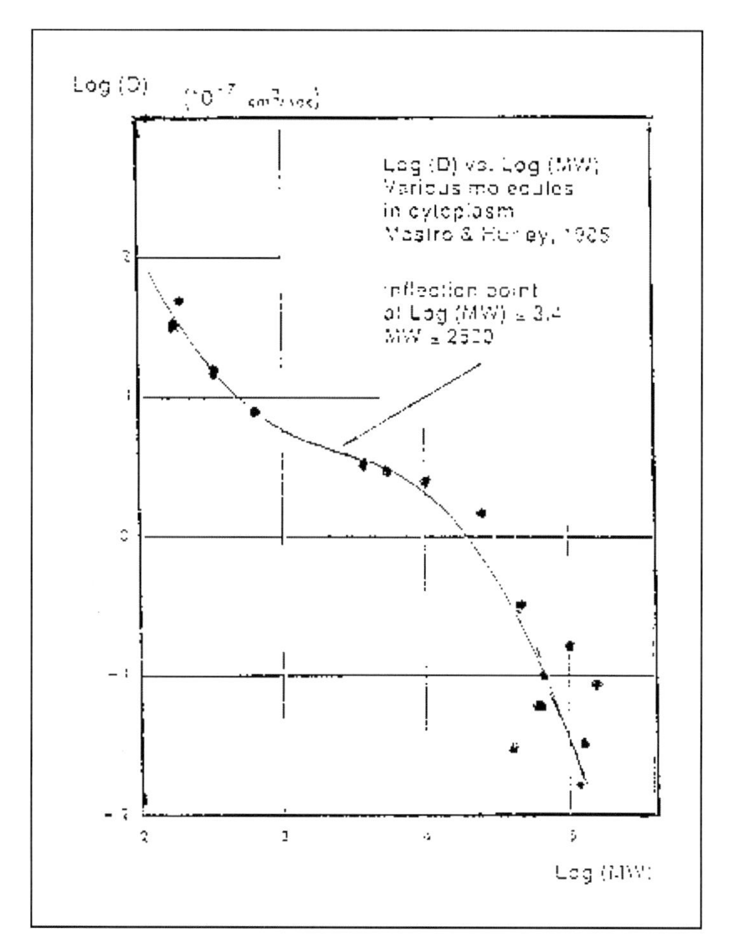

Figure 24. (A) Log (Diffusion coefficient) versus log (MW) for various molecules in the cytoplasm. (B) Product of viscosity and diffusion coefficient versus log (MW) for the molecules from Figure 24A. Data redrawn from Mastro and Hurley (1985)

3.9 Shear Rate Effects

VW is shear rate dependent and once destroyed by shear, the time to reform the vicinal hydration layers (when the shear stops) may be very long; of the order of minutes, hours or even a day. This finding has been corroborated many times. Kerr (1970) and Drost-Hansen (1976) studied the internal damping of a vibrating hairpin capillary (vibrating in a vacuum). The damping is caused primarily by the energy dissipation at the water/quartz interface of the hairpin element, showing a notable reduction at 30 °C, which we have ascribed to the breakdown of the extended interfacial water structures at the capillary surface at this temperature. In other words, the coupling between the water molecules and the confining surface

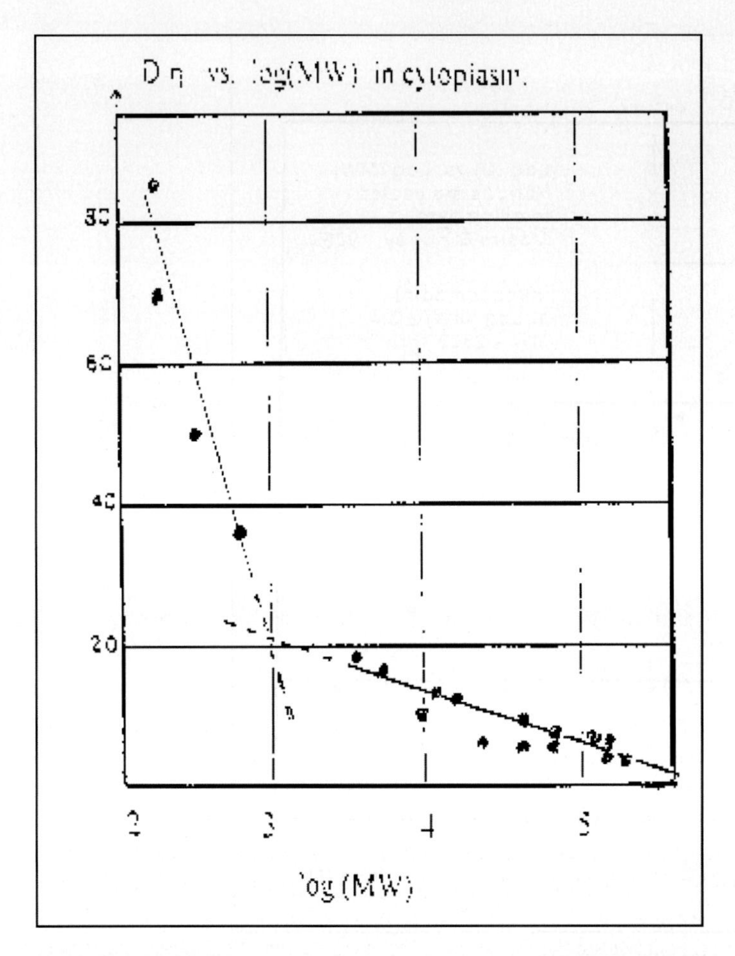

Figure 24. (Continued)

changes drastically at T_k. Kerr (1970) also noted that it was not possible to obtain reproducible results if another run was initiated immediately after completion of the previous run. The system had to be left undisturbed for as long as 24 h before reproducible results could be obtained. Likewise, Braun and Drost-Hansen (1976) carried out a number of DCS experiments with water in a porous silica gel. While each new run would show a distinct anomaly at T_k, such results could not be repeated unless the sample was allowed to sit at a lower temperature overnight. We return to the shear rate and temperature induced hysteresis in Section 3.11, which deals with rheological properties of aqueous macromolecular solutions, where the effects appear to be even more pronounced.

The effects of shear on some micellar systems were studied by Shephard et al. (1974), who measured the viscosity of various solutions prepared using some sulfonated alkanes [of the type considered for use in tertiary oil recovery

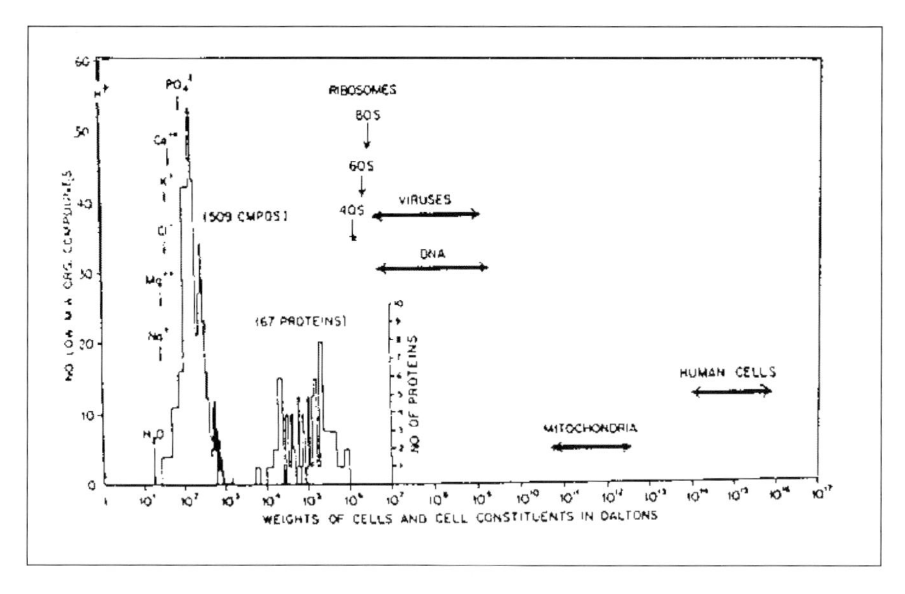

Figure 25. Solute concentrations in a typical eukaryotic cell versus MW of the solute. Data from Anderson and Green (1975), as presented in Clegg (1984a)

processes] in a brine solution (2.5% NaCl plus smaller amounts of $CaCl_2$ and $MgCl_2$). Figures 26A and B show some of their results, from which it is clear that the thermal anomalies near both 45 °C and ~60-62 °C rapidly become less prominent with increasing shear rate, consistent with the idea that the vicinal hydration is indeed sensitive to shear effects. It is interesting to speculate that the shear rate effect may ultimately have a bearing on the overall rheological properties of many, or most, aqueous suspensions, as well as macromolecular and micellar systems. Thus the transient properties of *VW* under conditions of shear may conceivably play a role in non-Newtonian behavior of such systems as dough; heavily hydrated clay deposits (as seen in mud slides and turbidity currents in marine environments), or the large micelles in such solutions as cetyltrimethyl ammonium salicylate (strongly elastic solutions, even at dilutions as high as 0.01%! – see also the mention of these micelles in the discussion of DCS studies of such solutions).

3.10 Possible Thermodynamic Implications of Vicinal Hydration

From the examples discussed above, it is clear that mechanical shearing notably affects the viscosities of macromolecular solutions. However, if a transport property is affected by shearing, it seems logical that thermodynamic properties may also be affected. Thus the proposal is made that in an osmometer with identical aqueous solutions (of a high MW polymer) on either side of a semi-permeable membrane, at least a transient osmotic pressure difference might be created if the contents on one side of the osmometer are suddenly stirred sufficiently vigorously. By

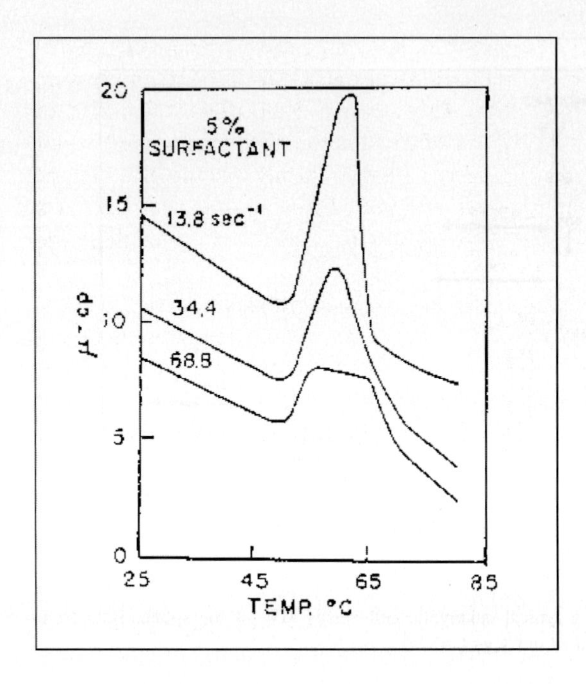

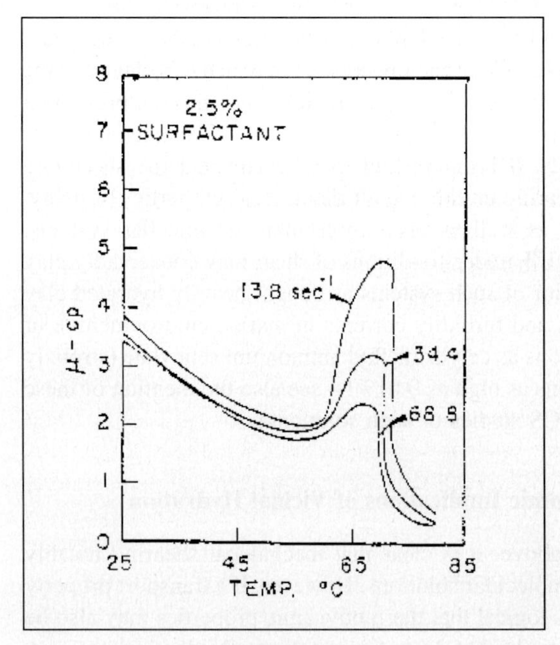

Figure 26. (A) Viscosity versus temperature of a surfactant dispersion (SS-5) in a dilute electrolyte solution, measured at 3 different shear rates. (B) Viscosity versus temperature of a surfactant dispersion (SS-2.5) in dilute electrolyte solution, measured at 3 different shear rates. Data from Shephard et al. (1974)

a similar argument, if a stagnant solution of a sufficiently concentrated macromolecular solution (or suspension of particles) is stirred, a (transient) vapor pressure difference should develop. Indeed, about 40 years ago, I did exactly this kind of experiment using 5 and 10% suspensions of montmorillonite: a (transient) vapor pressure increase was indeed been observed upon suddenly stirring the suspension! However, because this was a highly unexpected finding – and for lack of a deeper understanding of *VW* at that time – I was dissuaded from publishing the results.

Returning to transport phenomena: if the above arguments can be corroborated, one may conceivably expect differences as well in reaction rates involving solutes that are highly vicinally hydrated. As we have repeatedly shown, enzymes in solution are vicinally hydrated, as evidenced by the occurrence of thermal anomalies in Arrhenius graphs of enzymatic reaction rates (see Etzler and Drost-Hansen, 1979, 1983). Hence it is postulated that differences in rates of reaction involving sufficiently large enzymes may exist between reactions taking place in stagnant solutions compared to vigorously stirred systems. This may be particularly pronounced if the enzyme acts upon a relatively large substrate molecule with its own vicinal hydration.

In hemodynamics, it has long been known that shear may profoundly affect some of the dynamic properties of the circulating blood. One obvious site of such shear effects is the deformable particles of the blood, notably the platelets, leucocytes, and, in particular, the erythrocytes (Dintenfass, 1981; Uijttewaal, 1990). However, in the case of a more-or-less suddenly created stenosis, the locally increased shear rate due to increased velocity at the point of constriction may not only affect the geometric shapes of the deformable circulating cells, but might also affect the dynamic properties of some of the circulating macromolecules. Essentially 'stripped' of the 'protective' vicinal hydration hulls, due to the increased shear rate, some macromolecules may become more reactive and thus alter the homeostatic dynamic equilibria of the circulating blood.

3.11 Mechanistic Aspects

In considering slope changes in Arrhenius graphs of rate processes, it is important to recognize that distinct changes in slope may mean very significant differences in energies of activation for the processes (below and above a critical temperature, T_k). In other words, the energetics may differ notably – implying not merely a small change in the overall molecular dynamics, but sometimes possibly a very dramatic change in the underlying mechanisms of the reactions. This might be brought about by changes in a relatively small number of water molecules in the hydration shell, if these are very strongly bound, or, in the case where the energetics of the *VW* does not differ significantly from that of the bulk water, very *large* numbers of water molecules must be involved. The latter situation is by far the more probable.

3.12 Hysteresis

The fact that heating past one of the critical temperatures, T_k, destroys the vicinal hydration structure stable below that point, together with the very low rate of reforming of the hydration structures (after the temperature range has been decreased to a range below T_k) results in notable hysteresis. In addition, at constant temperature, shearing of the solutions will also destroy the vicinal hydration structures and again the time for reforming the appropriate hydration structure is long (minutes or hours). These facts, combined with the fact that many studies of polymer viscosities have been made in relatively high shear instruments, such as capillary viscometers – may offer an explanation why the effects described in this paper have rarely been noted or explored before. The formation of a 'stable' vicinal hydration structure will often be fairly fast: in a continuous heating run, once the temperature has exceeded T_k, the 'new' vicinal hydration structure stable between T_k and T_{k+1} may form sufficiently quickly so that another thermal transition is seen at T_{k+1}. However, in some cases of relatively fast heating – and depending on the nature of the solid interface or the specific type of polymer used – there may be insufficient time for the stable vicinal hydration structure to form, and in such cases only one (or perhaps two) of the thermal transitions may be seen in any given run.

3.13 Conformational Changes and Pre-denaturation

In view of the ubiquitous presence of *VW*, it may prove useful to review a large number of previously published data on biomacromolecular solutions. Studies where a parameter has been measured as a function of temperature might reflect previously overlooked influences of *VW* and the thermal anomalies. As an example, Lopez-Lacomba et al., 1989) used a DSC method to follow the thermal unfolding of myosin rod, light meromyosin and subfragment 2. An inspection of their data (notably their Figures 1-3) suggests that the thermal transitions of *VW* may play a role in the unfolding process, considering the events in the DSC curves that appear to occur near T_k. Likewise, the study by Urbanke et al. (1973) of conformational changes in $tRNA^{phe}$ (yeast) by a differential melting technique, suggests that some elements of transitions may be caused by – or at least be influenced by – the structural transitions of the *VW* of hydration of the nucleic acid. By the same token, the concept of 'pre-denaturation' may be influenced by – or simply reflect – the thermal transitions of the *VW*.

4. ROLE OF VICINAL WATER IN BIOPHYSICS AND CELL BIOLOGY

In view of the inevitable vicinal hydration of all solid surfaces and of large macromolecules, it is hardly surprising that *VW* plays a critical role in cell biology. Space limitations do not allow a review of the large amount of information now available on this subject, but below a number of physiological elements and processes are enumerated for which *VW* has been found to play an important – and sometimes a controlling – role. The reader is encouraged to check the various references listed, especially those of Clegg, Etzler, Mentre, Nishiyama, Wiggins, and the authors who contributed to the Symposium Proceeding: 'Cell-Associated Water' (Drost-Hansen and Clegg, 1979). More recently, the present author has discussed in some detail various aspects of the role of VW in cell biology (Drost-Hansen, 2001). Among the topics discussed in that paper from the point of view of *VW* and the thermal anomalies are:

- Enzyme kinetics and anomalies in Arrhenius graphs
- Membrane functioning
- Cell volume control
- Erythrocyte sedimentation rates
- Chromosome aberrations
- Seed germination
- Multiple growth optima
- Upper thermal limits for growth and Pasteurization temperature
- Body temperature selection in mammals and birds
- Thermal stress and hyperthermia therapy

5. SUMMARY AND CONCLUSIONS

It has long been apparent that solid surfaces induce *VW* structures. It now appears that sufficiently large macromolecules in aqueous solution are also vicinally hydrated. There exists a 'critical MW range' [MWc] above which all dissolved macromolecules are vicinally hydrated while molecules with lower MW are not vicinally hydrated: the transition appears to fall between 1,000 and 5,000 Daltons. Like solid surfaces, the vicinal hydration of macromolecules is essentially independent of the specific chemical details (the 'Substrate-Independent Effect' or 'Paradoxical Effect'). In the case of the hydration of solid surfaces, the geometric extent of the *VW* is probably in the range of several tens of molecular layers and possibly as many as 100 layers and *it is likely that the vicinal hydration of macromolecules is of the same order of magnitude.*

The most notable characteristics of vicinal hydration are the highly anomalous temperature dependencies: anomalous changes are seen over four narrow temperature ranges (T_k) centered around 15, 30, 45 and 60 °C. Apparently different geometric structures are stable between each of these intervals; heating a sample above any one of the characteristic temperatures destroys the vicinal hydration structure stable below the transition temperature and a new structure becomes stable. Upon lowering the temperature below the previous critical temperature, the vicinal hydration structures that are stable below the critical point become re-established. However, the reforming of the new structures takes time (depending on circumstances, from minutes to hours), and the properties may thus exhibit notable hysteresis.

The vicinal hydration structures are highly shear-rate dependent. This is clearly evidenced by the progressive decrease of change seen in the thermal anomalies (at T_k) as shear rate is increased. The reforming of the vicinal hydration structures may be slow, similar to the hysteresis observed upon heating and cooling a sample past any one of the critical temperatures, T_k. Furthermore, high pressure, as might be found in ultracentrifuge measurements, is also likely to reduce the amount of *VW* of hydration of macromolecules. In addition, if the temperature of a sample has exceeded one of the critical temperatures, T_k, just before an experiment is begun, then the intrinsic vicinal hydration may have been destroyed and may reform only very slowly. In other words, such parameters as intrinsic viscosities, diffusion coefficients and sedimentation coefficients will depend strongly on the experimental protocol. In view of these facts, it is perhaps not surprising that estimates of hydrodynamic radii (and other transport properties, as well as some thermodynamic properties) may differ, even notably, from one investigator to another.

Since vicinal hydration can occur at all solid interfaces (including membranes) and with all large macromolecules in solution, it is little wonder that all cellular systems (i.e., all living systems) also show the effects of VW. Any biophysical or molecular biology theory that does not allow for – and specifically includes – VW must be judged to be incomplete.

6. POSTSCRIPT

Even the most casual reading of this paper must surely suggest to the reader that thermal anomalies are an important and intrinsic characteristic of VW. Note that *their existence cannot be predicted from Ling's Association-Induction hypothesis for interfacial water* and in fact it does not appear that his hypothesis can ever accommodate, let alone explain, thermal anomalies. The *'Substrate Independence' effect is, of course, also inconsistent with Ling's theory, nor is the shear rate dependence a natural part of it.*

The Vicinal Water hypothesis, as it currently stands, is strictly empirical and does not allow for any detailed quantitative predictions or calculations. Progess in this direction will almost certainly have to await great advances in the statistical description of cooperative phenomena involving hydrogen bonding. On the other hand, Ling's Association-Induction hypothesis allows for semi-quantitative estimates of many parameters – even if some of these efforts may seem more like clever data-fitting than genuine theory. Furthermore, Ling's hypothesis critically depends on the existence of suitably located positive and negative charges to induce the alignment of water molecules, and the energetics of such geometries remain highly uncertain. However, is it possible that both Ling and I are 'partially blinded'? Do we merely describe 'different parts of the elephant'? In other words, is it possible that all interfaces of solids or large macromolecules induce VW (with all its attendant implications for density, heat capacity, entropy, transport properties, etc., and thermal anomalies), but within this framework of modified water structures the Association-Induction mechanisms may operate – assuming the presence of the requisite distribution of charges of opposite signs?

7. ACKNOWLEDGEMENTS

The author wishes to thank Drs. B. Prevel and Richard McNeer, who obtained a number of the viscosity and diffusion coefficient measurements, as well as Mr. Kip Vought, Mr. Julio Gamacho and Mr. Freddy Gonzales who contributed to much of the data collection.The author also gratefully acknowledges the support and hospitality of the US Air Force Clinical Investigation Directorate at Wilford Hall Medical Center, Lackland Air Force Base (San Antonio, Texas, USA), and in particular to the former Director of that Laboratory, Col John Cissik, PhD, and Lt Col Wayne Patterson, PhD.

I wish to thank Dr Denys Wheatley of BioMedES for preparation of the electronic form of this article, and for copy editing the proofs. The figures were taken from old drafts and were scanned in because access to the original data was unavailable. For this reason, please excuse the poor quality of many of them.

Finally, I would like to dedicate this paper to my friend, Dr Frank M Etzler, in recognition of his numerous and remarkable contributions to the field of vicinal water over the past thirty years. All of his experimental work has been of the highest quality and his insight into vicinal water is profound.

REFERENCES

Alpers W, Hühnerfuss H (1983) Molecular aspects of the system water/monomolecular surface film and the occurrence of a new anomalous dispersion regime at 1. 43 GHz. J Phys Chem 87:5251–5258

Anderson NG, Green JG (1975) The soluble phase of the cell. In: Roodyn (ed), Enzyme Cytology, Academic Press, New York, pp 475–490

Antonsen KP, Hoffman AS (1992) In: Harris JM (ed), Poly(ethylene glycol) Chemistry, Plenum Press, New York, pp 15–28

Bailey FE, Koleske JV (1976) Poly(ethylene oxide), Academic Press, New York

Braun CV Jr (1981) Calorimetric and dilatometric studies of structural properties and relaxations of vicinal water, MSc Thesis, Univ Miami at Coral Gables

Braun CV Jr, Drost-Hansen W (1976) A DSC study of the heat capacity of vicinal water in porous materials. In: Kerker M (ed), Colloid and Interface Science, Academic Press, New York, vol 3, pp 533–541

Brunner C (1847) Untersuchubg uberr die cohesion der flussigkeiten. Ann der Physik und Chem (Pogendorff's Annals) 70:481

Churev NV, Deryaguin BV (1985) Inclusion of structural forces in the theory of stability of colloids and films. J Colloid Interface Sci 103:542–553

Clegg JS (1979) Metabolism and the intracellular environment: the vicinal water network model. In: Drost Hansen W, Clegg JS (ed), Cell-associated Water, Academic Press, New York, pp 363–413

Clegg JS (1984a) Properties and metabolism of the aqueous cytoplasm and its boundaries. Am J Physiol 246:R133–R151

Clegg JS (1984b) Intracellular water and the cytomatrix: some methods of study and current views. J Cell Biol 99:167–171

Clifford J (1975) Properties of water in capillaries and thin films. In: Franks F (ed), Water – A Comprehensive Treatise Plenum Press, New York, pp 75–132

Daly JG, Hastings R (1981) Temperature dependence of Bragg scattering from crystallized suspensions of macroions. J Phys Chem 85:294–300

De Gennes PG (1997) Soft interfaces. The 1994 dirac memorial lecture. Cambridge University Press, Cambridge, p 117

Deryaguin BV, Physik Zeit. D Sowj. (1932) 4:431–432
Deryaguin BV, Physik Z (1933) 84:657–670
Deryaguin BV (1964) Recent research into the properties of water in thin films and in microcapillaries. The state and movement of water in living organisms, In 19th Symposium of Soc Exp Biol, Cambridge University Press, Cambridge, pp 55–60
Deryaguin BV (1975) Physik Zeit. D Sowj. 1932, also Zeitung für Physik 1933; 84:657–670 [For an extensive list of references, see list in 'Water – a comprehensive treatise' ('Water in disperse systems')], Plenum Press 5:335–336
Deryaguin BV (1977) Structural components of the disjoining pressure of thin layers of liquids. Croatica Chem Acta 50:187–195 (see also Clifford 1975)
Deryaguin BV, Churev NV (1987) Structure of water in thin layers. Langmuir 3:607–612
Dintenfass L (1981) Hyperviscosity in hypertension Pergamon Press, Sydney p 250
Dordick R, Drost-Hansen W (1981) High precision viscosity measurements. 2. Dilute aqueous solutions of binary mixtures of the alkali metal chlorides. J Phys Chem 85:1086–1088
Dordick R, Korson L, Drost-Hansen W (1979) High precision viscosity measurements on aqueous solutions of single and mixed electrolytes. 1. Alkali chlorides. J Colloid Interface Sci 72:206–214
Dorsey NE (1940) Properties of ordinary water substances, Rheinhold, New York
Dreyer G, Kahrig E, Kirstein D, Erpenbeck J, Lange F (1969) Structural anomalies of water. Naturwiss 56:558–559
Drost-Hansen W (1956) Temperature anomalies and biological temperature optima in the process of evolution. Naturwiss 43:512
Drost-Hansen W (1965) Aqueous interfaces – methods of study and structural properties. Part 1. Ind Eng Chem Res. March issue: 28–44, and Part 2 April issue:18–37
Drost-Hansen W (1965) The effects on biological systems of higher-order phase transitions in water. NY Acad Sci Ann, Art B 125:471–501
Drost-Hansen W (1968) Thermal anomalies in aqueous systems – manifestations of interfacial phenomena. Chem Phys Lett 2:647–652
Drost-Hansen W (1969) Structure of water near solid interfaces. Ind Eng Chem Res 61:10–47
Drost-Hansen W (1971) Role of water structure in cell-wall interactions. Fed Proc 30:1539–1548
Drost-Hansen W (1971) Structure and properties of water at biological interfaces. In: Brown HD (ed), Chemistry of the Cell Interface, Part B, Chapter 6, Academic Press, New York, pp 1–184
Drost-Hansen W (1972) Effects of pressure on the structure of water in various aqueous systems. The effect of pressure on organisms, vols XXVI, Symposia, Society for Experimental Biology, Cambridge University Press, pp 61–101
Drost-Hansen W (1976) The nature and role of interfacial water in porous media. Am Chem Soc, Div Petroleum Chem 21:278–280
Drost-Hansen W (1976) Structure and functional aspects of interfacial (vicinal) water as related to membranes and cellular systems. Colloq Internat du CRNS (L'eau et les systemes biologique) 246:177–186
Drost-Hansen W (1977) Effects of vicinal water on colloidal stability and sedimentation processes. J Colloid Interface Sci 58:251–262
Drost-Hansen W (1978) Water at biological interfaces – structural and functional aspects. Phy Chem Liquids 7:243–346
Drost-Hansen W (1981) Gradient device for studying effects of temperature on biological systems. J Wash Acad Sci 71:187–201
Drost-Hansen W (1982) The occurrence and extent of vicinal water. In: Franks F, Mathias S (ed), Biophysics of Water, Wiley and Sons, New York, pp 163–169
Drost-Hansen W (1985) Anomalous volume properties of vicinal water and some recent thermodynamic (DSC) measurements relevant to cell physiology. In: Pullman A, Vasilecu V, Packer L (ed), Water and Ions in Biological Systems, Plenum Press, New York, pp 289–294
Drost-Hansen W (1991) Temperature effects on erythrocyte sedimentation rates, cell volumes and viscosities in mammalian blood. Final grant report to USAF Office of Scientific Research, Bolling Air Force Base, Washington DC, p 29

Drost-Hansen W (1992) Rheological, biochemical and biophysical studies of blood at elevated temperatures Final grant report to US Office of Scientific Research, Boilling Air Force Base, Washington DC, p 40

Drost-Hansen W (1996) Biochemical and cell physiological aspects of hyperthermi Final grant report to US Office of Scientific Research, Boilling Air Force Base, Washington DC, p 20

Drost-Hansen W (1997) Long-range hydration of macromolecules in aqueous solutions. 2. 214th ACS National Meeting, Las Vegas, Nevada, Abstract 274

Drost-Hansen W (2001) Temperature effects on cell functioning – a critical role for vicinal water. In L'eau dans la cellule. J Cell Mol Biol 47:865–883

Drost-Hansen W, Neill H (1955) Temperature anomalies in the properties of liquid water. Phys rev, vols 100, Abstract 1800

Drost-Hansen W, Clegg JS (1979) Cell-associated water, Academic Press, New York, p 440

Drost-Hansen W, Lin Singleton J (1995) Our aqueous heritage: evidence for vicinal water in cells. In: Bittar EE, Bittar N (ed), Principles of Medical Biology, vol 4, JAI Press Inc, Greenwich, CN, pp 171–194; and 195–215

Drost-Hansen W, Braun CV Jr, Hochstim R, Crowther GW (1987) High precision dilatometry on aqueous suspensions: volume contraction upon settling. In: Ariman T, Nejat Veziroglu T (ed), Particulate and multiphase processes, vol 3, Hemisphere Publishing Corporation, Springer-Verlag, Berlin, pp 111–124

Etzler FM (1983) A statistical and thermodynamic model for water near solid surfaces. J Colloid Interface Sci 92:94–98

Etzler FM (1991) A comparison of the properties of vicinal water in silica gel, clay, wood, cellulose, and other polymeric materials. In: Levine H, Slade L (ed), Water Relationships in Foods, Plenum Press, New York, pp 805–821

Etzler FM, Drost-Hansen W (1979) A role for water in biological rate processes. In: Drost Hansen W, Clegg JS (ed), Cell-associated Water, Academic Press, New York, pp 125–164

Etzler FM, Drost-Hansen W (1983) Recent thermodynamic data on vicinal water and a model for their interpretation. Croatica Chem Acta 56:563–592

Etzler FM, Lilies TL (1986) Ionic selectivities by solvents in narrow pores: Physical and biophysical significance. Langmuir 2:797–800

Etzler FM, Fagundus D (1987) The extent of vicinal water. J Colloid Interface Sci 115:513–519

Etzler FM, White PJ (1987) Heat capacity of water in silica pores. J Colloid Interface Sci 94:98–102

Etzler FM, Conners JJ (1990) Temperature dependence of the heat capacity of water in small pores. Langmuir 6:1250–1253

Etzler FM, Conners JJ (1991) Structural transitions in vicinal water: pore size and temperature dependence of the heat capacity of water in small pores. Langmuir 7:2293–2297

Etzler FM, Conners JJ, Ross RF (1990) The structure and properties of vicinal water. In: Passeretti JD, Caulfield DF (ed), Materials Interactions Relevant to the Pulp, Paper and Wood Industries, Material Research Society Publishers

Etzler FM, Deanne R, Ibrahim TH, Burk TR, Neuman RD (2002) Direct adhesion measurements between pharmaceutical materials. Particles on Surfaces, VSP Utrecht, Netherlands, pp 7–16

Etzler FM, Ibrahim TH, Burk TR, Wiulling GA, Neuman RDv (2005) The Effect of the acid-base chemistry of lactose on its adhesion to gelatin capsules; conclusions from contact angles and other surface chemical techniques. Contact Angle, Wettability and Adhesion vol 2, VSP Utrecht

Falk M, Kell GS (1966) Thermal properties of water: discontinuities questioned. Science 154:1013–1015

Gekko K, Noguchi H (1971) Physicochemical studies of oligodextrans. 1. Molecular weight dependence of intrinsic viscosity, partial specific compressibility and hydrated water. Biopolymers 10:1513–1524

Henniker JC (1949) Rev Mod Phys 2:322–341

Hühnerfuss H (1987) Molecular aspects of organic surface films on marine water and the modification of water waves. La Chemica e L'Industria 107:97–101

Hurtado RM, Drost-Hansen W (1979) Ionic selectivities of vicinal water in pores of a silica gel. In: Drost Hansen W, Clegg JS (ed), Cell-associated Water, Academic Press, New York, pp 115–123

Johnson FH, Eyring H, Polisar MJ (1954) The kinetic basis of rheology, Wiley Sons, New York

Kaivarainen A (1995) Hierarchic concept of matter and field. Water, biosystems and elementary particles Privately published ISBN 0-9642557-0-7, p 483

Kerr J (1970) Relaxation studies on vicinal water. Dissertation, University of Miami, Coral Gables, FL

Korson L, Millero F, Drost-Hansen W (1969) Viscosity of water at various temperatures. J Phys Chem 73:34–38

Kurihara K, Kunitake T. (1992) Submicron-range attraction between hydrophobic surfaces in monolayer-modified mica in water. J Am Chem Soc 114:10927–10933

Lafleur M, Pigeon M, Pezelot M, Caille J-P (1989) Raman spectrum of interstitial water in biological systems. J Phys Chem 93:1522–1526

Lin GN (1965) The physical state of water in living cells and model systems. Ann New York Acad Sci, vols 125, pp 402–417 (article 2)

Ling GN (1962) A physical theory of the living state: the association-induction hypothesis, Blaisdell, Waltham, MA

Ling GN (1979) The polarized multilayer theory of cell water according to the adsorption-induction hypothesis. In: Drost Hansen W, Clegg JS (ed), Cell-associated Water Academic Press, New York, pp 261–269

Ling GN (1992) A revolution in the physiology of the living cell, Krieger Publishing Co, Malabar FL, p 378

Ling GN (2003) A new theoretical foundation for the polarized-oriented multilayer theory of cell water and for inanimate systems demonstrating long-range dynamic structuring of water molecules. Physiol Chem Phys Med NMR 35:91–130

Ling CS, Drost-Hansen W (1975) DTA study of water in porous glass. Adsortion at interfaces. ACS Symp Ser 8:129–156

Lopez-Lacomba JL, Gutzman M, Cortijo M, Mateo P, Aquirre R, Harvey SC, Cheung HC (1989) Differential scanning calorimetric study of the thermal unfolding of myosin rod, light meromyosin, and subfragment 2. Biopolymers 28:2143–22159

Low PF (1979) Nature and properties of water in Montmorillonite-water systems. Soil Sci Soc Am J 43:651–658

Lowe GDO (1987). Thrombosis and hemorheology. In: Cien S, Dormandy J, Ernst E, Matrai A Dordrecht (ed), Clinical hemorheology, Martinus Nijhof Publishers, pp 195–226

Mastro AM, Hurley DJ (1985) Diffusion of a small molecule in the aqueous compartment of mammalian cells. In: Welch R, Clegg JS (ed), Organization of cell metabolism, Plenum Press, New York, pp 57–74

Mentre P, Hui BH (2001) The effects of high hydrostatic pressures on living cells: A consequence of the properties of macromolecules and macromolecular associated water. Int Rev Cytol 201:1–84

Montejano JG, Hamann DD, Lanier TC (1983) Final strength and rheological changes during processing of thermally induced fish muscle gels. J Rheology 27:557–579

Montejano JG, Hamann DD, Lanier TC (1984) Thermally induced gelation of selected comminuted muscle systems – rheological changes during processing, final strengths and microstructure. J Food Sci 49:1496–1505

Nir S, Stein WD (1971) Two modes of diffusion. J Chem Phys 55:1598–1603

Okano M, Yoshida Y (1994) Junction complexes of endothelial cells in atherosclerosis-prone and atherosclerosis-resistant regions on flow dividers of brachiocephalic bifurcation in the rabbit aorta. Biorheology 31:155–169

Peschel G, Adlfinger KH (1969) Temperatur abhängigkeit der Viskosität sehr dünner Wasserschichten Quartzglasoberflächen. Naturwiss 58:558–559

Peschel G, Adlfinger KH (1970) Viscosity anomalies in liquid surface zones. 3. The experimental method. Ber Bunsen-Gesellschaft 74:351–357

Peschel G, Adlfinger KH (1971) Thermodynamic investigation of the liquid layers between solid surfaces. II. Water between entirely hydroxylated fused silica surfaces. Z Naturfor 26a:707–715

Peschel G, Adlfinger KH (1971) Viscosity anomalies in liquid surface zones. 4. The apparent viscosity of water in thin layers adjacent to hydroxylated fused silica. J Colloid Interface Sci 34:505–510

Peschel G, Belouschek P (1976) Eine neue Messmethode zur Untersuchung der Struktur dunner Elektrolytschichten zwischen Festkörperoberflächen. Prog Colloid Polym Sci 60:108–119

Peschel G, Belouschek P (1979) The problem of water structure in biological systems. In: Drost Hansen W, Clegg JS (ed), Cell-associated Water, Academic Press New York, pp 3–52

Phillips MC, Chapman D (1968) Biochim Biophys Acta 75:301

Rhykerd Jr CL, Cushman JH, Low PF (1991) Application of multiple-anlge-of-incidence ellipsometry to the study of thin films adsorbed on surfaces. Langmuir 7:2219–2229 [with references to Low's numerous earlier papers]

Sato M, Ohshima N (1994) Flow-induced changes in shape and cytoskeletal structure of vascular endothelial cells. Biorheology 31:143–155

Shoufle JA, Huang S-Y (1972) Tex J Sci 24:197 [see also J Geophys Res 1968; 73:3345]

Shoufle JA, Huang CT, Drost-Hansen W (1976) Surface conductance and vicinal water. J Colloid Interface Sci 54:184–202

Shephard J, Malmberg E, Logerot D (1974) Some rheological properties of an aqueous surfactant oil recovery agent. Preprint 48th Nat Colloid Symp, Austin TX, 191–196

Streekstra GJ (1990) The deformation of red blood cells in coquette flow. Dissertation, University of Utrecht, p 113

Uijttewaal W (1990) On the motion of particles in bounded flow: Applications in hemorheology. Dissertation, University of Utrecht, p 128

Urbanke C, Romer R, Maass G (1973) The binding of ethidium bromide to different conformations of tRNA: Unfolding or tertiary structure? Eur J Biochem 33:511–516

van Steveninck J, Paardekooper M, Dubbleman TMAR, Ben-Hur E, Leddeboer AM (1991) Anomalous properties of water in macromolecular gels. Biochim Biophys Acta 1115:96–100

Viani BE, Low PF, Roth CB (1983) Direct measurements of the relation between interlayer force and interlayer distance in the swelling of montmorillonite. J Colloid Interface Sci 96:229–234

Wiggins P (1975) Thermal anomalies in ion distribution in rat kidney slices and in a model system. Clin Exp Pharmacol Physiol 2:171–176

Wu MC, Lanier TC, Hamann DD (1985) Rigidity and viscosity changes of croaker actomyosin during thermal gelation. J Food Sci 50:14–19

Young TF (1966) Paper presented at Tetrasectional ACS meeting. Santa Fe, New Mexico

CHAPTER 10

THE LIQUID CRYSTALLINE ORGANISM AND BIOLOGICAL WATER

MAE-WAN HO[1,2,*], ZHOU YU-MING[1], JULIAN HAFFEGEE[1], ANDY WATTON[1], FRANCO MUSUMECI[3], GIUSEPPE PRIVITERA[3], AGATA SCORDINO[3] AND ANTONIO TRIGLIA[3]

[1] *Institute of Science in Society, PO Box 32097, London NW1 0XR, UK*

[2] *Biophysics Group, Department of Pharmacy, King's College, Franklin-Wilkins Bldg., London SE1 9NN, UK*

[3] *Dipartimento di Metodologie Fisiche e Chimiche per l'Ingegneria, Università di Catania, INFM Unità di Catania, Viale A. Doria 6, I-95125 Catania (Italy)*

Abstract: The organism is a dynamic liquid crystalline continuum with coherent motions on every scale. Evidence is presented that biological (interfacial) water, aligned and moving coherently with the macromolecular matrix, is integral to the liquid crystallinity of the organism; and that the liquid crystalline continuum facilitates rapid intercommunication throughout the body, enabling it to function as a perfectly coherent whole

Keywords: Liquid crystalline continuum, coherence, birefringence, nonlinear optics, delayed luminescence, bound water, free water, collagen, proton-conduction, intercommunication, body consciousness

1. THE LIQUID CRYSTALLINE ORGANISM

More than ten years ago, we discovered what living organisms look like under the polarized light microscope that geologists use for examining rock crystals (Ho and Lawrence, 1993; Ho and Saunders, 1994). They give brilliant dynamic liquid crystal displays in colours of the rainbow (see Figure 1). The colours depend on the coherent alignment of molecular dipoles in liquid crystal mesophases. But how can a living, breathing, squirming worm appear crystalline?

* Corresponding author.

G. Pollack et al. (eds.), Water and the Cell, 219–234.

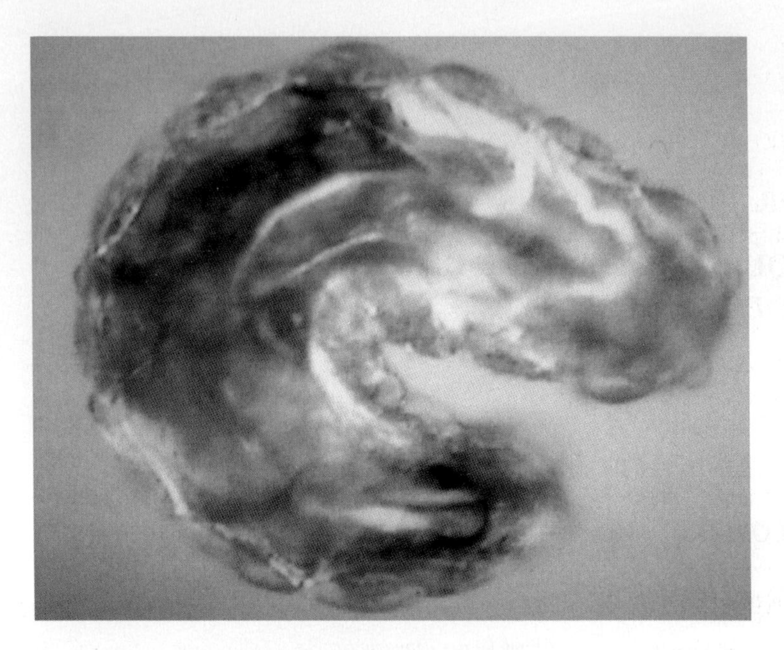

Figure 1. The rainbow worm: a freshly hatched fruitly larva

It is because all the molecular dipoles in the tissues are not only aligned, but also moving coherently together. Visible light vibrates at 10^{14} cycles per second, much faster than the coherent molecular motions in the organism, which is why the molecules look statically aligned and ordered to the light passing through.

Not only are the molecular dipoles in the tissues aligned, they are aligned in *all* the tissues, and aligned globally from head to tail. The antero-posterior axis is the optic axis, so when that axis is laid out straight at the correct angle (45°) to the optics, each tissue takes on a more or less uniform colour: blue, orange, red, or green. But when that axis is rotated 90°, blue changes to red, green to orange and *vice versa*, as characteristic of interference colours.

The fruit fly larva has neatly demonstrated the colour changes for us by making a circle with its flexi-liquid crystalline body. The most active parts of the organism have the brightest colours; the brighter the colours, the more coherent the molecular motions (see later).

One more thing about the rainbow worm; the colours are not just a function of the coherent motions of all the molecules in the tissues, they are the result of the accompanying coherent motions of the 70% by weight of biological water that enables the molecules to be mobile and flexible, which is why the worm, and we too, are flexible and mobile.

Imagine all the biological water dancing together with the molecules in the entire body, creating a quantum jazz of life that's improvised from moment to moment. This technique involves a small modification of that used for examining rock crystals which happens to greatly improve colour contrasts for the range of small birefringences found in biological liquid crystals, and can give high resolution images.

In the live recording of the adult brine shrimp, gently held in a cavity slide under a cover slip just heavy enough to prevent it darting about, you can see giant solitary waves or solitons, both stationary and mobile, passing down the gut.

The organism is a liquid crystalline continuum, coherent beyond our wildest dreams, perhaps even quantum coherent (Ho, 1993; 1998). We thought we made a new discovery but Joseph Needham had anticipated that and a lot more of what this paper is about in his book, *Order and Life* (1935). The properties of 'protoplasm' preoccupied many physical-minded biologists since the end of the 19th century, and fortunately for some of us, well into the 20th century (see Ling, 2001). Needham, who died a few years short of this century, proposed that all the remarkable properties of protoplasm could really be accounted for in terms of liquid crystals. Indeed, he suggested that living systems actually are liquid crystals.

2. RAINBOW WORM OPTICS

The optics of the rainbow worm is very straightforward. The organism is put between crossed polars in transmitted white light in series with a full wave plate that introduces a retardation of 560nm.

The colours are generated by interference when the plane-polarised light, doubly refracted by the birefringent crystals, is recombined on passing through the second polarizer; and depending on the wavelength of the light, there is either additive or destructive interference, so the white light becomes coloured.

The modification we introduced consist in placing the full wave plate at a very small angle of 7.5° to one of the polarizers, instead of 45° as is usually done.

We used the equation of Hartshorne and Stuart (1970) for two superposed birefringent crystals to represent the wave plate and the biological sample, to show why the colour contrast is so much better at the 7.5° angle than the 45° by plotting the intensity of the monochromatic wavelengths for red (700nm), blue (450nm) and green (560nm) respectively (Figure 2). The effect of the small angle is to reduce the contributions of red and blue relative to green, and to increase the difference between the maximum and minimum intensities of all the colours at the +45° and −45° angle of rotation.

For small birefringences (retardation less than 50 nm), the intensity of light is approximately linearly related to retardation. We derived equations to relate the difference between maximum and minimum intensity of the red or the blue light to the retardation of the sample. In practice, we determined the retardation of a sample relative to a mica standard for which the retardation has been measured exactly with monochromatic green light from a mercury lamp.

The next important result is mathematically quite involved and almost entirely the work of Zhou Yuming (2000). The detailed derivations are in a special chapter

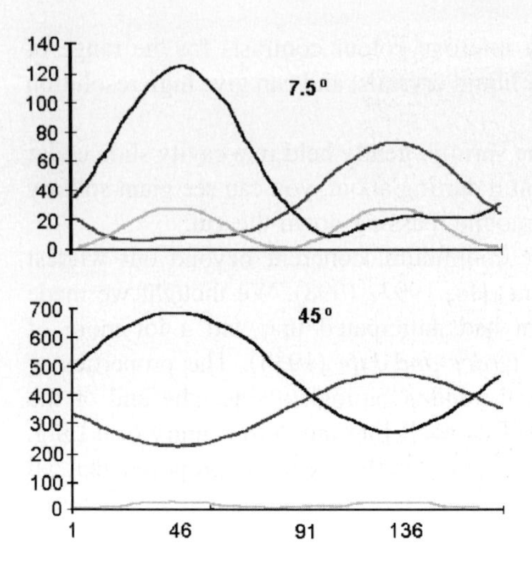

Figure 2. Intensity of red, green and blue light with angle of rotation (From Newton et al., 1995)

of his doctoral thesis. It says that birefringence is linearly related to the molecular alignment order parameter for nematic liquid crystals, which is a good first approximation to biological polymers. This is the reason for stating earlier that the most active parts of the organisms are the most coherent parts: the brightness of the colours is a direct measure of birefringence, and birefringence depends on coherence of molecular alignment. In fact, the linear relationship holds for both form and intrinsic birefringence, as Yuming showed in his thesis. But the distinction between form and intrinsic birefringence is extremely blurred; suggesting that biological water associated with proteins are all inseparably part of the intrinsic birefringence, and is perhaps the most important contribution to the liquid crystallinity of organisms.

These two main results – the linear variation in the intensity of the monochromatic red or blue light with birefringence, and the linear variation of birefringence with the molecular alignment order parameter – underpin a quantitative imaging technique that we have devised to determine the molecular alignment and the birefringence of liquid crystalline mesophases (see Figure 3).

The direction of the vector gives the direction of the molecular alignment at that point and its length is proportional to the retardation or the brightness. As can be seen, this technique is potentially very useful for working out the structure of soft tissues, bones, cartilage, other liquid crystalline composites, natural or self-assembled *in vitro*.

All sections and slides can be imaged directly in water without fixing or staining. As you shall see, water contributes a great deal to the birefringence. The details of this imaging technique and software are described in Ross et al., (1997).

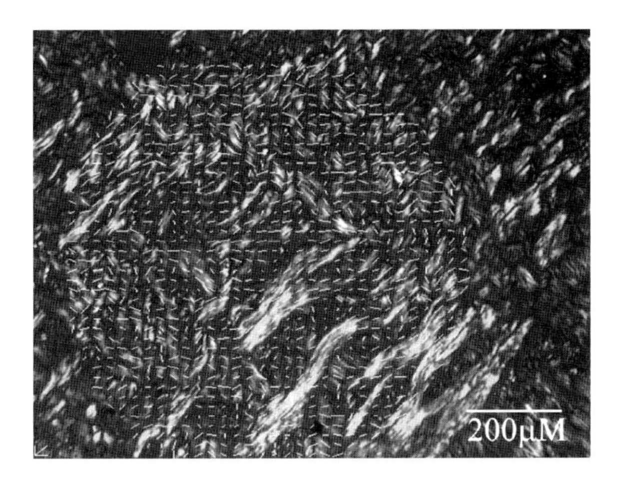

Figure 3. Section of pork skin with overlay of quantitative imaging

3. COLLAGEN AND THE LIQUID CRYSTALLINE LIVING MATRIX

Collagen is the most abundant protein in the organism, and is known to form liquid crystalline mesophases, at least, *in vitro*. It is the main protein in the extracellular matrix and connective tissues and may thus account for the liquid crystallinity of living organisms as a whole, facilitating intercommunication throughout the body.

Type I collagen is the archetype of all collagens as well as the most abundant. It is found in tendons, skin, and bone. The polypeptide chain is made up of the repeating tripeptide unit, Gly-X-Y, where X are Y are usually proline and hydroxyproline respectively. Three peptide chains are wound into a triple left-handed helix molecule, with the glycine in the middle and a stagger of one amino acid between neighbouring chains. The helix has a pitch of about 9.5Å and 3.3 units per turn.

The molecules are assembled into fibrils, fibrils into fibres, and many different hierarchical structures. What makes collagen most interesting is its associated biological water. We used our quantitative imaging technique to investigate how water contributes to the birefringence of the rat-tail tendon (Zhou, 2000). We looked at all organisms and sections in water, because it was the most convenient and also because it gives the highest birefringence. But is that due to *form birefringence*, which results from a difference in refractive index between the sample and the medium rather than *intrinsic birefringence* due to polarisability of the biological molecules?

The retardation of a fixed, unstained tendon 5 micron thick section of rat tail tendon was measured in both the air-dry state, and embedded in histomount. The relative retardation in histomount was between two to three-fold that of the dry section. As the refractive index of histomount is 1.29, while that of collagen is

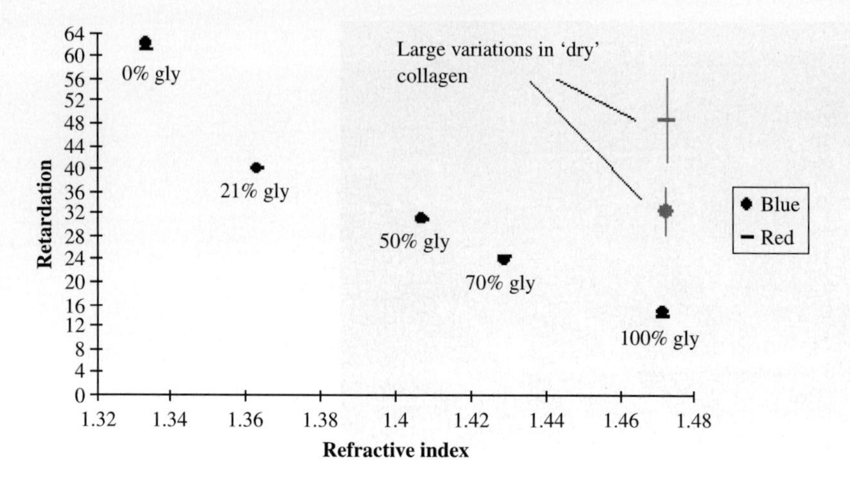

Figure 4. Relative retardation in increasing concentrations of glycerol in water (From Zhou, 2000)

around 1.47, the increase in relative retardation in histomount could be said to due to form birefringence, assuming that there is no interaction between histomount and collagen.

To investigate further, mixtures of water and glycerol were prepared in order to vary the refractive index between 1.333 for water and 1.471 for glycerol (see Figure 4).

As can be seen, increasing glycerol concentration increases the refractive index, but concentrations greater than 21% also begins to reduce the birefringence to below the level of 'dry collagen' (left overnight in a dessicator with silica gel). It is quite likely that this 'dry' collagen is actually hydrated to some degree (see next Section), and the effect of high concentrations of glycerol may involve dehydrating the collagen further.

It suggests that the role of water is not just an embedding medium, and introduces more than form birefringence. As water interacts extensively with collagen through hydrogen bonds, it would be expected to alter the intrinsic birefringence of the protein. One way to investigate the effects of hydrogen-bonding is to introduce solvents that perturb this hydrogen bonding.

Two series of mixtures of solvents were prepared, a glycerol in water mixture from 0% to 21%, and a series with matching refractive index of different concentrations of ethanol in methanol. The idea was that glycerol may be more 'water-like' in that it does not have the hydrophobic side-chain of the alcohols. The results are shown in Figure 5.

It can be seen that alcoholic solvents reduce birefringence by about 20% at all refractive index values.

The difference was more pronounced in unfixed tendon sections of the same thickness (see Figure 6). Here the birefringence in the aqueous solution was much stronger than in fixed sections, hence the difference between the aqueous and the

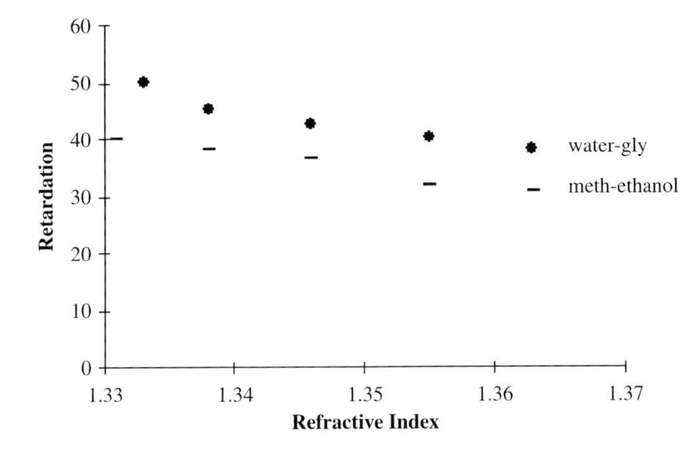

Figure 5. Retardation in water-glycerol mixtures vs methanol-ethanol mixtures in fixed rat tail tendon (From Zhou, 2000)

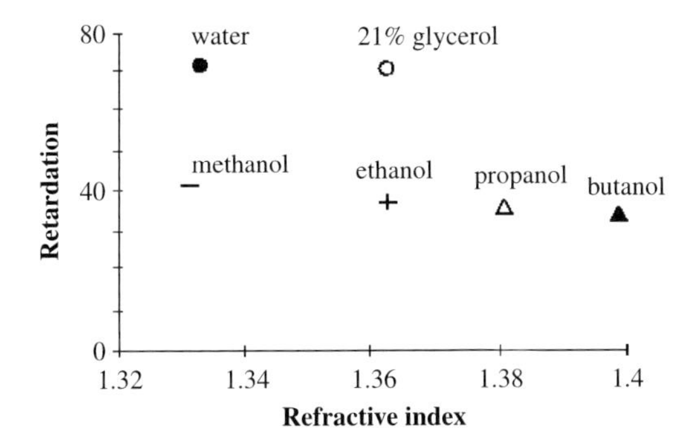

Figure 6. Retardation in water, 21% glycerol and various alcohols in fresh rat tail tendon (From Zhou, 2000)

alcohol solutions is correspondingly greater, while the change with refractive index was much less. This suggests that in unfixed sections, most of the total birefringence is intrinsic birefringence in both aqueous solutions and alcohol solutions. However, intrinsic birefringence also differs by more than 70% between the solvents, probably on account of large changes in molecular order, or protein conformation, or both.

These experiments show that biological water contributes a great deal to the intrinsic birefringence and liquid crystallinity of biological polymers.

4. COLLAGEN IS RICHLY HYDRATED

The hydration of proteins has been measured with dielectric relaxation and many other techniques.

In dielectric relaxation measurements, the sample is subjected to alternating electric fields of different frequencies. As the frequency of the applied field increases, the dipole moments of the molecules are unable to orient fast enough to keep up alignment with the applied electric field and the total polarization falls. This fall, with its related reduction of permittivity and energy absorption, is referred to as dielectric relaxation or dispersion. A complex permittivity ε^* describes the dielectric relaxation, the real part of which, ε'represents the permittivity of the medium and the imaginary component ε'' is the loss of the medium (Cole, 1975).

The frequency-dependent dielectric constant of the combined protein-water system can be written as a sum of four dispersion terms for the protein, bound water, free water and bulk water respectively (Pethig, 1992). The dielectric relaxation time for bulk water is about 8.3ps, for free water, 40ps, and for bound water 10 ns, compared to the typical protein myoglobin, which is 74ns.

Three populations of biological water have been identified in tendon, which is almost all type I collagen, by means of NMR (Peto et al., 1990), dielectric measurements and sorption experiments (Grigera and Berendsen, 1979). The most tightly bound fraction consists of 2 water molecules for every three amino acid residues and provides water bridges between the three strands of the collagen molecule, linking backbone carbonyl groups. This represents 0.125gwater/g collagen. A second, less tightly bound fraction is localized in the interstices of the quasi-hexagonal packing arrangement, which takes a further 0.35g/g, and consists of hydrogen-bonded chains of water molecules (Hoeve and Tata, 1978). A third population of more loosely bound water can be absorbed in the 'ground substance' in which the collagen fibrils are embedded. But considering the complicated hierarchical structure of most collagen structures, there are likely to be many different populations of water that are to varying extents 'restricted' in motion compared with bulk water (cf Fullerton, this volume).

Middendorf et al., (1995) reported that some 0.06g of water/g collagen remained in completely desiccated collagen, representing the most tightly bound fraction. This is about one molecule of water per triplet and is similar to that reported for the desiccated $(\text{Pro-Pro-Gly})_{10}$ peptide (Sakakibara et al., 1972), in which the water molecule forms a H-bond between glycine and the second proline in the triplet. Thus, the 'tightly bound' fraction of water probably consists of at least two populations of water molecules.

While the relaxation time of bulk water is about 8.3ps, the relaxation times of "free water" for collagen was reported to range from about 12 to 40 ps (Hayashi et al., 2002), indicating a rather uniform dynamical structure of water around the collagen triple helix.

Collagen is unusual among proteins in that there are very few direct H-bonds either within the chain or between chains. There is only one direct H-bond in each Gly-X-Y unit, the imide group of the Gly to the carbonyl group of the X residue in the adjacent chain. This leaves the carbonyl group of the glycine residues and the carbonyl of the Y residues with no amide H-bonding partner. In addition, the OH group of hydroxyproline points out from the triple helix and cannot directly H-bond to any other group within the molecule.

A detailed X-ray diffraction analysis was carried out by Bella et al., (1995) on the ordered water molecules in a collagen-like peptide with ten repeating units of Gly-Pro-Hyp and a single substitution of a Gly by an Ala residue in the middle of the peptide. The analysis showed that all available groups of the peptide backbone and the Hyp are involved in binding water molecules. In other words, most of the H-bonds in collagen structure are water mediated.

Water chains mediate H-bonding between carbonyl groups on the same chain as well as between different chains in the triple helix, and between the OH group of hydroxyproline with carbonyl groups in the same or different chains. The number of water molecules involved in bridging two groups appears to vary along the helix. On average, the carbonyl groups of Gly residues are bonded to one water molecule, while that of Hyp are bonded to two. The OH group of Hyp can bind two water molecules at two distinct sites, but not all positions are fully occupied (Brodsky and Ramshaw, 1997). Water bridges are also critical in connecting adjacent triple-helices and maintaining the molecular spacing (Bella et al., 1995). Local hydrogen-bonding network was observed in the interstitial waters. Some water molecules link up to four other water molecules, illustrating the three-dimensional hydrogen-bonded network of water around the collagens.

There is little or no direct contact between neighbouring collagen triple helices, suggesting that a uniform cylinder of water surrounds each triple helix.

There is disagreement over the role of Hyp and hydration in stabilizing the collagen triple-helix. In the computer simulation of the three-dimensional hydration structure of $(\text{Pro-Pro-Gly})_{10}$ (Gough et al., 1998) which has no Hyp, and consequently, no water bridge between a side-chain group and a backbone group, the triple helical structure of the peptide was found to be very similar to native collagen. The water bridges between the carbonyl groups are all interchain, and quite different from the results obtained by Bella et al., (1995). Instead, hydration is determined by the geometry of the backbone carbonyl groups and steric crowding surrounding them. Prolines on different chains are stacked against each other in the triple helix, regardless of whether the molecule is hydrated or not. These close contacts prevent hydration molecules from entering, and are the stabilizing factor in solution. All Hyp-containing triple-helices known to-date form direct hydrogen bonding interactions between Hyp hydroxyl groups of adjacent triple helices (Berisio et al., 2001).

A high resolution X-ray diffraction study of $(\text{Pro-Pro-Gly})_{10}$ (Berisio et al., 2002) found a thick cylinder of hydration, composed of as many as 352 water molecules surrounding the two triple helices in the asymmetric unit. These water sites occupy two hydration shells with equal population of the two shells. This is about 2 water molecules per amino acid, and includes the 'loosely bound' fraction identified from other studies.

5. PROBING BIOLOGICAL WATER OF COLLAGEN WITH DELAYED LUMINESCENCE

Bovine Achilles tendon has a very elaborate fractal structure of microfibrils, subfibrils, fibrils, fibres, and fibre-bundles (cf Fullerton, this volume); and we decided to use delayed luminescence to probe the different populations of biological water associated with it.

Delayed luminescence (DL) is the re-emission of ultraweak intensity light with delay time of milliseconds to minutes from all living organisms and cells on being stimulated with light.

Where does DL in living systems come from? When solid-state systems are excited by light, 'excitons' are generated which propagate within the system, some of which then decay radiatively back to the ground state over long time scales. A similar phenomenon occurs in the living system stimulated by light. The excitation is delocalized over the whole system, and cannot be assigned to specific 'chromophores', or specific molecular species that are excited. As distinct from stimulated emission from chromophores, DL from living cells and organisms typically covers a broad spectrum of frequencies, indicating the collective excitation of many coupled modes; all of which remarkably, decay hyperbolically back to the 'ground' according to the same hyperbolic decay equation (Musumeci et al., 1992),

$$I(t) = I_0/(1 + t/t_0)^m$$

where the parameters I_0, t_0 and m, are fitted using a non-linear least squares procedure. These parameters are very sensitive to the physiological states of the cell or organism, and have been used successfully to assess food quality, for example.

We were quite surprised to find that these parameters are also very sensitive to the degree of hydration (Figure 7).

We decided to use alternative parameters that enable us to relate the characteristics of DL more directly to the energy status of the system, as we have previously done

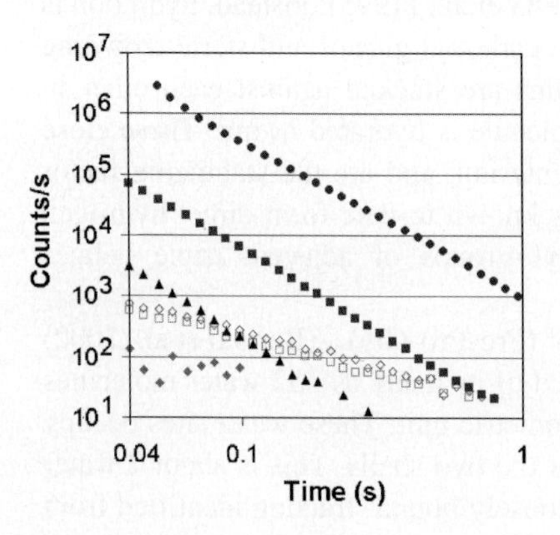

Figure 7. DL kinetics of bovine Achilles tendon at different hydration. (◇) Native sample, (□)1.6g/g, (♦)0.7g/g, (▲)0.4g/g. (■)0.2g/g, (•) fully dehydrated (From Ho et al., 2003)

(Ho, 1998; 2002). The total number of photon count, N, is connected to the total number of photons re-emitted, or the collective electronic levels excited that decay in a radiative way,

$$N = \int_{t_s}^{\infty} I(t)\, dt$$

where t_s is the start time of DL recording, and the probability of decay per excited level $P(t')$, expressed as,

$$P(t') = \frac{I(t')}{\int_{t'}^{\infty} I(t)\, dt}$$

In our condition, $t_0 \sim 0$, so the above equation becomes,

$$P_t(t') = \frac{I(t')}{\int_{t'}^{\infty} I(t)\, dt} = \frac{Rp}{t}$$

As the parameter N represents an extensive quantity, its values were normalized to the maximum value achieved by every sample in order to compare values from different samples. The resulting parameter is denoted the relative number of excited states Rn. Similarly, we use the parameter, Rp, the slope of $P(t)$ trend vs $1/t$.

When we plotted Rn and Rp against hydration levels expressed as g water/g dry collagen (dried to constant weight at 39C in a desiccator with activated silica gel), we identified what appear to be four states of hydration, each with distinctive values of Rp and Rn (see Figure 8). State 1, fully hydrated, has greater than 1.5g/g

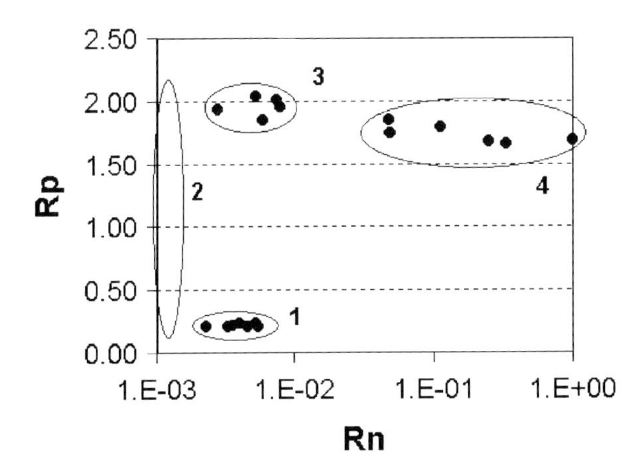

Figure 8. Plot of Rp versus Rn identifies four hydration states in bovine Achilles tendon (From Ho et al., 2003)

hydration (between 41 and 25.5 molecules of water per triplet, Gly-Pro-Pro, mw 305), and is characterized by low *Rn* and *Rp*; state 2, between 1.52g/g to about 0.53g/g (25.5 to 9 molecules of water per triplet) is characterized by such low levels of DL that it is not possible to calculate either *Rp* or *Rn* reliably; state 3, between 0.53g/g and 0.26g/g (9 to 4.5 molecule per triplet), is characterized by the highest *Rp* level while *Rn* levels remain almost as low as state 1; and finally, state 4, less than 0.26g/g (less than 4.5 molecule of water per triplet), has a broad range of high *Rn* values as well as a high *Rp*. The transitions between different states are apparently abrupt.

State 4 corresponds to the tightly bound water, state 3, the loosely bound, and states 2 and 1 must therefore be different populations of 'free water'. The loss of each population of water involves what seems like a global phase transition, each 'state' being maintained until nearly all of the population of water is lost. Fullerton et al., (this volume) have detected similar phase-transition like behaviour in their measurements of relaxation times of oriented bovine Achilles tendon using solid-state nmr – at hydrations of about 0.5, 0.26 and 0.06g/g – two of the values matching precisely those we obtain here.

Cell biologists have recently discovered a very interesting nonlinear optical phenomenon in collagen fibres, which enables the extracellular matrix and other collagenous material to be imaged without a fluorescent probe, using the multi-photon fluorescence microscope. Collagen, on absorbing simultaneously two low energy photons, generates second harmonic, frequency-doubled uv light, which stimulates it to fluoresce (Zipfel et al., 2003). We do not know whether the DL measured in bovine Achilles tendon is related to fluorescence, as we have been using a uv pulse laser (duration $\approx$ 5ns, $\lambda = 337.1$nm), to stimulate DL.

One mechanism suggested for DL is the formation and subsequent radiative decay of excited Davydov (1994) solitons (Brizhik et al., 2001), which may be especially applicable to collagen. Optical solitons, induced in collagen, could act as waveguides giving rise to other nonlinear effects such as second harmonic generation. Salguerio et al., (2004) have described a mechanism for generating second harmonic wave in a vortex soliton waveguide.

It is clear that collagen liquid crystal mesophases have very exciting properties that could be responsible for ultrafast intercommunication within the body.

6. WATER OF HYDRATION SUPPORTS JUMP CONDUCTION OF PROTONS

There have been many suggestions for years that interfacial water adsorbed onto the surface of proteins and membranes could support a special kind of jump conduction of protons.

As mentioned earlier, a complex permittivity describes dielectric relaxation, consisting of a real part representing the permittivity of the medium, the dielectric constant, and an imaginary component representing the loss of the medium, or conductivity. For a long time, it has been known that both the dielectric constant and conductivity for biological polymers tend to increase strongly with the degree of hydration.

Sasaki (1984) measured the dielectric dispersion of bovine Achilles tendon at several hydrations levels below 0.3 g water/g protein over the frequency range of 30 Hz to 100kHz. He found both the dielectric constant and conductivity increasing strongly with water content, especially in the lower frequency side. There was no dielectric absorption peaks within this range of frequencies. In the lower frequencies (< 1 kHz), ε'' varies as frequency, f, to the power $-n$.

$$\varepsilon'' \propto f^{-n}$$

This is apparently indicative of discontinuous jump of charge carriers between localized sites. The dielectric loss factor is proportional to the number of carrier jumps:

$$\varepsilon'' \propto \boldsymbol{J}_{\phi,f}$$

where $\boldsymbol{J}_{\phi,f}$ represents the number of jumps per unit volume performed by carriers within the period of oscillation of the external field and at the water content, ϕ. The charge carriers are presumed to be protons. There was a power-law relationship between conductivity σ and water content ϕ of the form,

$$\sigma(\phi) = X\phi^{Y}$$

where Y, the power of ϕ, is between 5.1 and 5.4, independent of the frequency of the electric field, and is thought to be related to the distance between ion-generating sites.

7. PROTON-NEURAL NETWORK

We have described some of the major findings suggesting that the main function of the liquid crystalline matrix in the body is to facilitate rapid intercommunication that makes organisms so perfectly coordinated, even organisms as large as whales or elephants. A substantial part of this intercommunication is associated with the biological water (Ho, 1998), of which proton currents are the best understood, although other nonlinear optical and phonon effects such as solitons may also be important as mentioned earlier.

Welch and Berry (1985) suggested that a proton-neural network is involved in regulating enzyme reactions within the cell, where metabolic reactions are predominantly of a redox nature. Proton currents may well flow throughout the extracellular matrix, and linked into the interior of every single cell through proton channels. Proton currents could flow from the most local level within the cell to the most global level of the entire organism. Protons (reducing power) give a boost of energy where it is needed.

Structural studies carried out on proton pumps such as bacteriorhodopsin and cytochrome oxidase within the past ten years show that they typically form a channel through the cell membrane which is threaded by a chain of hydrogen-bonded water molecules from one side of the membrane to the other. There is now evidence that protons can flow directly along the membrane within the interfacial water layer, from proton pump to ATP synthase, both of which are embedded in the membranes (see Ho, 2005).

A model of proton-conducting water chain or "proton-wire" has come from a further unexpected source: studies on carbon nanotubes.

Hummer et al., (2001) showed in computer simulations that a single-wall nanotube 13.4 Å long and 8.1Å in diameter rapidly filled up with water from the surrounding reservoir, and remained occupied by a chain of about 5 water molecules on average during the entire 66ns of simulation.

Water molecules not only penetrate into the nanotubes, but are also conducted through them. During the 66 ns, 1 119 molecules of water entered the nanotube on one side and left on the other, about 17 molecules per ns. The measured water flow through the twice as long channel of the transmembrane water-conducting protein aquaporin-1 is about the same order of magnitude. Water conduction occurs in pulses, peaking at about 30 molecules per ns, reminiscent of single ion channel activity in the cell, and is a consequence of the tight H-bond inside the tube.

There is a weak attractive van der Waals force between the water molecules and the carbon atoms, of 0.114 kcal per mol. Reducing this by 0.05kcal per mol (less than 5%) turns out to drastically change the number of water molecules inside the nanotube. This fluctuates in sharp transitions between empty states (zero water molecule) and filled states, suggesting that changes in the conformation of enzyme proteins may control the transport of water from one side to another in the cell membrane.

Do such water-filled channels conduct protons? The answer is yes. If there is an excess of protons on one side of the channel, positive electricity will spirit down fast, in less than a picosecond, some 40 times faster than similar conduction of protons in bulk water (Hummer, 2003).

Collagen in connective tissues has a special role to play in coordinating the activities of each and every cell throughout the body. Giant collagen fibres and especially their associated biological water may be jump-conducting cables linking distant sites with one another.

Ho and Knight (1998) proposed that the system of ramifying water channels along aligned collagen fibres may be the basis of the acupuncture meridian system of Traditional Chinese Medicine.

The liquid crystalline continuum provides rapid intercommunication throughout the body, enabling the organism to function as a perfectly coordinated whole. This "body consciousness" is common to all cells and organisms; it precedes the "brain consciousness" of the nervous system in evolution and works in tandem with it. It is, remarkably, nothing more than a guided matrix of biological water.

REFERENCES

Bella J, Brodsky B, Berman HM (1995) Hydration structure of a collagen peptide. Structure 3:893–906

Berisio R, Vitagliano L, Mazzarella L, Zagari A (2001) Crystal structure determination of the collagen-like polypeptide with repeating sequence Pro-Gyp-Gly: Implications for hydration. Biopolymers 56:8–13

Berisio R, Vitagliano L, Mazzarella L, Zagari A (2002) Crystal structure of the collagen triple helix model $[(\text{Pro-Pro-Gly})_{10}]_3$. Protein Sci 11:264–70

Brizhik L, Musumei F, Scordino A, Triglia A (2000) The soliton mechanism of the delayed luminescence of biological systems. Europhysics Lett 52:238–44

Brizhik L, Musumeci F, Scordino A, Triglia A (2001) Solitons an delayed luminescence. Phys Rev E 64:31902

Brodsky B, Ramshaw JAM (1997) The collagen triple-helix structure. Matrix Biol 15:545–54

Cole RH (1975a) Evaluation of dielectric behavior by time domain spectroscopy. I. Dielectric response by real time analysis. J Phys Chem 79:1459–69

Davydov AS (1994) Energy and electron transport in biological systems. In: MW Ho, F-A Popp, U Warnke (eds.), Bioelectrodynamics and Biocommunication. World Scientific, Singapore

Gough CA, Anderson RW, Bhatnagar RS (1998) The role of bound water the stability of the triple-helical conformation of $(\text{Pro-Pro-Gly})_{10}$. Journal of biomolecular structure & dynamics 15:1029–37. Journal code:8404176. ISSN:0739-1102. PubMed ID 9669549 AN 1998332212 MEDLINE

Grigera JR, Berendsen HJC (1979) The molecular details of collagen hydration. Biopolymers 18:47–57

Hartshorne NH, Stuart A (1970) Crystals and the Polarizing Microscope, Edward Arnold, London

Hayashi Y, Shinyashiki N, Yagihara S (2002) Dynamical structure of water around biopolymers investigated by microwave dielectric measurements via time domain reflectometry. J Non-Crist Solids 305:328–332

Ho MW The Rainbow and the Worm, The Physics of Organisms, World Scientific, 1993, 2nd ed 1998; reprinted 2000; 2001, 2003

Ho MW (2005) Positive electricity zaps through water chains. Science in Society (to appear)

Ho MW, Lawrence M (1993) Interference colour vital imaging: a novel noninvasive technique. Microscopy and Analysis, September:26

Ho MW, Saunders PT (1994) Liquid crystalline mesophases in living organisms. In: MW Ho, F-A Popp, U Warnke (eds.), Bioelectrodynamics and Biocommunication. pp 213–227 World Scientific Singapore.

Ho MW, Knight D (1998) Liquid crystalline meridians. Am J Chin Med 26:251–63

Ho MW, Musumeci F, Scordino A, Triglia A (1998) Influence of cations in extracellular liquid on delayed luminescence of *Acetabularia acetabulum*, J Photochem Photobiol Biol B 45:60–6

Ho MW, Musumeci F, Scordino A, Triglia A, Privitera G (2002) Delayed luminescence from bovine Achilles' tendon and its dependence on collagen structure. J Photochem Photobiol B, Biol 66:165–70

Ho MW, Haffegee JP, Privitera G, Scordino A, Triglia A, Musumeci F (2003) Delayed luminescence and biological water in collagen liquid crystalline mesophases (unpublished)

Hoeve CAJ, Tata AS (1978) The structure of water absorbed in collagen. J Phys Chem 82:1661–3

Hummer G (2003) Water and proton conduction through carbon nanotubes. Banff, April

Hummer G, Rasalah JC, Noworyta JP (2001) Water conduction through the hydrophobic channel of a carbon nanotube. Nature 414:188–90

Ling GN (2001) Life at the Cell and Below-Cell Level, Pacific Press, New York

Middendorf HD, Hayward RL, Parker SF, Bradshaw J, Miller A (1995) Vibrational neutron spectroscopy of collagen and model peptides. Biophys J 69:660–73

Musumeci F, Godlevski M, Popp FA, Ho MW (1992) Time behaviour of delayed luminescence in Acetabularia acetabulum. In FA Popp, KH Li, Q Gu (eds.), Advances in Biophoton Research World Scientific Singapore

Needham J (1935) Order and life Yale University Press, Mass

Newton R, Haffegee J, Ho MW (1995) Colour-contrast in polarized light microscope of weakly birefringetn biological specimens. J Microsc 180:127–8.

Pethig R (1992) Protein-water interactions determined by dielectric methods. Annu Rev Phys Chem 43:177–205

Peto S, Gillis P, Henri VP (1990) Structure and dynamics of wtaer in tendon from NMR relaxation measurements. Biophys J 57:71–84

Ross S, Newton R, Zhou Y-M, Haffegee J, Ho MW, Bolton JP, Knight D (1977) Quantitative image analysis of birefringent biological material. Journal of Microscopy 187:62–67

Sakakibara S, Kishida Y, Okuyama K , Tanaka N, Ashida T, Kakudo M (1972) Single crystals of $(Pro\text{-}Pro\text{-}Gly)_{10}$: a synthetic polypeptide model of collagen. J Mol Biol 65:371

Salgueiro JR, Carlsson AH, Ostrovskaya E, Kivshar Y (2004) Second-harmonic generation in vortex-induced waveguides. Optics Letters 29:503–505

Sasaki N (1984) Dielectric properties of slightly hydrated collagen: Time-water content superposition analysis. Biopolymers 23:1725–34

Welch GR, Berry MN (1985) Long-range energy continua and the coorindation of multienzyme sequences in vivo. In: GR Welch (ed.), Organized Multienzyme Systems Academic Press New York

Zhou Y-M (2000) Optical Properties of Living Organisms, PhD Thesis, Open University, United Kingdom, February.

Zipfel WE, Williams RM, Christie R, Nikitin AY, Hyman BT, Webb WW (2003) Live tissue intrinsic emission microscopy using multiphoton-excited native fluorescence and second harmonic generation. Proc Natl Acad Sci USA 100:7075–80.

CHAPTER 11

THE UNFOLDED PROTEIN STATE REVISITED

PATRICIO A. CARVAJAL[1,2,*] AND TYRE C. LANIER[1]

[1] *Department of Food Science, North Carolina State University, Raleigh NC 27695, USA*

[2] *Escuela de Alimentos, Pontificia Universidad Católica de Valparaíso. Av. Brasil 2950, Valparaíso, Chile*

Abstract: Most studies on proteins have centered on the conformation and stability of the folded state. The unfolded state has essentially been neglected because of its reputation of being devoid of biological function, and not well-defined. Recently the importance of unfolded segments, as part of the secondary structure of globular proteins and their role in the performance of biological functions, has become apparent. We also are beginning to realize that there may be a surprising simplicity to what previously appeared to be a heterogeneous disorder. Thus the unfolded state can be characterized as having, in part, the same conformation as that adopted by a single polypeptide chain of the collagen molecule, termed the polyproline II (PPII) conformation. This PPII conformation has emerged as an important member in both the globular protein secondary structure and the unfolded state. Additionally, the important role of water in the stabilization of this conformation is crucial being the major determinant of it

This overview compiles recent significant findings on the unfolded state and highlights the essential role of water to its structure. Furthermore, we extend these findings to suggest a possible mechanism on the structuring of water by the antifreeze glycoproteins

Keywords: Protein Hydration; Protein Unfolding; PPII, AFGP

1. INTRODUCTION

The term "unfolded state" is used to describe the collection of conformations populated under extreme non-native conditions, including high and low temperatures, high pressure, extremes of pH, and high concentrations of denaturant (Shortle, 1996). The conventional view of the unfolded state

* Corresponding author. P.O Box 7624, Raleigh NC 27695, USA; Fax: 919-515-7624; E-mail: Patricio@Rondanelli.com

G. Pollack et al. (eds.), Water and the Cell, 235–251.

is that denatured proteins and short peptides are featureless, random coil polymers, with seemingly little preference for any particular conformation. The observed intrinsic chain flexibility of unfolded proteins is thought to arise from internal motion around the peptide backbone. This view reflects the predictions of the random-coil model of Brant and Flory (1965), who treated an unfolded polypeptide like a synthetic flexible polymer. However, several lines of evidence accumulated over the past several years suggest that denatured protein chains in water may be far from random in their conformation.

2. RESIDUAL STRUCTURE

It is now becoming evident that when a globular protein unfolds, not all of its secondary structure is lost. Moreover, it has been shown that residual structure can exist even under the most severe denaturing conditions, such as high concentrations of strong denaturants. For example, in reduced unfolded hen lysozyme, six hydrophobic clusters are detected forming a network connected by cooperative interactions (Klein-Seetharaman et al., 2002). Recent investigations of barnase denatured by pH, urea, and temperature denatured suggest that some fraction of the unfolded state contains residual, nonrandom structure (Bond et al., 1997). The most remarkable case is bovine pancreatic trypsin inhibitor (BPTI), which remains practically intact in 8M urea (Chang and Ballatore, 2000).

The fact that residual secondary structure can prevail under such denaturing conditions suggests that unfolded proteins may be predisposed to adopt specific backbone conformations rather than that of a random coil (Uversky and Fink, 2002).

3. NATIVELY UNFOLDED PROTEINS

The assumption that a globular protein requires a folded conformation to have specific biological function has pervaded protein science and related fields for over 100 years. It is now recognized that a large number of proteins and protein domains contain little or no ordered secondary structure (α-helix, β-sheet or β-turns) or tertiary structure and yet do have specific biological functions under physiological conditions (Dunker et al., 2002). These proteins are characterized by an extended conformation with a high intramolecular flexibility due to a high degree of exposure of the peptide backbone to the solvent. Such residues have been referred to by different terms, such as natively denatured (Schweers et al., 1994) intrinsically unstructured (Weinreb et al., 1996), intrinsically disordered (Wright, 1999), or most recently, natively unfolded proteins (Dunker, 2001). The terms reflect the idea that these proteins or protein segments, while having biological function, behave as random coils.

The current list of such natively unfolded proteins contains more than 100 entries and includes proteins with a range of functions such as signal transduction (Kay et al., 2000), transcription (Giesemann et al., 1999), cell motility (Mahoney et al., 1997), and immune response (Jardetzky et al., 1996). However, no enzymatic activity has yet been assigned to any of these. Some of the most studied natively unfolded proteins in the last past 5 years include the casein milk proteins (Farrell et al., 2002), the gluten cereal proteins (Blanch et al., 2003), protease inhibitors (calpastatin, and Bowman-Birk types) (Smyth et al., 2001), the brain synuclein proteins (Maiti, 2004), and the microtubule-associated tau protein (Linder et al., 2000).

A distinctive feature of this group of proteins, which sets them apart from other secondary structures is their high hydration capacity. A recent solid-state NMR study (Bokor et al., 2005) revealed that the activation energy obtained for the dynamic of the most strongly bound part of the hydration shell was 50% larger for natively unfolded proteins than for globular types.

Interestingly, these natively unfolded proteins have amino acid compositional bias, being substantially depleted in Trp, Cys, Phe, Ile, Tyr, Val, Leu, and enriched in Glu, Lys, Arg, Gln, Ser, Ala, Pro, and Gly (Dunker et al., 2002; Tompa, 2002). At a glance, this compositional bias denotes a low overall hydrophobicity and large net charge.

It has further been suggested that interaction with water molecules is favored due to the extended nature of these proteins, with the backbone carbonyl (CO) and amide (NH) groups pointed out from the helical axis into the solvent in a strategic manner. Alanine and residues with long, flexible side chains (such as Glu, Lys, Arg and Gln) seem not to occlude the backbone from water access, or do so to a limited extent, while bulky branched or aromatic residues, such as Leu, Ile, Val, Trp, Phe , Tyr and Trp, are not favored because they occlude the peptide backbone from access to solvent (Shah et al., 1996; Chellgren and Creamer, 2004).

There has been recent emphasis to characterize the structure of the natively unfolded proteins. The use of spectroscopy techniques such as circular dichroism (CD), vibrational circular dichroism (VCD), and Raman optical activity (ROA) has revealed that such structures present some more regular type of conformational order. ROA spectroscopy has shown its ability to both probe the conformation of polypeptide backbone and to distinguish different elements of the secondary structure, and the loops and turns linking these (Barron et al., 2002).

For example, the ROA spectra of caseins, ω-glutens, brain protein synucleins and microtubule-associated tau proteins were found to be very similar, being dominated by a strong positive band centered at $\sim 1318\,\text{cm}^{-1}$ (Syme et al., 2002). This band has been identified as that of the left-handed poly(L-proline) II (PPII) helix, a well-known secondary element that exists in collagen. Additionally, DSC measurements on these proteins revealed no evidence for high temperature thermal transition, a typical result for extended conformations such as PPII.

In relation to globular proteins, several statistical surveys of structures in the Protein Data Base (PDB) show also that PPII is a commonly occurring conformation (Adzhubei and Sternberg, 1993; Sreerama and Woody, 1994; Stapley and Creamer, 1999). It is estimated that up to 10% of individual amino acid residues that are not assigned to regular secondary structures are PPII. However, these PPII helices tend to be short, no more than 5 or 6 residues long (Stapley and Creamer, 1999). It is interesting to note that all PPII structures have been found on the protein surface where they can maximize their interaction with water.

4. POLYPROLYNE II CONFORMATION

Collagens are built up of three polypeptide chains structured in a PPII-helix conformation twisted about each other. This regular structure arises because the collagen polypeptide chains consist largely of the repeated proline-rich sequences $(Gly\text{-}X\text{-}Y)_n$, with proline residues at the X positions and hydroxyproline residues at the Y positions. Furthermore, their conformation resembles essentially those adopted by homopolymers of 3 or more proline residues in water (Schweitzer-Stenner et al., 2003). The main difference is that homopolymers of proline form helices that remain in no apparent contact with each other.

Proline is unique among the 20 amino acids in having a cyclic side chain that includes its backbone nitrogen atom. This limits rotation about the peptide backbone $N\text{-}C_\alpha$ bond (ϕ torsion angle), allowing only two types of geometric structures (Reiersen and Rees, 2001). Furthermore, the lack of a hydrogen substituent on its imide nitrogen prevents backbone residues from engaging in the usual hydrogen-bonding observed in α-helices or β-sheets.

In aqueous solution, homopolymers of proline have a strong preference for the left-handed, all trans, extended helix; i.e., the PPII conformation (Figure 1). In the PPII conformation, the peptide bonds adopt average backbone dihedral angles of $(\phi, \psi) = (-78°, +149°)$, corresponding to a region of the Ramachandran map slightly to the right of the β region. The PPII conformation adopted by homopolymers of proline is very stable at temperatures as high as 90 °C (Kelly et al., 2001) and ionic interactions (pH and salt effects) cannot disrupt it to any great extent (Rucker and Creamer, 2002). In a non-aqueous environment, however, polyprolines assume a more compact right-handed, all cis helix with backbone dihedral angles of $(\phi, \psi) = (-83°, +158°)$ (Schweitzer-Stenner et al., 2003). This conformation is properly called the polyproline I (PPI) conformation (Figure 1). It is stabilized by van der Waals forces on the interior of the helix (Counterman and Clemmer, 2004).

The important role of water in the stabilization of the PPII conformation is demonstrated by its mutarotation dependence upon the polarity of its solvent. When

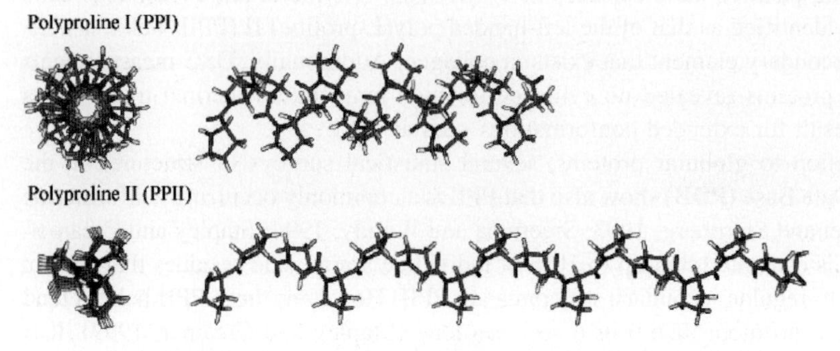

Figure 1. Representation of structures for a 15-residue poly-L-proline. The PPI helix contains all cis-residues; PPII contains all trans-residues (Counterman and Clemmer, 2004)

polyproline in the PPI configuration is introduced into water it will mutarotate to the PPII configuration. Likewise polyproline in the PPII conformation will mutarotate to PPI if introduced into aliphatic alcohol such as propanol or methanol (Counterman and Clemmer, 2004; Kakinoki et al., 2005).

Additionally, a recent study on the gas phase conformations of varying lengths of polyproline ions demonstrated that while the PPI conformation is maintained in the gas phase, the PPII conformation is not. The authors suggest that as the aqueous phase was removed from the PPII-structured polyproline during an electrospray process, the loss of water destabilized the PPII helix. Although it was not clear what conformations were formed from PPII polyproline in the gas phase, a mixture of cis- and trans-proline was evident (Counterman and Clemmer, 2004). This study also clearly demonstrated the critical importance of water in stabilizing the PPII helix.

In the case of the collagen triple-helix, it is stabilized in part by hydrogen bonding that occurs only every third residue, mainly between the backbone NH of glycine and the backbone CO of the residue in the X position of the adjacent chain (Figure 1a) (Brodsky and Ramshaw, 1997). The remaining two backbone CO groups in each tripeptide, as well as any backbone NH groups of non-proline X and Y residues, are not involved in either intra- nor inter-chain hydrogen bonds with other groups. In addition, the hydroxyl groups of the hydroxyproline residues point outward from the triple helix and therefore cannot directly hydrogen bond to any other groups within the molecule (Brodsky and Ramshaw, 1997).

The manner by which the hydrogen bonding potential of this conformation is satisfied was revealed by the determination of the structure of a triple-helical collagen-like molecule $(GlyProHyp)_{10}$ by X-ray crystallography (Bella et al., 1994; Bella et al., 1995). This first high-resolution structure of a collagen-type triple helix revealed an ordered and thick cylinder of hydration surrounding the triple helix (Figure 2a). Water molecules bridge hydrogen bonds between the hydroxyl groups of hydroxyproline and the peptide backbone CO and NH (if available) groups both within each chain and between different chains (Figure 2b). The number of water molecules involved in bridging two groups appears to vary along the molecule, such that two, three, four, or even five water molecules may form a chain linking the two groups (Bella et al., 1995). Additionally, all side chains, as well as the backbone CO group of the glycine in all three chains, are found on the outside of the triple helix molecule and in contact with water molecules.

It is worth noticing that crystallographic studies have also been carried out on $(Gly\text{-}Pro\text{-}Pro)_{10}$ sequences. In comparison with the $(Gly\text{-}Pro\text{-}Hyp)_{10}$ peptide, both structures demonstrate very similar molecular conformation and analogous hydration patterns involving carbonyl groups. Differences among the structures occur primarily in the extended water structure (Kramer et al., 1998; Berisio et al., 2002). It was concluded that while hydroxyproline is not necessary for hydration, its presence adds stability and interconnectivity to the water network that are probably necessary in the packing assemble of triple-helices.

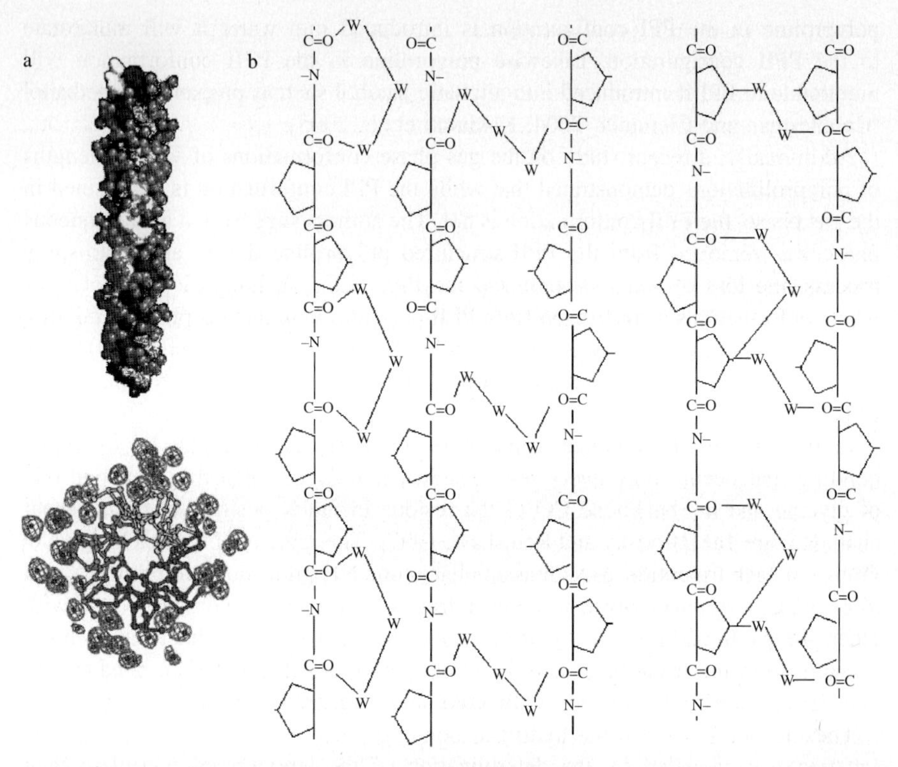

Figure 2. (a) Overall view of the hydration layer surrounding collagen molecule and the electron density map of the water molecules. (Bella et al., 1995; Kramer et al., 1998). (b) A schematic drawing illustrating the types of water hydrogen bonding patterns found in the triple-helix:water mediated hydrogen linking carbonyl groups; and water mediated hydrogen bonding linking hydroxyproline OH group and carbonyl groups (Bella et al., 1995)

5. PPII CONFORMATION OF NON-PROLINE PROTEINS

Although originally defined for the conformation adopted by polymers of proline, the PPII helical conformation can be adopted by amino acid sequences other than those based on proline.

Tiffany and Krimm first observed such phenomena almost 40 years ago. They noticed that the CD spectra of charged or unfolded (denatured) forms of polyglutamic acid and polylysine structures resembled that of homopolymers of proline or collagen. Indeed, they emphasized that the similarities of the random coil spectrum and that of polyprolines were too strong to correspond to a true random coil (Tiffany and Krimm, 1968).

Later the authors pointed out that electrostatic interaction is not the only driven force that can give rise to extended PPII structure (Tiffany and Krimm, 1972) since the same PPII type of CD spectrum can be displayed for systems in which electrostatic interaction is not a factor (Tiffany and Krimm, 1973). In the same paper the authors also propose that water should play a special role in stabilizing the PPII structure, by hydrogen bonding to exposed carbonyl groups.

Since this time, a number of groups have examined this hypothesis using a variety of peptides systems and several biophysical methods. Each of these groups has found evidence in support of Tiffany and Krimm's hypothesis.

6. HOST-GUEST STUDIES

The tendency of each amino acid to form the PPII conformation has been quantified using a host guest model system in several studies (Shah et al., 1996; Creamer and Campbell, 2002; Rucker et al., 2003; Chellgren and Creamer, 2004). Such experiments consist of introducing guest residues into the center of a polyproline-based peptide. The tendency toward the PPII conformation can then be determined by CD spectroscopy. In particular, PPII-typical CD spectra are characterized by a negative band near 200 nm and by a weaker positive band at about 220 nm (Figure 3a) (Chellgren and Creamer, 2004).

Figure 3b illustrates the effect of adding Glu (Q), Asn (N), Ala (A) and Val (V) to a host system P_3XP_3. It is observed that alanine does not significantly disrupt the PPII conformation, while valine strongly disfavors it (Chellgren and Creamer, 2004).

Such host-guest experiments have demonstrated that each residue possesses its own propensity to induce the PPII conformation, with proline being the most stabilizing in this respect. Charged residues (Glu, Lys, Arg and Asp) were among the most stabilizing of this conformation after proline, with Ala and Gln also falling into this category. Note that this list has a striking resemblance to the list of amino

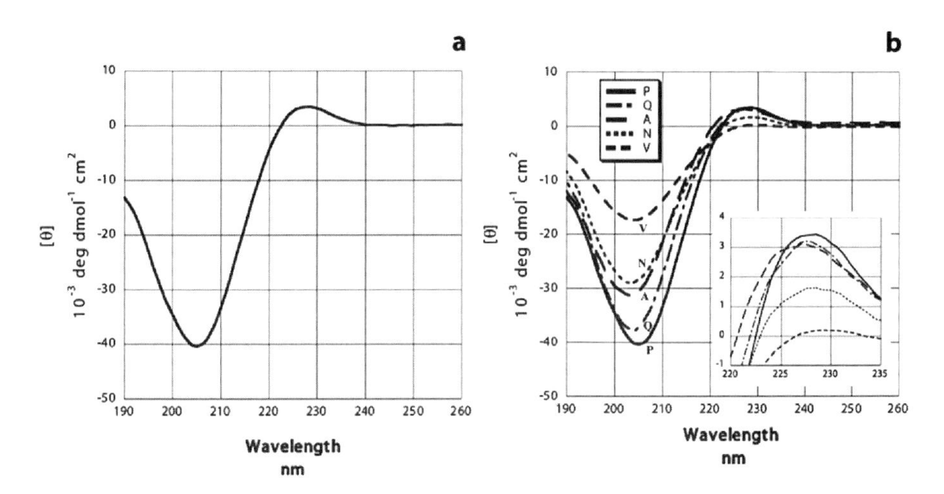

Figure 3. CD spectra of (Pro)X(Pro) at 5 °C. (a) $(Pro)_7$ peptide (b) $(Pro)_7$ peptide (solid line) with Gln (long- and short-dashed line), Ala (long-dashed line), Asn (short-dashed line), and Val (Medium-dashed line) single guest residue peptides. Inset shows maxima (Chellgren and Creamer, 2004)

acids typically found in natively unfolded proteins discussed above. Next lower in stability was a group including the nonpolar residues Leu, Phe, Ile, Val and Met, along with Asn, Ser, His, Thr and Cys. Tyr and Trp residues seemed highly destabilizing (Shah et al., 1996; Rucker et al., 2003; Chellgren and Creamer, 2004).

As we have previously established, the hydration of the peptide backbone seems crucial to maintain the PPII conformation and stability of homopolymers of proline. The incorporation of a guest non-proline residue into this rigid structure would mainly alter the water network around the peptide backbone. As mentioned above, β-branched and bulky residues seem to be the most disrupting types.

7. SHORT PEPTIDES STUDIES

In the past 5 years, one of the most important and influential studies on the unfolded state has been achieved with short model peptides such as polyalanine. These studies have altered dramatically the understanding of the nature and dynamics of the random coil peptides and unfolded proteins.

Combined evidence from theoretical computer modeling studies of short peptides (too short to form any detectable α-helix or β-sheet) in aqueous solution and from a variety of spectroscopic studies, including CD (Rucker and Creamer, 2002), NMR (Poon et al., 2002), two-dimensional vibrational spectroscopy (Woutersen and Hamm, 2001), VCD (Keiderling et al., 1999), and vibrational Raman spectroscopy (Blanch et al., 2000), reveal that the PPII helix is the dominant conformation in a variety of these short peptides.

For example, using NMR and CD spectroscopy, Shi et al. (2002) recently demonstrated that a seven-residue alanine peptide is predominantly in the PPII conformation in aqueous solution. Following up on this result, Eker et al. (2004) showed, using a variety of spectroscopic techniques, that the acidic and basic tripeptides, triglutamate, triaspartate and trilysine, adopted a distorted PPII conformation. Interestingly, a comparison of structures obtained from the spectra measured at acid, neutral and alkaline pH strongly suggested that the structural preference of all these peptides does not depend on the protonation states of the residues. Earlier, Rucker and Creamer (2002) showed similar results for a seven residue lysine peptide which retained PPII helical CD signals over a range of pH levels. They concluded that PPII helices must be preferred conformations for the polypeptide backbone and that electrostatic repulsion is not a driving force for PPII helix formation (Rucker and Creamer, 2002; Eker et al., 2004). In contrast, tripeptides such as trivaline and triserine only adopt an extended β-strand conformation (Eker et al., 2002; Eker et al., 2003).

Much larger alanine peptides have also been studied. For instance, Asher et al. (2004), using UV Raman spectroscopy, examined the melting of a 21-residue, mainly alanine, peptide (containing three arginines to confer solubility). The peptide was mainly in an α-helix conformation at 0 °C and melted to a PPII conformation as the temperature increased. Above room temperature the peptide existed mainly as a PPII helix. The authors did not observe evidence of any other significantly populated intermediates.

The recent reports by several laboratories that PPII is the major backbone conformation present in short alanine peptides has motivated an interest in finding the cause of this preference. There is general agreement that solvation is probably an important factor. For instance, the unsolvated PPII conformation in polyalanine is not stable in the gas phase, but it is stable in water (Drozdov et al., 2004).

Experiments in water have also shown that alanine peptides fluctuate between a PPII and an extended β-strand conformation (Shi et al., 2002; Eker et al., 2002; Schweitzer-Stenner et al., 2004), while valine and proline peptides exist only as β-strand and PPII conformations, respectively. At low temperatures, and as the number of alanine residues in the peptide increases, the PPII fraction substantially increases. This last observation has been interpreted as indicating that the addition of an alanine residue changes the hydration shell of the peptide in a way that stabilizes the peptide-solvent interaction and its PPII conformation (Schweitzer-Stenner et al., 2004). Using a molecular dynamic approach, Garcia supported this hypothesis, establishing that a peptide segment comprising four alanine residues is needed for the formation of a strongly hydrated groove around the peptide backbone which stabilizes the PPII conformation (Figure 4a) (Garcia, 2004).

Poon et al., (2002) hypothesized that bridging water molecules are responsible for an alanine dipeptide adopting the PPII conformation. They argue that the PPII helix is better configured for stability than other forms since both the CO and NH units to which water dimers bind are coplanar, permitting nearly linear hydrogen bond

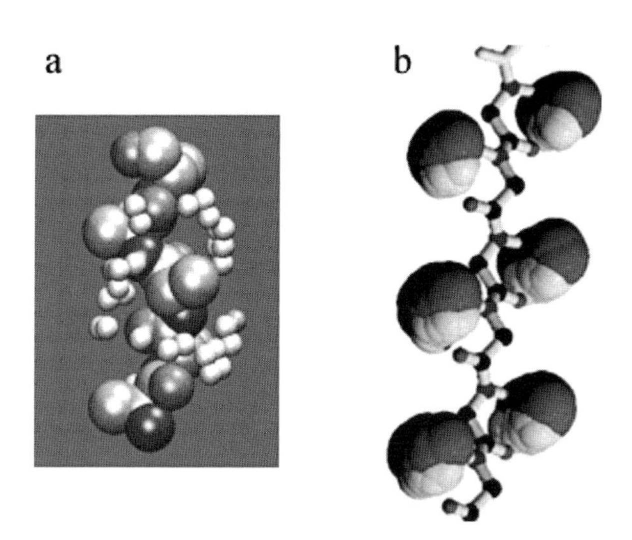

Figure 4. (a) alanine peptide in PPII conformation showing a hydrated groove around the peptide backbone (Garcia, 2004). (b) An alanine β-strand on which clustering waters (big balls) from multiple simulation have been superimposed. Some proximate water molecules are not hydrogen-bonded to the peptide (Mezei et al., 2004)

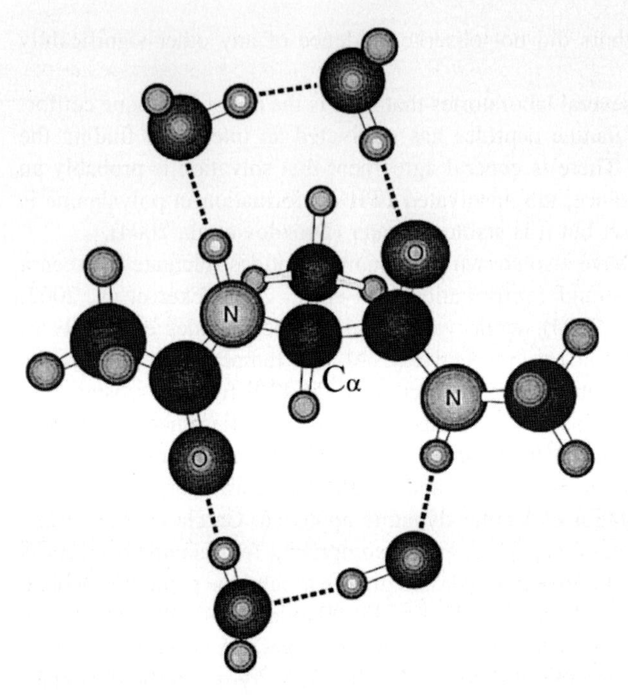

Figure 5. Polyproline II geometry in an alanine residue showing two double-water bridges (Poon et al., 2002)

angles through bridges (Figure 5). Such a bridging structure suggests that effects of cooperative hydrogen bonding may be quite important in forming a stable hydration groove around the peptide backbone. A similar mechanism has been considered for the collagen hydration discussed above.

A recent Monte Carlo simulation of the interaction between water and a 12-residue alanine peptide has complemented this later finding (Mezei et al., 2004). The simulation included water molecules (explicit), and it was possible to examine the hydrogen bonding interactions made between water and the alanine peptide as an α-helix, β-strand, or PPII conformation. This study found that the apparent enthalpic interaction between water and the alanine peptide is significantly stronger in the PPII conformation than in the extended β conformation or α-helix. This suggests that β strands induce formation of entropically disfavored (ordered) water, reminiscent of the hydrophobic effect (Figure 4b).

Thus, the PPII structure fully utilizes the hydrogen bonding capacity of the CO and NH groups, maximizes peptide-water cooperativity, and leaves the first solvation layer of hydration able to participate in further hydrogen bonding with the next solvation shell. PPII helices produce a less disruptive effect on surrounding water organization as compared to β strands (Kentsis et al., 2004; Mezei et al., 2004).

These studies strongly suggest that "random coil" peptides have definitive backbone conformations and water plays a major role in determining the conformation of proteins in the unfolded state. The concept of a denatured protein as a structureless random chain is no longer valid when backbone conformations of individual residues are considered (Baldwin, 2002).

8. ANTIFREEZE GLYCOPROTEIN MECHANISM

Many Arctic and Antarctic fish species secrete high concentrations of antifreeze glycoproteins (AFGPs) in their body fluid when in a subzero temperature environment. These AFGPs are responsible for the observed freezing point depression, inhibition of ice nucleation and crystal growth, and morphological changes of the ice crystals in the immediate vicinity of the AFGPs (Yeh and Feeney, 1996; Ben, 2001; Harding et al., 2003). AFGPs induce these effects by strongly influencing water-self organization; structuring vicinal water molecules, thereby inhibiting the transition of water into the ice state (Pollack, 2002).

AFGPs are able to depress the freezing point temperatures to a level more than 200–300 times on a molal basis than that which would be expected for ordinary cryoprotectants (sugars and polyols) or salts (DeVries et al., 1970). Such an impressive freezing point depression of AFGPs does not, however, significantly alter the melting temperature of the same solution. The freezing and melting temperature differential is referred to as thermal hysteresis and is often taken as a primary manifestation of antifreeze activity by AFGPs.

A typical AFGP is composed of a repeating tripeptide unit $(\text{Ala-Ala-Thr})_n$ in which the secondary hydroxy group of the threonine residue is glycosylated with the disaccharide β-D-galactosyl-(1,3)-α-D-*N*-acetylgalactosamine (Figure 6a) (Yeh and Feeney, 1996). Eight distinct isomers of AFGP, ranging in molecular mass from 2.7 to 32 KDa, have been isolated from the blood serum and tissues of polar fishes. These primarily differ in the number of tripeptide repeating units, which range from four to fifty. Minor differences are found in the amino acid composition for the low molecular weight AFGPs where Ala is occasionally substituted for Pro and/or Thr residues substituted with Arg.

The exact mechanism whereby these molecules inhibit ice crystal growth at the molecular level remains a source of intense debate. Some researchers have long proposed that the binding of AFGPs to the ice surface likely involves hydrogen bonding between the hydroxyl groups of the disaccharide residue and the ice surface. However, this hypothesis has been challenged on several levels. Of note is the fact that substitution of Arg for Thr removes the disaccharide from one of the tripeptides units, but this structural modification does not affect antifreeze activity (Schrag et al., 1982; Burcham et al., 1986).

Another consistent problem on attempting to elucidate this mechanism is that the water-ice interface has not been well characterized. In fact, the interface itself is probably not an abrupt transition as typically represented (Ben, 2001). In fact, recent evidence shows the loss of organized ice structure at the interface as being fairly gradual, occurring over approximately 10 angstroms (Harding et al., 2003).

Based on the short alanine peptide studies discussed above, we propose an alternative mechanism of action by antifreeze glycoproteins. We suggest that the freezing point depression/thermal hysteresis depends on a conformational rearrangement of the AFGPs that optimizes the PPII conformation in the supercooling regime.

This hypothesis is strongly supported by the work of Bush and Feeney (1986) who performed many variable-temperature ^{1}H- and ^{13}C-NMR studies, as well as NOE (Nuclear Overhauser Effect) experiments. Based on these experiments and CD measurements they suggested that low-molecular-weight glycoproteins exist in solution as PPII helices at low temperatures, whereas at higher temperatures the structure becomes more like a flexible coil. A recent molecular dynamic simulation (Nguyen et al., 2002) and a NMR measurement of a synthetic AFGP trimer [(AlaAlaThr*)$_3$] (Figure 6b) (Tachibana et al., 2004a) have confirmed these earlier findings.

Interestingly, this later study also showed that a single synthetic tripeptide (monomer) of AFGP was able to influence the ice conformation, but not the thermal

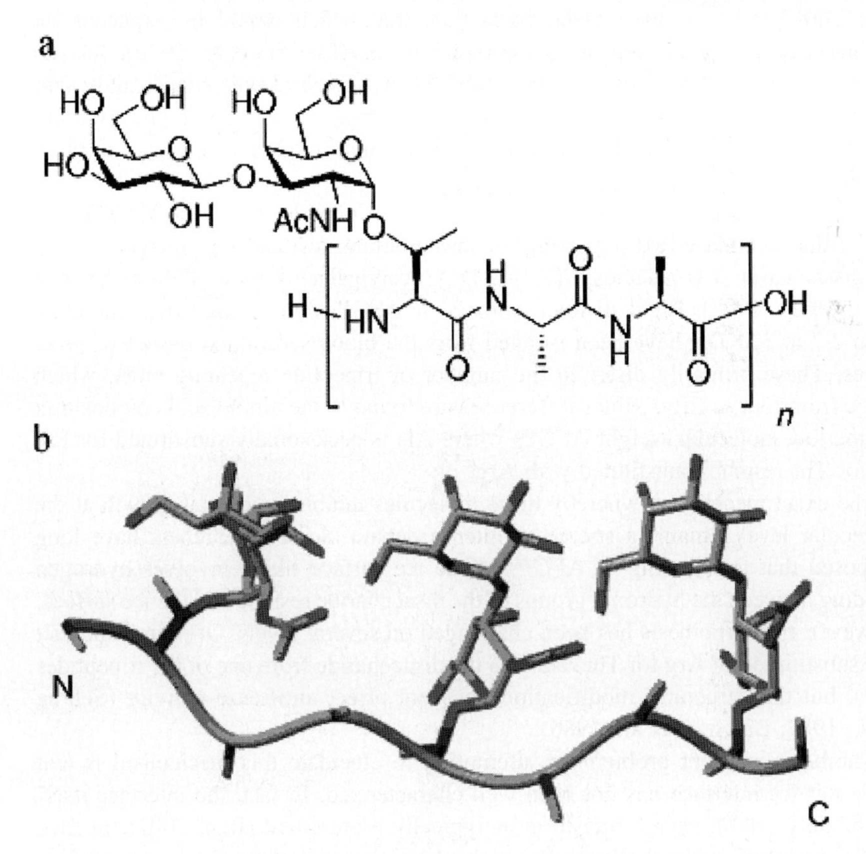

Figure 6. (a) A general AFGP repeat (Ben, 2001) (b) and the aligments of sugars and hydrophobic side chains on a PPII backbone conformation (Tachibana et al., 2004a)

hysteresis, of the AFGP solution. A dimer seemed enough to form the characteristic PPII conformation and confer as appreciable a level of antifreeze activity as did the larger AFGP species. This finding supports Garcia's modeling study that hypothesizes that a minimum of four alanine residues are necessary for the formation of a strongly hydrated groove around the peptide backbone to stabilize the PPII conformation (Garcia, 2004).

Structure-functional studies have now shown conclusively that both the acetamide group (AcNH) on the sugar moiety and the methyl of the threonine residue play key role on the antifreeze activity. Early Mimura et al., (1992) hypothesized that the NH of the acetamide group makes a stabilizing hydrogen bond to the backbone Thr carbonyl oxygen, thereby stabilizing the PPII conformation. This non-covalent stabilization mimics how proline fixes the backbone torsional angle by its covalent structure in proline-rich peptides. Synthesized glycoproteins with deleted amide proton at the galactose residue showed neither strong nor negligible positive shoulder around 220 nm in its CD spectra (Tachibana et al., 2004a,b). This suggests a partial structural collapse of the PPII structure.

The uniqueness of the PPII conformation of AFGPs is that it allows all backbone groups to be positioned on the same side of the molecule facing in close proximity the disaccharide group, while the relatively hydrophobic Ala groups occupy the primary position on the opposite side (Mimura et al., 1992). This arrangement is crucial since the use of a glycosylated serine residue in place of threonine, which has a side-chain hydroxyl group but no methyl group, was unable to confer thermal hysteresis (Tachibana et al., 2004a). Furthermore, this arrangement adopted an α-helical form. It has been also found that in the glycosylated serine substituted glycopeptides that the acetamide groups assumed more apical positions with the peptide backbones distanced from the sugar, allowing more rotational freedom around the o-glycosidic bond (Naganagowda et al., 1999; Kindahl et al., 2000).

Since the hydrogen bonding between the NH proton of AcNH and the CO bond of threonine stabilizes the carbohydrate group against the backbone (Mimura et al., 1992), the most likely mechanism of stabilization by the disaccharide residues is the solute exclusion theory of Timasheff (Bolen and Baskakov, 2001). It mimics the same effect as a solution of high sugar concentration stabilizes proteins. We also consider that the sugars add stability and interconnectivity in the extended water structure created by the peptide backbone. This effect is similar to how the hydroxyl groups of the hydroxyproline on the collagen molecule stabilize the water layers.

It should be understood that the proline and arginine residues found in lower molecular weight AFGP species could act to add to stabilize the PPII conformation since both residues do not restrict the access of water to the peptide backbone and have a high propensity to form PPII helices.

Interestingly, ice nucleation proteins (INPs), which represent the antithesis of AFPs in that INPs promote the formation of ice, have been suggested to form β-helices (Graether and Jia, 2001).

9. CONCLUSIONS

Recent studies have shown a radically different picture of the unfolded state. In this new view, unfolded proteins have a more limited range of conformations than was formerly appreciated. It provides a basis for understanding not only the nature of the unfolded state but also the earliest event that occurs during folding.

The critical role of water has also emerged as a factor to condition protein conformation. In this new model, the optimal bridging of water with the peptide backbone groups (carbonyl and amide) determines a well known conformation termed polyproline II (PPII). The specific role of the side chains is to modulate conformations by interfering to certain degree with the solvation of the peptide backbone.

The molecular mechanism of action describing how biological antifreeze glycoproteins alter the water structuring can be resolved by invoking the PPII conformation function as a main role.

This new understanding has potentially broad reaching implications, particularly with respect to modeling the unfolded state and understanding the determinants of protein stability.

ACKNOWLEDGEMENTS

Funded by a grant from USDA/NRIGP Project 2002-0891.

REFERENCES

Adzhubei A, Sternberg M (1993) Left-handed polyproline II helices commonly occur in globular proteins. J Mol Biol 229:472–493

Asher SA, Mikhonin AV, Bykov S (2004) UV Raman demonstrates that alpha-helical polyalanine peptides melt to polyproline II conformations. J Am Chem Soc 126:8433–8440

Baldwin R (2002) A new perspective on unfolded proteins. Adv Protein Chem 62:361–367

Barron LD, Blanch EW, Hecht L (2002) Unfolded proteins studied by raman optical activity. Adv Prot Chem 62:51–90

Bella J, Eaton M, Brodsky B, Berman HM (1994) Crystal-structure and molecular-structure of a collagen-like peptide at 1.9-Angstrom resolution. Science 266:75–81

Bella J, Brodsky B, Berman HM (1995) Hydration structure of a collagen peptide. Structure 3:893–906

Ben RN (2001) Antifreeze glycoproteins – Preventing the growth of ice. Chembiochem 2:161–166

Berisio R, Vitagliano L, Mazzarella L, Zagari A (2002) Crystal structure of a collagen triple helix model $[(\text{Pro-Pro-Gly})]_3$. Protein Sci 11:262–270

Blanch EW, Morozova-Roche LA, Cochran DA, Doig AJ, Hecht L, Barron LD (2000) Is polyproline II helix the killer conformation? A Raman optical activity study of the amyloidogenic prefibrillar intermediate of human lysozyme. J Mol Biol 301:553–563

Blanch EW, Kasarda DD, Hecht L, Nielsen K, Barron LD (2003) New insight into the solution structures of wheat gluten proteins from Raman optical activity. Biochemistry 42:5665–5673

Bokor M, Csizmok V, Kovacs D, Banki P, Friedrich P, Tompa P, Tompa K (2005) NMR relaxation studies on the hydrate layer of intrinsically unstructured proteins. Biophysical J 88:2030–2037

Bolen DW, Baskakov IV (2001) The osmophobic effect: Natural selection of a thermodynamic force in protein folding. J Mol Biol 310:955–963

Bond CJ, Wong KB, Clarke J, Fersht AR, Daggett V (1997) Characterization of residual structure in the thermally denatured state of barnase by simulation and experiment: Description of the folding pathway. Proc Natl Acad Sci USA 94:13409–13413

Brant DA, Flory PA (1965) Configuration of random polypeptide chains 2: Theory. J Am Chem Soc 87:2791–2800

Brodsky B, Ramshaw JAM (1997) The collagen triple-helix structure. Matrix Biol 15(8-9):545–554

Burcham TS, Osuga DT, Rao BNN, Bush CA, Feeney RE (1986) Purification and primary sequences of the major arginine-containing antifreeze glycopeptides from the fish eleginus-gracilis. J Biol Chem 261:6384–6389

Bush CA, Feeney RE (1986) Conformation of the glycotripeptide repeating unit of antifreeze glycoprotein of polar fish as determined from the fully assigned proton NMR-spectrum. Int J Pept Protein Res 28(4):386–397

Chang JY, Ballatore A (2000) The structure of denatured bovine pancreatic trypsin inhibitor (BPTI). Febs Lett 473:183–187

Chellgren BW, Creamer TP (2004) Short sequences of non-proline residues can adopt the polyproline II helical conformation. Biochemistry 43:5864–5869

Counterman AE, Clemmer DE (2004) Anhydrous polyproline helices and globules. J Phys Chem B 108:4885–4898

Creamer TP, Campbell MN (2002) Determinants of the polyproline II helix from modeling studies. Adv Protein Chem 62:263–282

DeVries AL, Komatsu SK, Feeney RE (1970) Chemical and physical properties of freezing point-depressing glycoproteins from Antarctic fishes. J Biol Chem 245:2901–2908

Drozdov AN, Grossfield A, Pappu RV (2004) Role of solvent in determining conformational preferences of alanine dipeptide in water. J Am Chem Soc 126:2574–2581

Dunker AK, Lawson JD, Brown CJ, Williams RM, Romero P et al. (2001) Intrinsically disordered protein. J Mol Graph Model 19:26–59

Dunker AK, Brown CJ, Lawson JD, Lakoucheva LM, Obradovic Z (2002) Intrinsic disorder and protein function. Biochemistry 41:6573–6582

Eker F, Cao X, Nafie LA, Schweitzer-Stenner R (2002) Tripeptides adopt stable structures in water. A combined polarized visible Raman, FTIR, and VCD spectroscopy study. J Am Chem Soc 124:14330–14341

Eker F, Griebenow K, Schweitzer-Stenner R (2003) Stable conformations of tripeptides in aqueous solution studied by UV circular dichroism spectroscopy. J Am Chem Soc 125:878–885

Eker F, Griebenow R, Cao X, Nafie LA, Schweitzer-Stenner R (2004) Tripeptides with ionizable side chains adopt a perturbed polyproline II structure in water. Biochemistry 43:613–621

Farrell HM, Qi PX, Wickham ED, Unruh JJ (2002) Secondary structural studies of bovine caseins: Structure and temperature dependence of beta-casein phosphopeptide (1-25) as analyzed by circular dichroism, FTIR spectroscopy, and analytical ultracentrifugation. J Protein Chem 21:307–321

Garcia AE (2004) Characterization of non-alpha helical conformations in Ala peptides. Polymer 45:669–676

Giesemann T, Rathke-Hartlieb S, Rothkegel M, Bartsch JW, Buchmeier S, Jockusch BM, Jockusch H (1999) A role for polyproline motifs in the spinal muscular atrophy protein SMN – Profilins bind to and colocalize with SMN in nuclear gems. J Biol Chem 274:37908–37914

Graether SP, Jia Z (2001) Modeling Pseudomonas syringae ice-nucleation protein as a β-helical protein. Biophys J 80:1169–1173

Harding MM, Anderberg PI, Haymet ADJ (2003) 'Antifreeze' glycoproteins from polar fish. Eur J Biochem 270:1381–1392

Jardetzky TS, Brown JH, Gorga JC, Stern LJ, Urban RG, Strominger JL, Wiley DC (1996) Crystallographic analysis of endogenous peptides associated with HLA-DR1 suggests a common, polyproline II-like conformation for bound peptides. Proc Natl Acad Sci USA 93:734–738

Kakinoki S, Hirano V, Oka M (2005) On the stability of Polyproline I and II structures of proline oligopeptides. Polymer Bull 53:109–115

Kay BK, Williamson MP, Sudol M (2000) The importance of being proline: the interaction of proline-rich motifs in signaling proteins with their cognate domains. FASEB J 14:231–241

Keiderling TA, Silva RA, Yoder G, Dukor RK (1999) Vibrational circular dichroism spectroscopy of selected oligopeptide conformations. Bioorg Med Chem 7:133–141

Kelly MA, Chellgren BW, Rucker AL, Troutman JM, Fried MG (2001) Host-guest study of left-handed polyproline II helix formation. Biochemistry 40:14376–14383

Kentsis A, Mezei M, Gindin T, Osman R (2004) Unfolded state of polyalanine is a segmented polyproline II helix. Prot Struct Funct Bio 55:493–501

Kindahl L, Sandstrom C, Norberg T, Kenne L (2000) H-1 NMR studies of hydroxy protons of Asn- and Ser-linked disaccharides in aqueous solution. J Carbohydrate Chem 19:1291–1303

Klein-Seetharaman J, Oikawa M, Grimshaw SB, Wirmer J, Duchardt E et al Long-range interactions within a nonnative protein. Science 295:1719–1722

Kramer RZ, Vitagliano L, Bella J, Berisio R, Mazzarella L, Brodsky B, Zagari A, Berman HM (1998) X-ray crystallographic determination of a collagen-like peptide with the repeating sequence (Pro-Pro-Gly). J Mol Biol 280:623–638

Linder S, Hufner K, Wintergerst U, Aepfelbacher M (2000) Microtubule-dependent formation of podosomal adhesion structures in primary human macrophages. J Cell Sci 113:4165–4176

Mahoney NM, Janmey PA, Alamo SC (1997) Structure of the profiling-poly-L-proline complex involved in morphogenesis and cytoskeletal regulation. Nat Struct Biol 4:953–960

Maiti NC, Apetri MM, Zagorski MG, Carey PR, Anderson VE (2004) Raman spectroscopic characterization of secondary structure in natively unfolded proteins: alpha-synuclein. J Am Chem Soc 126:2399–2408

Mezei M, Fleming PJ, Srinivasan R, Rose GD (2004) Polyproline II helix is the preferred conformation for unfolded polyalanine in water. Prot Struct Funct Bio 55:502–507

Mimura Y, Yamamoto Y, Inoue Y, Chujo R (1992) NMR-study of interaction between sugar and peptide moieties in mucin-type model glycopeptides. Int J Biol Macromol 14:242–248

Naganagowda GA, Gururaja TL, Satyanarayana J, Levine MJ (1999) NMR analysis of human salivary mucin (MUC7) derived O-linked model glycopeptides: comparison of structural features and carbohydrate-peptide interactions. J Pept Res 54:290–310

Nguyen DH, Colvin ME, Yeh Y, Feeney RE, Fink WH (2002) The dynamics, structure, and conformational free energy of proline-containing antifreeze glycoproteins. Biophysical J 82:2892–2905

Pollack GH (2002) The cell as a biomaterial. J Mater Sci Mater Med 13:811–821

Poon CD, Samulski ET, Weise CF, Weisshaar JC (2002) Do bridging water molecules dictate the structure of a model dipeptide in aqueous solution? J Am Chem Soc 122:5642–5643

Reiersen H, Rees AR (2001) The hunchback and its neighbours: proline as an environmental modulator. Trends Biochem Sci 26:679–684

Rucker AL, Creamer TP (2002) Polyproline II helical structure in protein unfolded states: Lysine peptides revisited. Protein Sci 11:980–985

Rucker AL, Pager CT, Campbell MN, Qualls JE, Creamer TP (2003) Host-guest scale of left handed polyproline II helix formation. Protein Struct Funct Genet 53:68–73

Schrag JD, Ogrady SM, DeVries A (1982) The relationship of amino acid composition and molecular-weight of antifreeze glycopeptides to non-colligative freezing point depression. Biochim Biophys Acta 717:322–326

Schweers O, Schönbrunn-Hanebeck E, Marx A, Mandelkow E (1994) Structural studies of tau protein and Alzheimer paired helical filaments show no evidence for ß-structure. J Biol Chem 269:24290–24297

Schweitzer-Stenner R, Eker F, Perez A, Griebenow K, Cao X, Nafie LA (2003) The structure of tri-proline in water probed by polarized Raman, Fourier transform infrared, vibrational circular dichroism, and electric ultraviolet circular dichroism spectroscopy. Biopolymers 71:558–568

Schweitzer-Stenner R, Eker F, Griebenow K, Cao X, Nafie L (2004) The conformation of tetra-alanine in water determined by polarized Raman, FT-IR and VCD spectroscopy. J Am Chem Soc 126:2768–2776

Shah NK et al. (1996) A host-guest set of triple-helical peptides: Stability of Gly-X-Y triplets containing common nonpolar residues. Biochemistry 35:10262–10268

Shi ZS, Olson CA, Rose GD, Baldwin RL, Kallenbach NR (2002) Polyproline II structure in a sequence of seven alanine residues. Proc Natl Acad Sci USA 99:9190–9195
Shortle D (1996) The denatured state (the other half of the folding equation) and its role in protein stability. Faseb J 10:27–34
Smyth E, Syme CD, Blanch EW, Hecht L, Vasak M, Barron LD (2001) Solution structure of native proteins with irregular folds from Raman optical activity. Biopolymers 58:138–151
Sreerama N, Woody RW (1994) Poly(Pro) II helices in globular proteins: identification and circular dichroism analysis. Biochemistry 33:10022–10025
Stapley BJ, Creamer TP (1999) A survey of left-handed polyproline II helices. Prot Sci 8:587–595
Syme CD, Blanch EW, Holt C, Jakes R, Goedert M, Hecht L, Barron LD (2002) A Raman optical activity study of rheomorphism in caseins, synucleins and tau – New insight into the structure and behaviour of natively unfolded proteins. Eur J Biochem 269:148–156
Tachibana Y, Fletcher GL, Fujitani N, Tsuda S, Monde K, Nishimura SI (2004a) Antifreeze glycoproteins: Elucidation of the structural motifs that are essential for antifreeze activity. Angew Chem Int Ed Engl 43:856–862
Tachibana Y, Monde K, Nishimura SI (2004b) Sequential glycoproteins: Practical method for the synthesis of antifreeze glycoprotein models containing base labile groups. Macromolecules 37:6771–6779
Tiffany ML, Krimm S (1968) New chain conformations of poly(glutamic acid) and polylysine. Biopolymers 6:1379–1383
Tiffany ML, Krimm S (1972) Effect of temperature on circular-dichroism spectra of polypeptides in extended state. Biopolymers 11:2309–2316
Tiffany ML, Krimm S (1973) Extended conformations of polypeptides and proteins in urea and guanidine hydrochloride. Biopolymers 12:575–587
Tompa P (2002) Intrinsically unstructured proteins. Trends Biochem Sci 27:527–533
Uversky VN, Fink AL (2002) The chicken egg scenario of protein folding revisited. Febs Lett 515:79–83
Weinreb PH, Zhen W, Poon AW, Conway KA, Lansbury PT (1996) NACP, a protein implicated in Alzheimer's disease and learning, is natively unfolded. Biochemistry 35:13709–13715
Woutersen S, Hamm P (2001) Isotope-edited two-dimensional vibrational spectroscopy of trialanine in aqueous solution. J Chem Phys 114:2727–2737
Wright PE, Dyson HJ (1999) Intrinsically unstructured proteins: re-assessing the protein structure-function paradigm. J Mol Biol 293:321–331
Yeh Y, Feeney RE (1996) Antifreeze proteins: Structures and mechanisms of function. Chemical Reviews 96:601–617

CHAPTER 12

SOME PROPERTIES OF INTERFACIAL WATER: DETERMINANTS FOR CELL ARCHITECTURE AND FUNCTION?

FRANK MAYER[1,*], DENYS WHEATLEY[2] AND MICHAEL HOPPERT[1]

[1] *University of Goettingen, Institute for Microbiology and Genetics, Grisebachstrasse 8, D-37077 Goettingen, Germany*

[2] *University of Aberdeen, Cell Pathology Division, Hilton MG7, Aberdeen AB24 4FA, UK*

Abstract: Interfacial water is the water moiety in a living cell located in the immediate vicinity of particulate cell components of any kind, of which most will be proteins and membranes. It is often called vicinal water. According to theoretical considerations and experimental findings, interfacial water exhibits structural organizations that differ from 'free', i.e. 'bulk' water. These specific structural properties may be influenced by structural properties of the cellular particulate components. The deviation of vicinal/interfacial water from that of 'free' water implies, but does not necessarily ensure, that functional differences exist between them. Hence, there exists a range of mutual interdependences of the properties of the particulate cell components regarding the potential for structuring of water, water structure, the functional properties of interfacial water, and the physiologically important properties of the particulate cell components surrounded by interfacial water. In this communication, simulated examples are described that illustrate some of the possible consequences of these mutual interdependences for the architecture and function of prokaryotic and eukaryotic cells at the cellular and macromolecular levels, which may not be trivial. In addition, examples typical for the structural organization of prokaryotic and eukaryotic cells are described that support the notion of mutual interdependences

Keywords: Vicinal water; Interfacial water; Model systems; Reversed micelle; Enzyme kinetics; Cell architecture

---

* Corresponding author. Tel.: +49-(0)-4141-45866; *E-mail address*: *fmayer12@gmx.de (F.Mayer); mhopper@gwdg.de (M.Hoppert); wheatley@abdn.ac.uk (D N Wheatley).

G. Pollack et al. (eds.), Water and the Cell, 253–271.

1. INTRODUCTION

The living cell comprises a highly organized and regulated dynamic system composed of a multitude of building units organized into functional units of various degrees of complexity. Water is the irrevocable prerequisite for the working of the system.

A wealth of data on the properties of liquid water has been reported (Muller 1988; Kusalik and Svishchev, 1994) that illustrate why water can play some special roles in the living cell (Cho et al., 1996; Clegg and Wheatley, 1991; Drost-Hansen, 2001; Etzler and Drost-Hansen, 1983; Malone and Wheatley, 1991; Wheatley, 1991, 1993a,b; Wheatley and Clegg, 1994; Wheatley and Malone, 1993; Wheatley et al., 1984). The importance of this area of research is illustrated by the fact that a number of conferences were held (e.g., Gordon Research Conference – GRC – on Physics and Chemistry of Water and Aqueous Solutions 1994, a GRC on Macromolecular Organization and Cell Function 1998, and a GRC on Interfacial Water in Cell Biology 2004). 'Bulk' or 'free' water is only one of the various ways that liquid water can be organized. 'Dense' water (low degree of hydrogen bonding), and 'less dense' ('expanded') water (high degree of hydrogen bonding – extreme in ice with straight hydrogen bonds) were proposed to exist under specific conditions (Wiggins, 1990, 1995, 2001). In a closed system, these kinds of water moieties are balanced: deviation of water density from that of bulk water in one partial volume of the system is presumably compensated by an inverse deviation of density of the water moiety in some neighbouring partial volume to maintain a balance. Exactly defined borderlines between the different water densities obviously do not exist, and it should be kept in mind that partitioning of cell water into water moieties of different density cannot be static, but rather a procession of transient states of high dynamic activity.

2. MODEL SYSTEMS FOR ANALYSIS

2.1 The Reversed Micelle

In vitro model systems for closed systems, especially reversed micelles, were used for studies on the characteristic features of these modifications of the organisation of water (Hoppert et al., 1994; Hoppert and Mayer, 1999b; Khmelnitsky et al., 1989; Strambini and Gonnelli, 1988). In a reversed micelle, liquid water exists as nanoscale microdroplets in an organic solvent as a matrix. These microdroplets are built up in solutions consisting of appropriate surfactants or surfactant/co-surfactant combinations in an organic solvent, such as alkanes or benzenes. At appropriate surfactant/solvent/water ratios, reversed micellar solutions (water-in-oil microemulsion) form, where droplets of water (typically of 5 to 25 nanometer in diameter (i.e., structures a few millionths of a millimetre across) are surrounded by surfactant molecules. These surfactant molecules are directed with their polar or ionic head groups to the water pool and their non-polar parts to the outside (organic solvent). Figure 1 represents the different water moieties and their distribution in a reversed micelle. For comparison, a conventional micelle is also depicted.

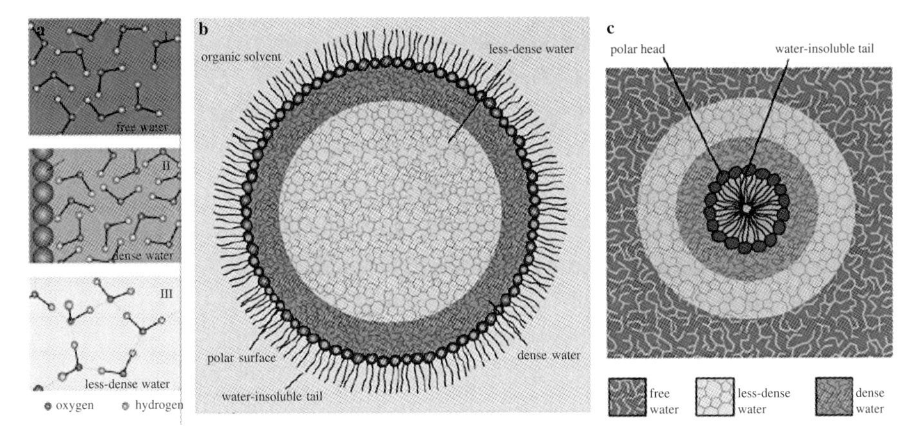

Figure 1. Structural organization of water. (a) in comparison with water molecules in free ('bulk') water (I), water molecules close to polar 'charged' surfaces lie very near one another (II). Molecules in this so-called dense water generally form fewer hydrogen bonds (dashed lines) than there are seen in free water. To offset the effects of dense water, some water molecules are arrayed in a configuration that is actually less dense than free water (III). These molecules form many hydrogen bonds, and the resulting structure resembles that of ice. These structures are adapted from models developed by Philippy Wiggins (see references therein). (From Hoppert and Mayer, 1999b). (b) a reversed micelle. These compartments enclose, rather than exclude, water. Amphiphilic molecules – that is, molecules containing both water-soluble (usually polar) and water-insoluble portions – form their outer boundary such that the polar heads face the interior water, while the water-insoluble tails project into the medium, which is an uncharged organic solvent. In this configuration, dense and less-dense water layers would be expected to form inside the compartment. (From Hoppert and Mayer, 1999b). (c) a conventional micelle. Micelles are formed by amphiphilic molecules (see Figure 1b). When placed in an aqueous environment, certain amphiphilic molecules become arrayed such that the polar heads contact the aqueous medium, and the water-insoluble tails are tucked inside the compartment, away from the water. A theory proposes that the water layer closest to the polar heads is dense water, followed by a layer of less-dense water, followed by free water. (From Hoppert and Mayer, 1999b)

Are reversed micelles with entrapped proteins appropriate systems for the collection of data on enzymes and structural proteins in the living cell? A comparison of small reversed micelles, with entrapped protein molecules (Figure 2a), with the known situation in the interior of a bacterial cell regarding the presumably very low amount of free water and the presence of 'surfaces' with polar or ionic groups in both systems (Figure 3) show some obvious similarities. 'Molecular crowding' (Eggers and Valentine, 2001; Ellis, 2001; Minton, 2001; Van den Berg et al., 1999; Zimmermann and Minton, 1993) as postulated for the interior of a bacterial cell – leaving only minor spaces for water – appears to be more closely simulated by enclosure of an enzyme in a reversed micelle, i.e., in a system with similar restrictions regarding the amount of water and the presence of 'surfaces' in the microenvironment of the protein, as compared to conventional aqueous buffer systems.

Individual protein molecules and high molecular weight protein complexes have dimensions similar to those of small reversed micelles. Therefore, for studies aimed

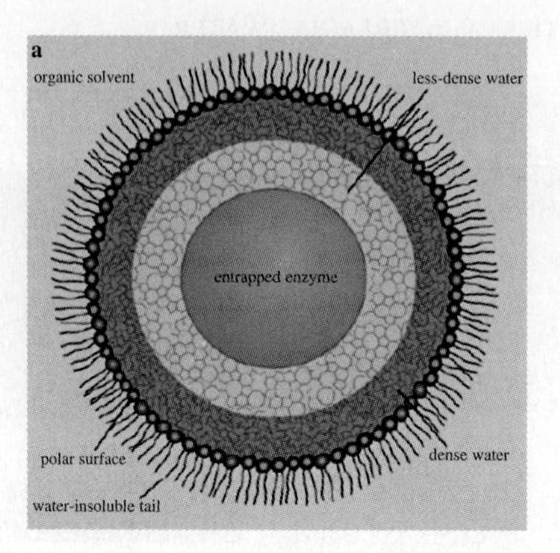

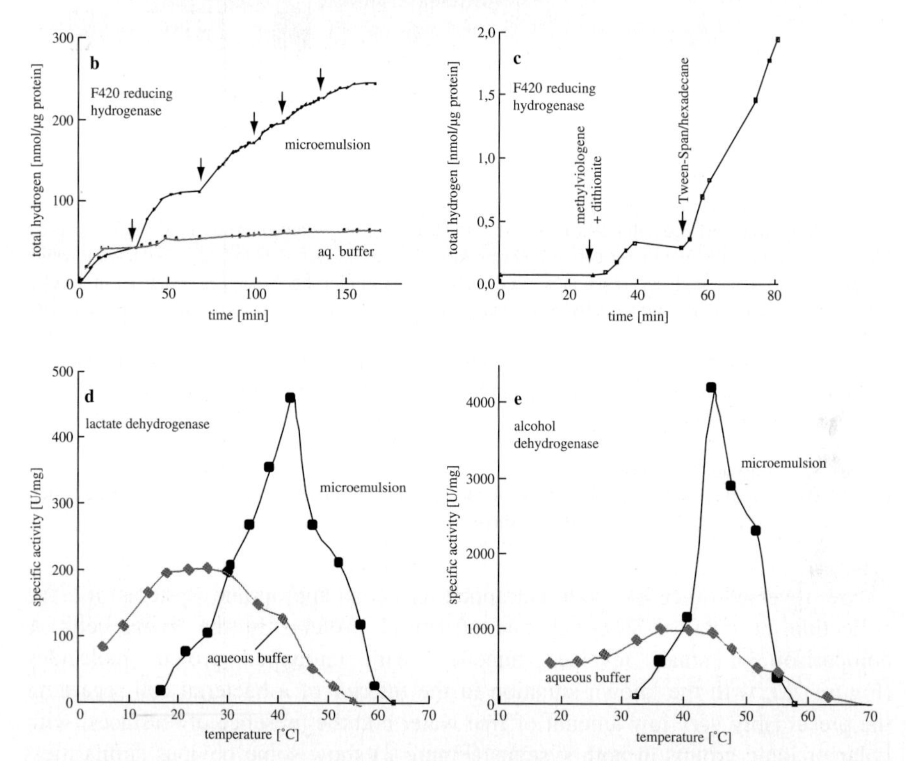

Figure 2. Enzymes trapped inside reversed micelles. (a) an enzyme molecule is placed inside a reversed micelle as described in Figure 1c. It has been proposed that water structure may influence the behaviour of enzymes and proteins in general. Enzymes placed inside reversed micelles are assumed to display specific activities similar to those inside living cells (see Figures 2b to e). (From Hoppert and Mayer, 1999b). b to e, behaviour of enzymes in reversed micelles ('microemulsion') and aqueous buffer, respectively. (b) hydrogen production of the F420-hydrogenase at 60 °C in microemulsion and aqueous buffer solution. The arrows indicate the points of time of methylviologene additions to both assays. (From Hoppert et al., 1994). (c) recovery of F420-hydrogenase activity after inactivation by incubation in aqueous buffer solution (12 h at 60 °C). Addition of substrate, reducing agent and Tween-Span/hexadecane as indicated. (d) (e) thermal stability of enzymes (d, lactate dehydrogenase; e, alcohol dehydrogenase) is increased in microemulsion as compared to aqueous buffer (From Hoppert et al., 1994)

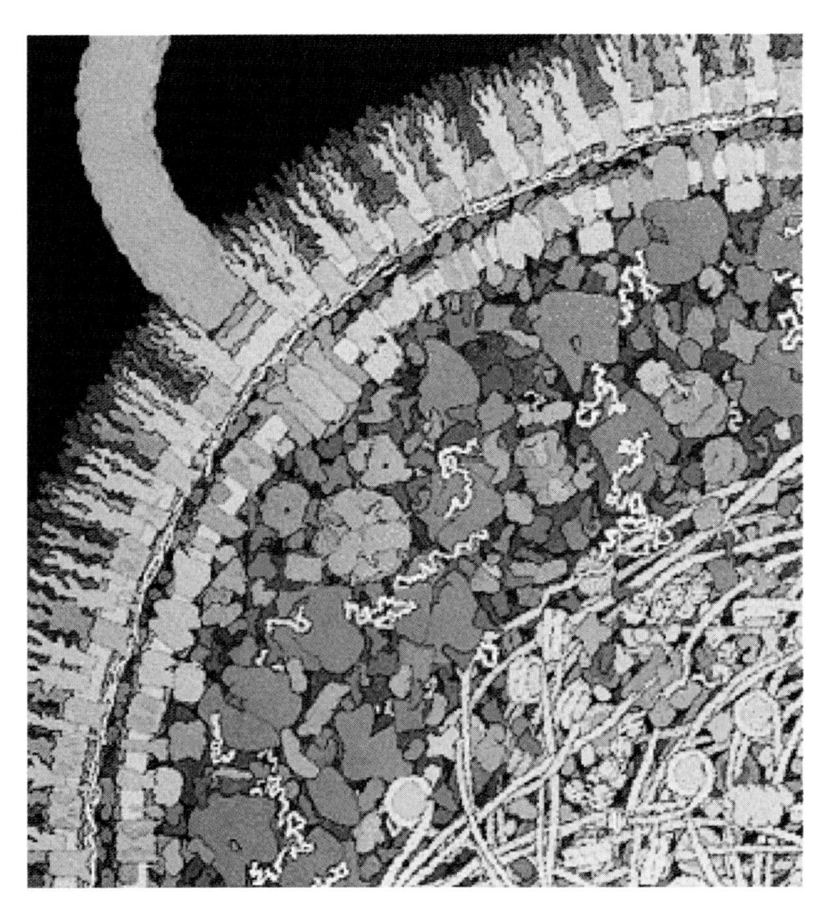

Figure 3. Cellular and macromolecular architecture of a typical eubacterial cell *(Escherichia coli)*: 'Macromolecular crowding'. The cell is surrounded by two membranes (see Figure 5a) enclosing a – narrow – periplasmic compartment that is used for capturing an sorting nutrients and wastes. At the center of the cell, densely packed DNA strands, with bound proteins, are folded into a compact nucleoid. The cytoplasm occupies the remaining portions of the cell, and is filled with ribosomes and many different enzymes and multiprotein complexes. The averaged distance from protein to protein is assumed to be less than the diameter of an average protein molecule. 'Molecular crowding' is the consequence, with a very low amount of free water. Note: the drawing does not show the bacterial cytoskeleton (see Figure 7h). (From Hoppert and Mayer, 1999b; drawing: David S.Goodsell, The Scripps Institute)

at measuring the effects of the availability of only very restricted amounts of water in very small spaces – as it is the case in a reversed micelle and in the living (bacterial) cell – on stability and on specific enzyme activity, reversed micelles were prepared with, on average, just *one* enzyme molecule entrapped per micelle. Under these conditions, hydrophilic proteins are known to be located in the center of reversed micelles, surrounded by a thin layer of water (Figure 2a). Enclosure of only one protein molecule per reversed micelle might appear to be not realistic when a comparison of the reversed micelle system with the situation in a cell is intended. However, this approach allows much better quantitative evaluations of the measured data compared to reversed micelles with a higher – and probably not precisely known – number of entrapped protein molecules per reversed micelle.

2.2 Kinetic Analysis of Enzymes in Reversed Micelles

These studies have revealed remarkable differences in the measured parameters when compared with respective data determined for these enzymes in the free state, i.e., in bulk water (Figures 2b-e; Hoppert et al., 1994). Hydrophilic enzymes showed a bell-shaped course of activity depending on the size of the micelle. At optimized micellar sizes, specific activities were 2- to 10-fold higher than in conventional aqueous buffer solutions, and the temperature optima of various enzymes were shifted by 10° to 16 °C.

Differences of this magnitude, measured in optimized reversed micellar systems, would not be too surprising when enzyme complexes are investigated that are associated with the membrane in order to perform their function in the living cell, e.g., enzymes involved in energetics (Gerberding and Mayer, 1993, and below). The *in vivo* state of such enzymes would certainly not be appropriately simulated after removal of the enzymes from the membrane or disorganization of the membrane, with subsequent transfer in bulk water/buffer. It is notable that these differences in specific activity could be measured for bacterial enzymes that had been characterized, by independent techniques, to be 'soluble' (Hoppert and Mayer, 1999b, Hoppert et al., 1994, 1997; Mayer, 1993a,b). In principle, high specific enzyme activities may be assumed to better reflect the natural state of an enzyme than low specific activities. Accepting this conclusion, these results indicate that for soluble enzymes as well, specifically structured water in their microenvironment may be of importance.

2.3 Solubility, Stability and Membrane Association

Additional studies were carried out with liposomes to which enzymes (again enzymes that are known to be 'soluble', e.g., bacterial α-amylase, guanylate kinase from *Saccharomyces cerevisiae*) were coupled from the outside by a histidine-tag or *via* a strep-tag (Figure 4; Wichmann et al., 2003). Although enzymes in this situation are not placed in a closed system, it could be shown that stability and specific activity of these enzymes were simultaneously modified after coupling and were especially influenced by the lipid used for the liposome assembly. From these *in vitro* data, one can conclude that for high degrees of stability and specific activity of enzymes *in vivo*, location of enzymes in narrow spaces is not an irrevocable prerequisite. Appropriate structuring of water may also occur in open systems, especially close to the surface of cellular membrane systems due to exposure of polar or ionic groups. In these cases, balancing of partial volumes of differently structured water moieties as mentioned above for the small reversed micelle is not obvious.

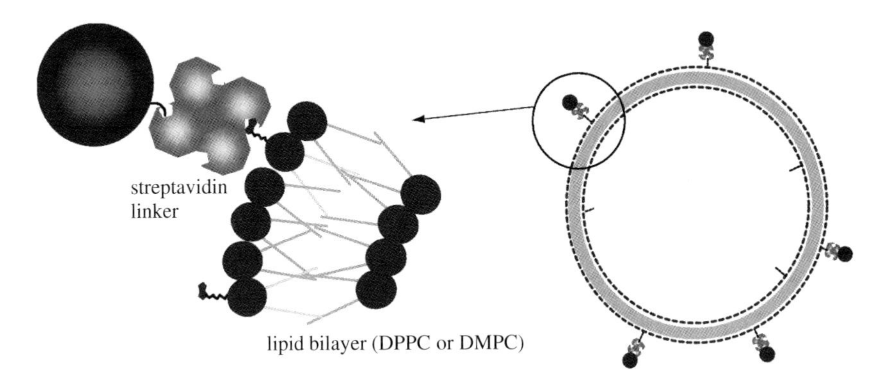

Figure 4. Assembly of enzyme-liposome complexes. At the right-hand side, a schematically drawn liposome with attached enzymes is depicted. Part of the diagram (circled) is drawn, at the left-hand side, with more details: a Streptavidin linker connects an enzyme molecule with the surface of the liposome. The size of the linker determines the distance of the enzyme from the liposome. A biotin tag not occupied by a linker is also shown attached to the outside of the liposome. DPPC, 1,2-dipalmitoyl-*sn*-glycero-3-phosphocholine; DMPC, 1,2-dimyristoyl-*sn*-glycero-3-phosphocholine (From Wichmann et al., 2003)

Probably, exposure of polar or ionic groups causes a gradient of water structure in the outside water, similar to that depicted in Figure 1c for a conventional micelle, into which the catalytic sites of the enzymes are properly placed by adjustment, mediated by a linker, of a certain distance of the enzyme from the 'surface'.

High specific activity of an enzyme can be taken as an indicator for an appropriate microenvironment of the enzyme. High specific activity could either be brought about by the fact that unfolding of the enzyme polypeptide is reduced in comparison with – artificial – unfolding in bulk water (Hoppert et al., 1994, 1997), or by the fact that the specific features of the microenvironment allow or support defined conformational changes of a protein, needed for the catalyzed reaction to take place or step forward. Provided this conclusion holds some validity, protection against unfolding of a polypeptide would be brought about by a 'stabilized' structure of the surrounding water *per se*, i.e., water with a high degree of hydrogen bonding ('low density water'). In fact, experimental evidence is available (Hoppert et al., 1994) that shows that even partially unfolded polypeptides may be folded back into the optimal conformation when entrapped in a reversed micelle, as deduced from an increase in specific activity when these polypetides are transferred from bulk water/buffer into the reversed micelle. On the other hand, water exhibiting a very low degree of hydrogen bonding ('high density water') would favour the flexibility of a polypeptide. When such flexibility is needed for an enzyme to perform its catalytic function, high-density water would be favourable. Hence, it appears feasible to assume that both high- and low-density water may be modifications that favour different enzymes, activities or specific properties depending on the circumstances.

As a first conclusion, it may be assumed that enzymes may have, *in vivo*, specific activities similar to those values determined in reversed micelles or after attachment to surfaces of lipid vesicles. The low values found in free water/buffer appear to be artefacts. Hence, the approaches used for simulation of the state of enzymes in the living cell – reversed micelles, surfaces of lipid vesicles – may be appropriate simulation systems. This may have interconnected reasons: one is that cells very often exhibit narrow spaces in which the catalytic sites of enzymes are located; this can be observed both in bacteria and in eukaryotic cells (see below). Such a design of the interior of a cell has the consequence that the structure of the water moiety adjacent to the 'surfaces' enclosing the narrow spaces is different from 'bulk' water (Parsegian and Rau, 1984). As indicated by the results of experiments where enzymes were attached to the outer surface of lipid vesicles, such localization may well be optimal and appears to fulfil the prerequisite functions of catalytic activity. Seemingly, it is of advantage for the living cell to place the catalytic sites of enzymes in such a way that they are 'embedded' in the most appropriate kind of structured water, independent of the way that the structuring of the water in the microenvironment is achieved.

As mentioned above, micelle-enzyme systems and lipid vesicles as used for the experiments described above are characterized by the fact that the enzyme molecules are located very close to 'surfaces' exhibiting polar or ionic groups that modify the structure of the water moiety in the immediate vicinity. Besides surfaces of membranes, such 'surfaces' can very well also be exposed surfaces of proteins or of structures other than proteins, e.g., nucleic acids (Brown et al., 1999; Rupley and Careri, 1991; Schneider et al., 1979; Sunnerhagen et al., 1998; Swaminathan et al., 1997; Timasheff, 1993).

3. CELLULAR ARCHITECTURE OF PROKARYOTIC AND EUKARYOTIC CELLS

3.1 'Narrow Spaces' ('nano spaces') and Polar or Ionic 'Surfaces' in Cells

Here, we present an overview regarding the occurrence of narrow spaces ('nano spaces') and exposed 'surfaces' in prokaryotic and eukaryotic cells, and the positioning of various enzymes and other proteins, both 'soluble' and 'membrane-bound' or 'membrane-associated', within prokaryotic and eukaryotic cells (Figures 5-7). On the basis of these data, we can begin to discuss some existing mutual interactions between water structure, enzyme localization, cell architecture, working of the normal cell, and their derangements in the workings of injured cells.

3.1.1 The prokaryotic cell

The envelope of prokaryotic cells is characterized by stratified layers surrounding the cytoplasm, with variations depending on the complexity of the organism. The innermost layer is the cytoplasmic membrane. In typical Gram-negative eubacteria (Figure 5a), the next layer is the peptidoglycan or murein layer embedded in the periplasmic space, followed by the outer membrane. Gram-positive bacteria lack the outer membrane; often, their outermost wall layer is a monolayer ('surface layer') consisting of complexes of protein molecules (Figure 5b). In principle, such a layout also constitutes a kind of periplasmic space (Beveridge, 1995). In Archaea, various modifications of the envelope structure can be found. In general, even in cases where the only layer additional to the cytoplasmic membrane is a surface layer, water moieties very restricted in space are formed within the cell wall. This can be achieved by a cup-like design of the individual surface layer complexes (Figure 5c). In conclusion, the typical prokaryotic cell envelope provides various kinds of 'narrow spaces' and 'surfaces'.

In the cytoplasm proper of prokaryotic cells, most obvious structural differentiations visible in phototrophic bacteria (and in the non-phototrophic, nitrate-producing bacterium *Nitrosococcus oceanus*; see Mayer, 1999) are invaginations of the cytoplasmic membrane, forming membrane vesicles or stacks of membranes (Figure 5d). Other kinds of differentiation are the chlorobium vesicles in certain anaerobic phototrophic bacteria. In all these cases, lumina are created with very restricted inner space, defined by 'surfaces' with considerable lateral extension. One might claim that enlargement of surfaces is needed to provide sufficient surfaces for a large amount of photopigments and reaction centers for photosynthesis. Nevertheless, it should be stated that the interiors of cells of many prokaryotic species exhibit a high degree of substructural organization, providing 'narrow spaces' and 'surfaces'.

3.1.2 Eukaryotic cells

It is no surprise that also in the eukaryotic cell (Figure 6a) structural differentiations can be found that are very similar to those observed in prokaryotes. After all, chloroplasts and mitochondria of recent eukaryotic cells have been derived, during evolution, from precursors of today's prokaryotes that were incorporated into precursors of today's eukaryotic cells by the process of endocytosis. In addition to these organelles, typical eukaryotic cells are characterized by a multitude of other organelles ('compartments'), such as vacuoles, the Golgi membrane stacks, vesicles of the dictyosomes, the endoplasmic reticulum, and the lumen within the two membranes surrounding the nucleus. Such a compartmentalization is interpreted to be the prerequisite for the ordered and regulated working of a eukaryotic cell. However, a possible interconnection between such a cell architecture and implicit formation of microenvironments, as far as water structures are concerned, is seldom mentioned in textbooks as if it was of no importance.

3.2 Common Principles of Localization of Functional Proteins to Specific Cellular Sites

3.2.1 The prokaryotic cell

A number of enzyme complexes and other functional proteins in a prokaryotic cell, especially those involved in cell energetics, are localized to membranes. The basic construct of such an enzyme is that it consists of a membrane-integrated part and one or two (hydrophilic) parts extending either into the cytoplasm or

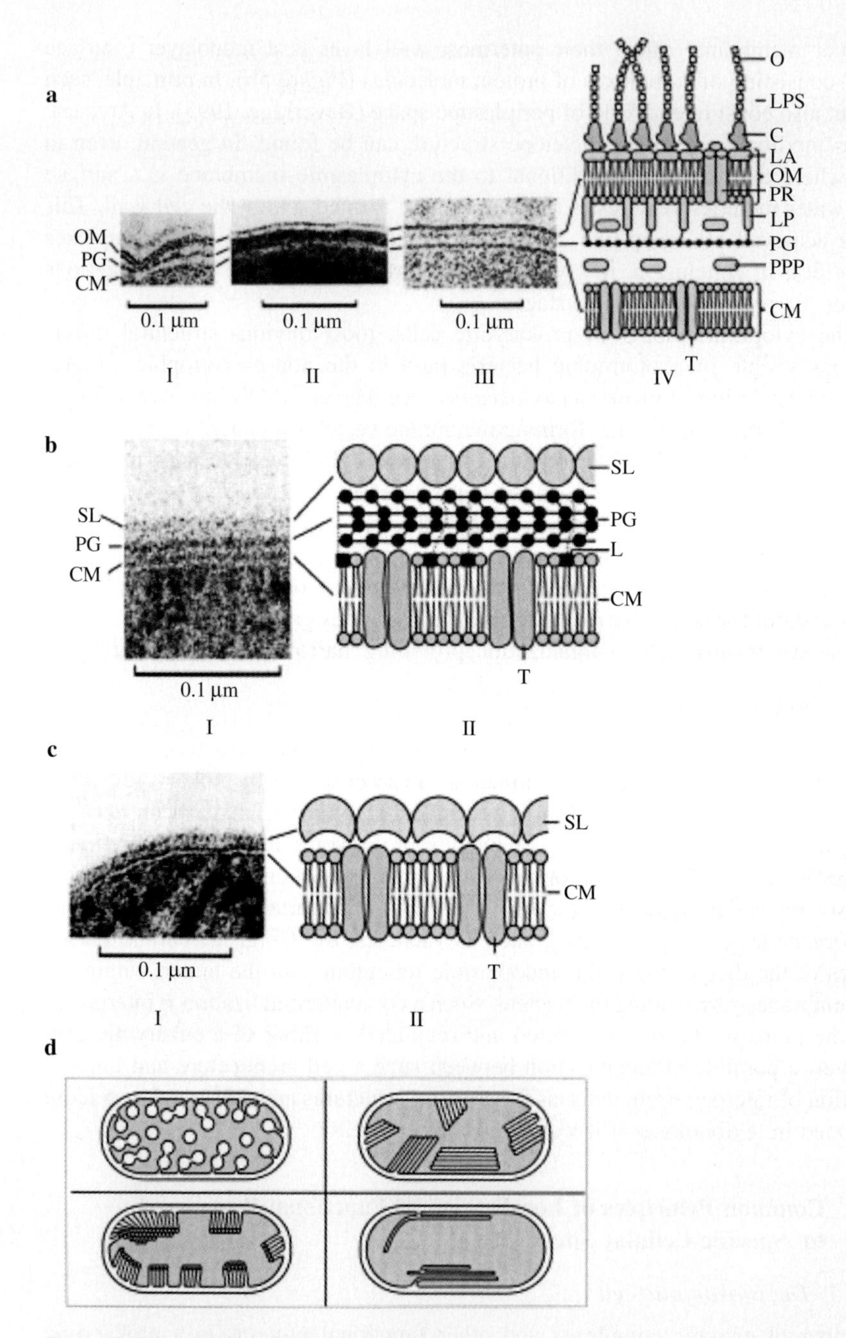

Figure 5. Cell envelopes and intracytoplasmic membranes in prokaryotes. (a) the cell envelope of Gram-negative eubacteria: (I) conventionally chemically fixed, dehydrated, resin-embedded, ultrathin-sectioned cell (*Acinetobacter* spec. strain MJT/F5); note the wavy appearance of the outer membrane, shrinkage by loss of material, and "empty" appearance of the space between the cytoplasmic and the

into the opposite direction (Stupperich et al., 1993). Such a distribution of masses is not surprising in cases where the overall reaction catalyzed by the enzyme complex needs interaction of compounds or ions present inside with those outside of the cytoplasm, and involvement of membrane-integrated components. Typical examples are the F_0F_1 ATPase and the A_0A_1 ATPase (Figure 7a) (Mayer et al., 1987; Reidlinger et al., 1994), and the functional units of phototrophic bacteria (and chloroplasts) comprised of light harvesting complexes and the reaction center (photosystems). Figure 7b depicts a photosystem II complex. A structural principle common to these partially membrane-integrated complexes is the fact that their hydrophilic part, usually carrying the catalytic sites, is kept at a distance from the membrane surface that usually is in a quite narrow but distinct range, from 1 to 6 nm. In the group of F_0F_1 and A_0A_1 ATPases, this is achieved by two 'stalks' of defined length. A comparison with the situation inside a reversed micelle (see above) reveals that this range is very similar to that of the distance of an entrapped enzyme molecule in an optimized reversed micelle (Figure 2a).

Surprisingly, a number of typical 'soluble' or 'cytosolic' enzymes in bacteria were also membrane-associated (Hoppert and Mayer, 1999a). At first sight, there is no obvious reason for such localization in these cases. After all, the reactions catalyzed by these enzymes are restricted either to the cytosol or to the exterior of the

◄———

Figure 5. outer mebrane, which seems to contain only the peptidiglycan layer. (II) cryosection of an *Alcaligenes eutrophus* (new name: *Ralstonia eutropha*) cell; stabilization of the section with methylcellulose, contrasted with uranyl acetate. Note the compact appearance of the cell envelope. The peptidoglycan present between the cytoplasmic and outer membrane is not discernible as a distinct layer. (III) cryosection of a frozen-hydrated, not chemically fixed, unstained *Escherichia coli* cell. The layers of the cell envelope are clearly visible. In the space between the cytoplasmic and the outer membrane, only the peptidoglycan can be discerned. Because of the lack of contrast, the other components present in the periplasmic space (see II, and Figure 3) cannot be seen. (IV) diagrammatic view of the macromolecular architecture; the lines indicate the respective layers visible in the ultrathin sections depicted in I to III Abbreviations: C, core region of the lipopolysaccharide; CM, cytoplasmic membrane; LA, lipid A; LP, lipoprotein; LPS, lipopolysaccharide (sugar); O, O-specific side chains of lipopolysaccharide; OM, outer membrane; PR, porin; PG, peptidoglycan (murein); PPP, periplasmic protein; T, transmembrane protein (b), the cell envelope of Gram-positive eubacteria: (I) conventionally chemically fixed, dehydrated, resin-embedded, and ultrathin-sectioned cell (*Thermoanaerobacterium thermosulfurogenes* EM1) exhibiting a cytoplasmic membrane (CM), thick peptidoglycan layer (PG), and a surface layer (SL) (see also Figure 7h). (II) diagrammatic view of the macromolecular architecture; the lines indicate the respective layers visible in the ultrathin section (I). The lipoteichoic acid molecules (L) extend into the cell wall. T, transmembrane protein. (c) the cell envelope of an archaebacterium (Archaeon (I), conventionally chemically fixed, dehydrated, resin-embedded, and ultrathin – sectioned cell (*Methanogenium marisnigri*), exhibiting an envelope of very low structural complexity. (II) diagrammatic view of the macromolecular architecture. The cell envelope is composed of the cytoplasmic membrane (CM), which contains proteins (T) and carries a surface layer (SL). The structural units of the surface layer exhibit a cup-like shape. Note that this view is only one of the variations of the structural organization of the cell envelope of the Archaea. (d) scheme of the structure and arrangement of intracytoplasmic membranes (ICM) bearing the photosynthetic apparatus in phototrophic bacteria: I, vesicle type; II, tubuli; III, flat, thylakoid-like membranes in regular stacks; IV, large thylakoids, partially stacked, partially irregularly arranged. (a – d, from Mayer, 1999)

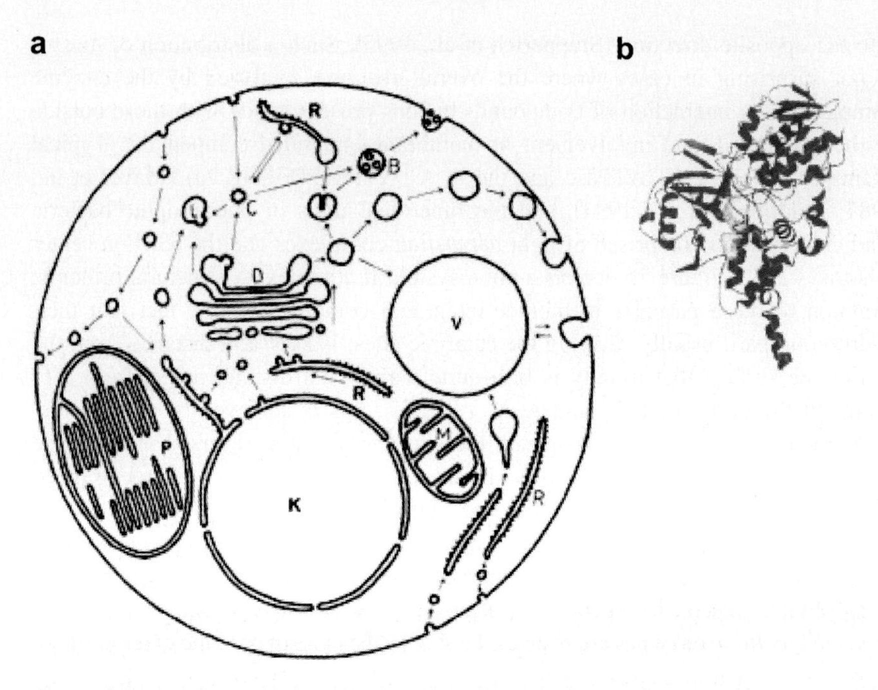

Figure 6. Structural organization of eukaryotic cells, and of a protein attached to mitochondria. (a) diagrammatic view of an idealized eukaryotic cell, exhibiting membrane-enclosed – narrow-compartments and their interactions by membrane flow and membrane transformation. Note: chloroplasts and mitochondria are not involved in these dynamic processes. Abbreviations: B, C, L, vesicles; D, dictyosome; K, nucleus; M, mitochondrium; P, chloroplast; R, endoplasmic reticulum; V, vacuole. (b) ribbon diagram of human monoamine oxidase B (MAO B). The protein has a single transmembrane helix that anchors it to the outer membrane of the mitochondrium. Nevertheless, the protein is considered monotopic because the bulk of the 520 residues, including the active side, is outside of the membrane (From Binda et al., 2002)

cell. Examples are the methyl-CoM methyl reductase, organized as the 'methanoreductosome', in the methanogenic archaeon, *Methanothermobacter thermoautotrophicus* (former name: *Methanobacterium thermoautotrophicum*) (Figure 7c) (Hoppert and Mayer, 1990; Mayer, 1993a; Mayer et al., 1988; Ossmer et al., 1986), the F420 reducing hydrogenase in the same archaeon (Figure 7d) Braks et al., (1994), and the α-amylase in the eubacterium, *Thermoanaerobacterium thermosulfurogenes* (Figures 7e,f) (Antranikian et al., 1987; Mayer, 2003b; Specka et al., 1991). This latter enzyme may be exposed to the 'outside' where it attacks the high molecular weight substrate starch, or it is released into the culture medium. During the experiments performed for the characterization of this enzyme, we observed that the amount of 'free' enzyme measured in the culture supernatant did increase when starch and phosphate were present in limited concentrations. Two features were observed: one that the cells lost their wall under these fermentation conditions and the enzyme attached to the outer surface of the cytoplasmic membrane was exposed to the medium. The other was that even the 'free' enzyme present in the culture supernatant was attached to membrane vesicles (Figure 7f inset). The distance of the 'head part' of the enzyme (carrying the catalytic site) was again localized to the surface of the membrane vesicle at a distance of about 3 to 4 nm. A distance around this value was also measured for the F420 hydrogenase mentioned above (Figure 7d). For the methano-reductosome (Figure 7c; it carries the catalytic sites in its 'head part'), a distance in the range of about 8 to 15 nm was measured. In all these cases, the enzymes were kept at their place by specific 'linkers' of defined length (Figure 7g). No indications could be found for a function of these linkers other than keeping the enzymes at their place.

A feature discovered only recently in bacteria might be important for a better understanding of the working of prokaryotes. Not only eukaryotes but also prokaryotes possess cytoskeletons (Hegermann et al., 2002; Mayer, 2003a,b; Mayer et al., 1998). Figure 7h illustrates several aspects of the structural organization of the cytoskeleton of a typical eubacterium. It became clear that the prokaryotic cytoskeleton might have a number of variations. The common principle of these modifications is that the cytoskeleton comprises structural elements located close to the inner face of the cytoplasmic membrane, and fibrillar structures crossing the cytoplasm. In the case of the 'basic' or 'primary' cytoskeleton in *Escherichia coli* (Mayer, 2003a,b), a close interaction of ribosomes/polysomes with the cytoskeleton could be observed. Such an interaction of ribosomes with elements of the cytoskeleton is also known for the eukaryotic cell (Hesketh and Pryme, 1991). In principle, cytoskeletal elements ought to be viewed as an additional kind of 'surface' within the cell, able to modify the structure of water in their immediate vicinity. The observed interactions of ribosomes with cytoskeletal elements, both in eukaryotes and in prokaryotes, might be of functional importance, and specific modifications of the water close to the cytoskeletal elements might play a role in these functions. It appears feasible to assume that not only ribosomes, but also other cellular components might interact with cytoskeletal elements not only in eukaryotes but also in prokaryotes.

3.2.2 *The eukaryotic cell*

As mentioned above, the presence of chloroplasts and mitochondria in the eukaryotic cell can be explained on the basis of the endosymbiont theory. Hence, the above comments on the localization of proteins specific for bacteria, to 'surfaces' in chloroplasts and mitochondria in the eukaryotic cells need no further discussion. Also, a possible water-organizing role of cytoskeletal elements in the eukaryotic cell has been sufficiently commented above.

There are dynamic structural and functional components that are specific for the eukaryotic cell. Many of them interact by membrane flow and membrane transformation (Figure 6a). For example, the eukaryotic cell harbours the endoplasmic reticulum (R in Figure 6a). A great deal of reactions important for the working of the eukaryotic cell takes place at 'surfaces' of the membranes making up this system, and within the narrow space 'inside' the membrane system. Hence, the endoplasmic reticulum, together with other membrane systems in the eukaryotic

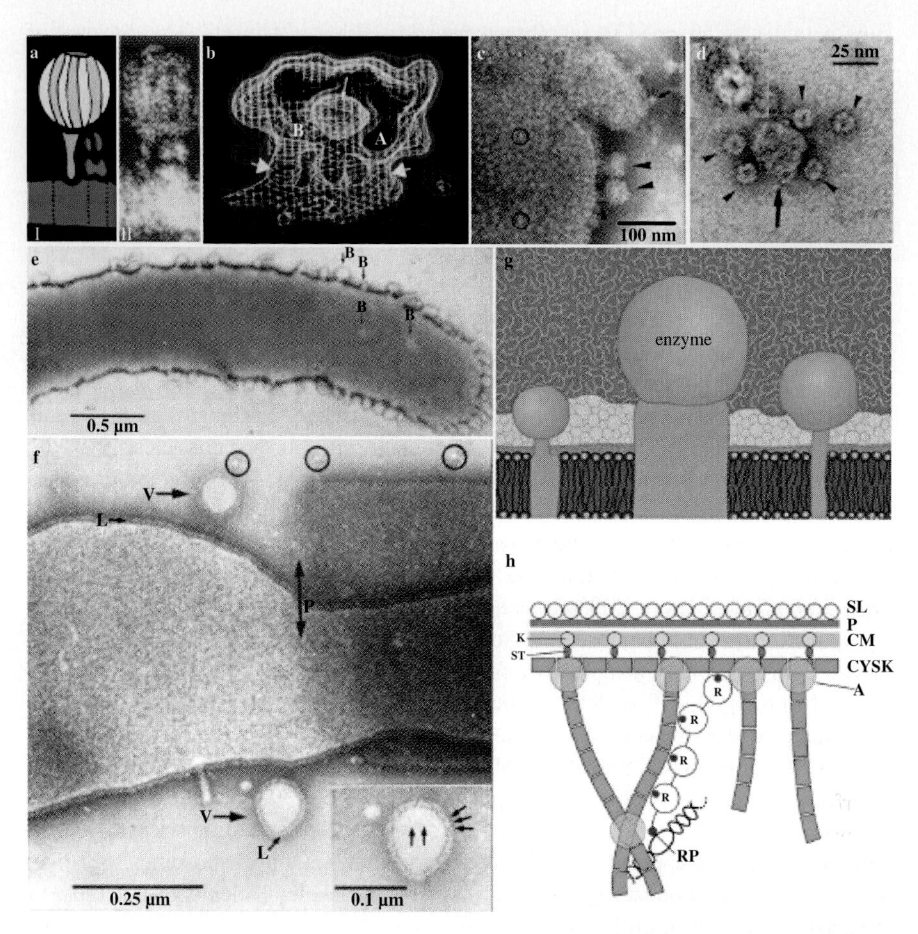

Figure 7. Structural organization and membrane attachment of bacterial protein complexes. (a) A_0A_1 ATPase; right: electron micrograph; left: model. (From Lingl et al., 2003). (b) the photosystem II complex (functionally intact structure occurring in cyanobacteria and plant chloroplasts; shown here: spinach chloroplast membrane with adhering photosystem II complex). A, B, stromal domains; thin arrow, stain-filled cavity after negative staining; large arrows, expected position of the outer (stromal) leaflet of the lipid bilayer (From Holzenburg et al., 1993). (c) methano-reductosomes (arrowheads), attached to the cytoplasmic side of the (artificially reversed) cytoplasmic membrane of *Methanococcus voltae*. Bar= 100 nm. (From Hoppert and Mayer, 1990). (d) F420-reducing hydrogenase complexes, attached to the cytoplasmic side of the (artificially reversed) cytoplasmic membrane of *Methanobacterium thermoautotrophicum*. Inset: higher magnification. (From Braks et al., 1994). (e) cell remnant of *Thermoanaerobacterium thermosulfurogenes* EM1 after removal of the wall layers by growth, under starch limitation, in continuous culture. The cell still exhibits its elongated shape; the cytoplasmic membrane is exposed to the environment; formation of blebs can be seen (B, arrows). Dimension is given in μm. (f) as e, but prior to the complete loss of the peptidoglycan layer. Circles, free enzymes (amylase/pullulanase); vesicles, originating from surface blebs, can be seen (V and inset); blebs and vesicles are densely covered by membrane-attached enzyme molecules (L, and arrows in the inset). Dimensions are given in μm.(e and f, From Antranikian et al., 1987). (g) diagrammatic view of enzyme complexes attached, at defined distances determined by linkers, to a membrane surface, exposing their

cell, may be envisaged as prominent structural components of the eukaryotic cell that may be involved in the working of the cell not only by creating compartments, but also by an ability to modify water structure in a sense discussed above for membranes in the prokaryotic cell (Luby-Phelps, 2000).

A specific protein in the eukaryotic cell, monoamine oxidase (MAO A, MOA B; Figure 6b) shall be described and discussed that appears to be a suitable example for a principally 'soluble' (monotopic) enzyme in the eukaryotic cell that is, nevertheless, membrane-associated (Binda et al., 2002; see above for similar cases in prokaryotes). This enzyme degrades amine neurotransmitters, such as dopamine, norepinephrine, phenylethylamine, and serotonin. It is a well-known target for antidepressant and neuroprotective drugs. Mutation in the genes coding for MOA A or MOA B results in monoamine oxidase deficiency, or Brunner syndrome. The protein is known as a prominent marker enzyme for the outer mitochondial membrane. A dimer of monomers of the protein is attached to that face of the membrane that is directed towards the cytosol by linker structures that are parts of the amino acid chain of the monomers. Though being a marker enzyme for the outer membrane of mitochondria, this enzyme is by no means a typical mitochondrial enzyme. Its function is not part of other functions of the mitochodrium, and the outer mitochondrial membrane is assumed, according to the endosymbiont theory, to be derived from the cytoplasmic membrane of the eukaryotic cell. Regarding the basic topology of the attachment of the enzyme to the mitochondrium, actually the enzyme is formally attached to the former cytoplasmic face of the eukaryotic cell's plasma membrane. As long as no experimental data are available regarding the specific reasons why this principally 'soluble' enzyme is attached to an intracellular 'surface', only speculations can be discussed. In the light of the data presented above on the influence of 'surfaces' on water structure, one idea might be to suggest the specific location of the enzyme to a membrane to be just as favourable, if not more so, than compared with the enzyme's location 'freely' in the cytosol of the cell.

4. CONCLUSION AND PERSPECTIVES

Detailed experimental analyses of possible roles of water structure in microenvironments in the living cell have been widely neglected (Cho et al., 1996; Wheatley, 1991) and should be extended. After all, indications are cumulating that a living

◄

Figure 7. catalytic sites to specifically structured water moieties. It is proposed that two (dense water, less dense water; s. Figure 1a) of the three different water structures depicted in the diagram are brought about by the properties of the membrane. (From Hoppert and Mayer, 1999b). (h) diagrammatic view of the interaction of the cytoplasmic membrane (CM) of a typical eubacterium (depicted here is a Gram-positive bacterium with a thick peptidoglacan layer, P, and a surface-layer, SL) with the bacterial cytoskeleton (CYSK); ST, stalks with terminal knobs (K), connecting the CM with the CYSK; A, additional proteins stabilizing the contact of cytoskeletal fibrils. Note: the diagram also illustrates an interaction of ribosomes (R) with cytoskeletal elements; RP, RNA polymerase. (From Mayer, 2003a,b)

cell is more than just the sum of 'parts' accessible for today's instruments, and that the component 'water' may be of prominent importance. We expect more data supporting our view on the importance of the properties of interfacial water for cell architecture and working. Of special interest might be analyses directed towards disorders and malfunctions in the working of cells that might be connected to so far completely unknown factors influencing the structure of water moieties in the microenvironment of functionally important polypeptides. As already mentioned above, properties such as the state of folding of a protein may be significantly influenced by this microenvironment. It can be speculated that even major known conversions of protein conformation (in prion proteins) might be caused, initiated, mediated or supported by alterations of water structure in the protein's microenvironment. Alterations could be envisaged to take place by – even transient – changes of kinds or concentrations of ions in immediate vicinity of the polypeptide, or by translocation of the protein into a different microenvironment. One might speculate that conversions of this kind could also take place in cases where an energy barrier has to be overcome that protects the protein from being converted. One could also envisage that a protein of this kind is principally 'conditioned' for conversion; conversion does not, however, take place so long as the energy barrier cannot be overcome. Changes, even of a transient kind, in the structure of the interfacial water forming the microenvironment for the protein could lower the energy barrier. This could have the effect that the probability of conversion increases. Once converted, the protein would then not be able to return to its original state of folding due to the energy barrier. Cases are known where conversion of a minor part of a protein population can play the role of a 'seed' for the conversion of the whole population of this kind of protein. It was suggested that this is caused by a 'domino effect' that takes place by induction of an altered water structure around the next protein in very close vicinity of the first one, and so on. We expect that pathological changes in cells might finally be attributed to or mediated by alterations of the structure of interfacial water.

ACKNOWLEDGEMENTS

We thank Mohamed Madkour and Carolin Wichmann for the preparation of several of the figures.

REFERENCES

Antranikian G, Herzberg C, Mayer F, Gottschalk G (1987) Changes in the cell envelope structure of *Clostridium* sp. strain EM1 during massive production of α-amylase and pullulanase. FEMS Microbiol Lett 41:193–197

Beveridge TJ (1995) The periplasmic space and the periplasm in Gram-positive and Gram-negative bacteria. ASM News 61:125–130

Binda C, Newton-Vinson P, Hubalek F, Edmonson DE, Mattevi A (2002) Structure of human monoamine oxidase B, a drug target for the treatment of neurological disorders. Nature Struct Biol 9:22–26

Braks IJ, Hoppert M, Roge S, Mayer F (1994) Structural aspects and immunolocalization of the F420-reducing and non-F420-reducing hydrogenases from Methanobacterium thermoautotrophicum Marburg. J Bacteriol 176:7677–7687

Brown MP, Grillo AO, Boyer M, Royer CA (1999) Probing the role of water in the tryptophan repressor-operator complex. Protein Sci 8:1276–1285

Cho CH, Sing S, Robinson GW (1996) Liquid water and biological systems: The most important problem in science that hardly anyone wants to see solved. Faraday Discuss 103:19–27

Clegg JS, Wheatley DN (1991) Intracellular organization: evolutionary origins and possible consequences of metabolic rate control in vertebrates. Am Zool 31:504–513

Drost-Hansen W (2001) Temperature effects on cell functioning – A critical role for vicinal water. Cell Mol Biol 47:465–483

Eggers DK, Valentine JS (2001) Crowding and hydration effects on protein conformation: A study with sol-gel encapsulated proteins. J Mol Biol 314:911–922

Ellis RJ (2001) Macromolecular crowding: Obvious but underappreciated. Trends Biochem Sci 26:597–604

Etzler FM, Drost-Hansen W (1983) Recent thermodynamic data on vicinal water and a model for their interpretation. Croatia Chemica Acta 56:563–592

Gerberding H, Mayer F (1993) Interaction and compartmentalization of the components of bacterial enzyme systems involved in cell energetics. Z Naturforsch 48c:535–541

Hegermann J, Herrmann R, Mayer F (2002) Cytoskeletal elements in the bacterium *Mycoplasma pneumoniae*. Naturwissenschaften 89:453–458

Hesketh JE, Pryme IF (1991) Interaction between mRNA, ribosomes and the cytoskeleton. Biochem 277:1–10

Holzenburg A, Bewley MC, Wilson FH, Nicholson WV, Ford RC (1993) The three-dimensional structure of photosystem II. Nature 363:470–472

Hoppert M, Mayer F (1990) Electron microscopy of native and artificial methyl-reductase high-molecular weight complexes in strain Göl and Methanococcus voltae. FEBS Lett 267:33–37

Hoppert M, Mayer F (1999a) Principles of macromolecular organization and cell function in bacteria and archaea. Cell Biochem Biophys 31:247–284

Hoppert M, Mayer F (1999b) Prokaryotes. Am Sci 87:518–525

Hoppert M, Braks IJ, Mayer F (1994) Stability and activity of hydrogenases of *Methanobacterium thermoautotrophicum* and *Alcaligenes eutrophus* in reversed micellar systems. FEMS Microbiol Lett 118:249–254

Hoppert M, Mlejnek K, Seiffert B, Mayer F (1997) Activities of microorganisms and enzymes in water-retricted environments: biological activities in aqueous compartments at μm-scale. Instruments, Methods, and Missions for the Investigation of Extraterrestrial Microorganisms, SPIE Proceed. Series 3111:501–509

Khmelnitsky YL, Kabanov AV, Klyachko NL, Levashov AV, Martinek K (1989) Enzymatic catalysis in reversed micelles. In: Structure and reactivity in reversed micelles pp 230–261 Elsevier Amsterdam

Kusalik PG, Svishchev IM (1994) The spatial structure of liquid water. Science 265:1219–1221

Lingl A, Huber H, Stetter KO, Mayer F, Kellermann J, Müller V (2003) Isolation of a complete A_1A_0 ATPsynthase comprising nine subunits from the hyperthermophile *Methanococcus jannaschii*. Extremophiles 7:249–257

Luby-Phelps K (2000) Cytoarchitecture and physical properties of cytoplasm: Volume, viscosity, diffusion, intracellular surface area. Int Rev Cytol 192:189–221

Malone PC, Wheatley DN (1991) Diffusion: a bigger can-of worms? Nature 349:343

Mayer F (1993a) "Compartments" in the bacterial cell and their enzymes. ASM News 59:346–350

Mayer F (1993b) Principles of functional and structural organization in the bacterial cell: "Compartments" and their enzymes. FEMS Microbiol Rev 104:327–346

Mayer F (1999) Cellular and subcellular organization of prokaryotes. In: Lengeler JW, Drews G, Schlegel HG (eds), Biology of the Prokaryotes. Thieme Stuttgart, pp 20–46

Mayer F (2003a) Cytoskeletons in prokaryotes. Cell Biol Int 27:429–438

Mayer F (2003b) Das bakterielle Cytoskelett – Ein aktuelles Problem der Zellbiologie der Prokaryoten. Naturwiss.Rundsch 56:595–605
Mayer F, Jussofie A, Salzmann M, Lübben M, Rohde M, Gottschalk G (1987) Immunoelectron microscopic demonstration of ATPase on the cytoplasmic membrane of the methanogenic bacterium strain Göl. J Bacteriol 169:2307–2309
Mayer F, Rohde M, Salzmann M, Jussofie A, Gottschalk G (1988) The methano-reductosome: A high molecular weight enzyme complex in the methanogenic bacterium strain Göl that contains components of the methylreductase system. J Bacteriol 170:1438–1444
Mayer F, Vogt B, Poc C (1998) Immunoelectron microscopic studies indicate the existence of a cell shape preserving cytoskeleton in prokaryotes. Naturwissenschaften 85:278–282
Minton AP (2001) The influence of macromolecular crowding and macro-molecular confinement on biochemical reactions in physiological media. J Biol Chem 276:10577–10580
Muller N (1988) Is there a region of highly structured water around non-polar solute molecules? J Solution Chem 17:661–672
Ossmer R, Mund T, Hartzell PL, Konheiser U, Kohring GW, Klein A, Wolfe RS, Gottschalk G, Mayer F (1986) Immunocytochemical localization of component C of the methylreductase system in *Methanococcus voltae* and *Methano-bacterium thermoautotrophicum*. Proc Natl Acad Sci USA 83:5789–5792
Parsegian VA, Rau DC (1984) Water near intracellular surfaces. J Cell Biol 99:191–200
Reidlinger J, Mayer F, Müller V (1994) The molecular structure of the Na^+-translocating F_1F_0ATPase of *Acetobacterium woodii* as revealed by electron microscopy resembles that of H^+-translocating ATPases. FEBS Lett 356:17–20
Rupley JA, Careri G (1991) Protein hydration and function. Adv Protein Chem 41:37–172
Schneider AS, Middaugh CR, Oldewurtel MD (1979) Role of bound water in biological membrane structure: Fluorescence and infrared studies. J Supramol Struct 10:265–275
Specka U, Spreinat A, Antranikian G, Mayer F (1991) Immunocytochemical identification and localization of active and inactive α-amylase and pullulanase in cells of *Clostridium thermosulfurogenes* EM1. Appl Environ Microbiol 57:1062–1069
Strambini GB, Gonnelli M (1988) Protein dynamical structure by tryptophan phosphorescence and enzymatic activity in reverse micelles: 1. Liver alcohol dehydrogenase. J Phys Chem 92:2850–2853
Stupperich E, Juza A, Hoppert M, Mayer F (1993) Cloning, sequencing and immunological characterization of the corrinoid-containing N^5-methyltetrahydromethanopterin: coenzyme M methyltransferase from *Methanobacterium thermoautotrophicum*. Eur J Biochem 217:115–121
Sunnerhagen M, Denisov VP, Venu K, Bonvin AM, Carey J, Halle B, Otting G (1998) Water molecules in DNA recognition I: Hydration lifetimes of trp operator DNA in solution measured by NMR spectroscopy. J Mol Biol 282:847–858
Swaminathan R, Hwang CP, Verkman AS (1997) Photobleaching and anisotropy decay of green fluorescent protein GFP-S65T in solutions and cells: Cytoplasmic viscosity probed by GFP translational and rotational diffusion. Curr Genet 40:2–12
Timasheff SN (1993) The control of protein stability and association by weak interactions with water: How do solvents affect these processes? Annu Rev Biophys Biomol Struct 22:67–97
Van den Berg B, Ellis RJ, Dobson CM (1999) Effects of macromolecular crowding on protein folding and aggregation. EMBO J 18:6927–6933
Wheatley DN (1991) Water: Biology's forgotten molecule. Biologist 38:45–49
Wheatley DN (1993a) Water in life. Nature 366:308
Wheatley DN (1993b) Diffusion theory and biology: Its validity and relevance. J Biol Educ 27:181–188
Wheatley DN, Malone PC (1993) Heat conductance, diffusion theory and intracellular metabolic regulation. Biol Cell 79:1–5
Wheatley DN, Clegg JS (1994) What determines the basal metabolic rate of vertebrate cells in vivo? BioSystems 32:83–92
Wheatley DN, Inglis MS, Clegg JS (1984) Dehydration of HeLa S-3 cells by osmotic shock: I. Effects on volume, surface, morphology and proliferate capacity. Mol Physiol 6:163–181

Wichmann C, Naumann PT, Spangenberg O, Konrad M, Mayer F, Hoppert M (2003) Liposomes for microcompartmentation of enzymes and their influence on catalytic activity. Biochem Biophys Res Commun 310:1104–1110

Wiggins PM (1990) Role of water in some biological processes. Microbiol Rev 54:432–449

Wiggins PM (1995) High and low-density water in gels. Prog Polymer Sci 20:1121–1163

Wiggins PM (2001) High and low density intracellular water. Cell Mol Biol 47:735–744

Zimmermann SB, Minton AP (1993) Macromolecular crowding: Biochemical, biophysical, and physiological consequences. Annu Rev Biophys Biomol Struct 22:27–65

CHAPTER 13

DONNAN POTENTIAL IN HYDROGELS OF POLY(METHACRYLIC ACID) AND ITS POTASSIUM SALT

ALEXANDER P. SAFRONOV[1,*], TATYANA F. SHKLYAR[2], VADIM S. BORODIN[2], YELENA A. SMIRNOVA[1], SERGEY YU. SOKOLOV[2], GERALD H. POLLACK[3] AND FELIX A. BLYAKHMAN[2]

[1] *Chemistry Department, Urals State University, 620083, Lenin St. 51, Yekaterinburg, Russia*
[2] *Physics Department, Urals State University, 620083, Lenin St. 51, Yekaterinburg, Russia*
[3] *Department of Bioengineering, Box 357962, University of Washington, Seattle, WA, 98195, USA*

Abstract: Donnan potentials have been measured in polyelectrolyte hydrogels gels of poly(methacrylic acid) and their potassium salts in water, using Ag/AgCl microelectrodes at 298 K. The Donnan potential varied from −80 to −40 mV as a function of gels' cross-link density and the fraction of potassuim methacrylate monomer units. Negative values of the potential increase with the decrease in cross-link density of the gel. Gels with an increasing fraction of potassium methacrylate yield less negative values. The results are discussed from the viewpoint of the Donnan theory initially developed for membrane potential. The theory is qualitatively consistent with observed dependencies of the potential. However several quantitative differences are present, whose sources are analyzed

Keywords: Donnan potential, polyelectrolyte hydrogels

1. INTRODUCTION

In 1911 Donnan studied the conditions under which equilibrium is established between two electrolyte solutions separated by a semi-permeable membrane that prevents the transfer of at least one ionic species. The equilibrium is characterized by

* Corresponding author. Chemistry Department, Urals State University, 620083, Lenin St. 51, Yekaterinburg, Russia; Fax: 007-3432 615 978. *E-mail address*: alexander.safronov@usu.ru.

G. Pollack et al. (eds.), Water and the Cell, 273–284.

an unequal distribution of diffusible ions, which results in a measurable difference in the electric potential between the solutions on each side of the membrane. The nature of the equilibrium and the existence of the potential have both become associated with Donnan's name (Donnan, 1924). Later, it was found that this phenomenon is not restricted only to the presence of semi-permeable membranes, and IUPAC now defines (Compendium of Chemical Terminology, 1997) Donnan equilibrium and potential as the product of any kind of restraint, such as gelation, gravitation, etc., that prevents some ionic components from moving from one phase to the other, but allows other components to do so.

In recent years it has been shown that the Donnan potential exists in hydrogels of synthetic polymers (Guelch et al., 2000). In anionic gels of poly(acrylic acid) and its alkali salts it was shown that the interior of the gel has a negative electric potential relative to the exterior, while in cationic gels of poly(dimethyl,diallyl ammonium chloride) this potential was positive (Guelch et al., 2000; Gao et al., 2003). The existence of such potential is interesting both for applied chemical engineering and fundamental polymer science. In the former case the electrical activity in aqueous gels might be a useful property of 'smart' materials for sensors and other practical applications (Guelch et al., 2000; Osada et al., 2002).

From the fundamental point of view, electrical potentials make gels promising models for some types of living cell phenomena (Pollack 2001). One of the most interesting examples is the membrane potential, which is involved in such vital functions as the nerve impulse and muscle contraction (Alberts et al., 1994). Although such potential is not commonly ascribed to a Donnan potential, it clearly arises from restraints on free ion exchange between exterior and interior of the cell. It was shown that the propagated potential change along the membrane, commonly known as the action potential, is accompanied by a swelling of the cytoskeletal gel beneath the membrane (Tasaki, 1999a). This feature stimulated the hypothesis that the electrical activity of the cell is tightly linked with the fundamental properties of the gel interior, including phase transitions that take place under certain conditions (Tasaki, 1999a, 1999b, 2002).

However, the few experimental works cited above dealt mainly with the electric potential's existence, rather than with the influence of properties of the gel on the potential. In some cases (Guelch et al., 2001) the polyelectrolyle gel was not characterized, i.e., neither its network density, nor the degree of ionization. In other cases (Gao et al., 2003) the potential was measured between gel and water solution of alkali-metal salt. Even though the potential is a manifestation of a Donnan potential, no comparison with predictions of classic Donnan theory (Donnan, 1924) was made.

In the present study we put forward and answer some simple but fundamental questions concerning Donnan potential in polyelectrolyle gels. We study the influence of network density and charge density on the value of Donnan potential and compare the results with predictions of the Donnan theory.

2. METHODS

Gels of poly(methacrylic) acid (PMAA) were made by free-radical polymerization of partially neutralized methacrylic acid with N,N'-methylene-diacrylamide as a cross-linker in aqueous solution. All reagents were purchased from Merck (Schuchardt, Hohenbrunn). In order to provide the series of gels with varying electric charge density, 0, 25, 50, or 75% of methacrylic acid monomers were neutralized by the required amounts of potassium hydroxide before polymerization. In each case the overall monomer (methacrylic acid and its potassium salt) concentration was 2.7 M, while the cross-linker to monomer concentrations were set at 1:100, 1:300, 1:500 in order to make gels with varying network density. Potassium persulphate (0.5 g/l) was used as initiator. Polymerization was carried out in PE probe tubes 10 mm in diameter for 1 h at 80°C and then 3 h at 50°C. After polymerization, gel samples were washed in distilled water to remove the sol fraction and free salts. Water was renewed every day for 3 weeks. The degree of swelling of gel samples was determined by measuring the net weight of the residue after drying.

Since methacrylic acid and its potassium salt have different activity in free radical polymerization, the actual molar fraction of potassium methacrylate monomer units in the gels is different from that in monomer mixture. The former was determined by element analysis. It gave 0, 13, 30, and 47% of potassium methacrylate monomer units relative to 0, 25, 50, and 75% of potassium methacrylate in the monomer mixture.

The experimental equipment for potential measurement was designed around an optical microscope; it contained a thermostatic bath for the gel sample, two micromanipulators for two identical Ag/AgCl glass-micropipette electrodes 1 μm in tip diameter filled with 3 M KCl, one of which penetrated into the swollen gel, the other placed in water outside. (Identical electrodes were used to avoid the possibility of potential shift due to a difference in electrodes.) The potential difference between microelectrodes was measured using an instrumental amplifier on the base of an integrated circuit "INA 129" ('Burr-Brown', USA). The main amplifier parameters are: input impedance – 10^{10} Ohm, frequency bandwidth –0...107 Hz, gain – 50. To reduce the influence of electromagnetic interference on the potential difference measurement, special wire shields were provided around the unit. Typically, the peak to peak noise of the electrical potential was approximately 5 mV.

No salt was added to the system in potential measurements, so as to avoid unanticipated effects on the gel.

3. RESULTS AND DISCUSSION

Figure 1 shows the degree of swelling (α) of gels of different cross-link density, as a function of the fraction (γ) of potassium methacrylate monomer units. We took

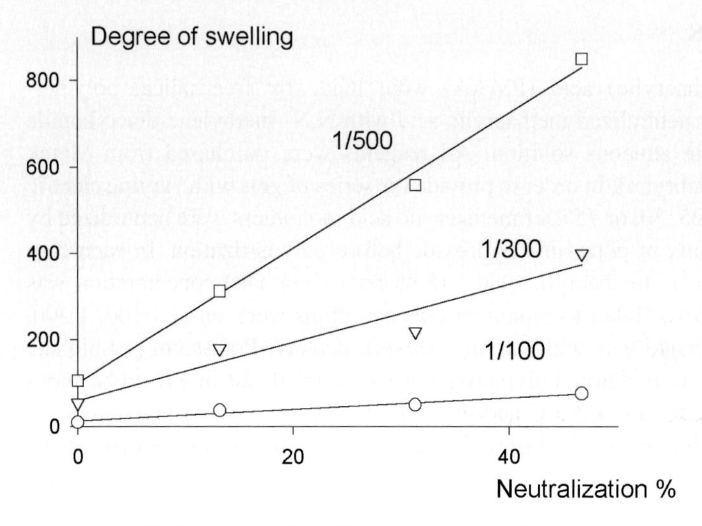

Figure 1. Equilibrium degree of swelling of aqueous gels of poly(methacrylic acid) of different network density on the molar fraction of neutralized residues

cross-linker to monomer ratio as a parameter (ω) of relative network density. The results fit fairly well following linear regressions:

$$\omega = 1/500: \quad \alpha = 85 + 9.91\gamma$$

$$\omega = 1/300: \quad \alpha = 52 + 4.31\gamma$$

$$\omega = 1/100: \quad \alpha = 11 + 0.86\gamma$$

As might be anticipated, the degree of swelling increases with the decrease in cross-link density. It also increases with the fraction of salt residues in the gel network.

When placed in water, both methacrylic acid and potassium methacrylate monomer units tend to dissociate. However, the former is very weak electrolyte: less than 1% of such monomer units dissociate. Meanwhile, all potassium methacrylate monomer units dissociate, providing negative electric charge on the polymer network arising from the carboxylate groups and positively charged potassium counter-ions in the surrounding solution. Therefore, gels of poly(methacrylic acid) were assumed essentially uncharged, while gels with the increasing fraction of potassium methacrylate monomer units had increasing degrees of ionization. The actual charge density on the polymer network cannot be readily evaluated, as it depends not only on the fraction of carboxylate groups, but on the number of condensed counter-ions (i.e., absorbed on the negatively charged chains) which is an implicit function of gel concentration and network density. To be definite, in the present study we considered the fraction of potassium methacrylate monomer units as the relative parameter of the electric charge of the gel network.

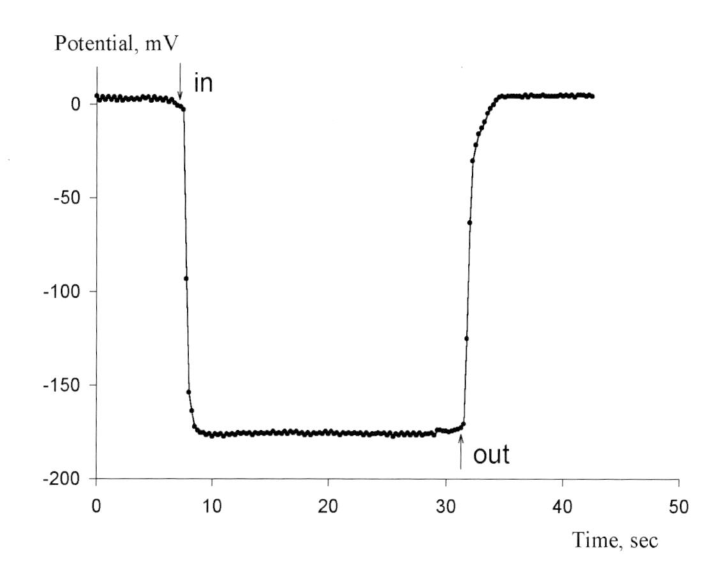

Figure 2. Donnan potential in uncharged PMAA gel with 500 monomer units per cross-link

The increase of the degree of swelling with γ shown in Figure 1 is conventionally known as polyelectrolyte swelling. It stems from multiple reasons, among which the repulsion of the negatively charged chains, osmotic pressure of counter-ions, and enhancement of interaction between water molecules and charged chains are usually mentioned.

Figure 2 shows typical experimental plot of potential measurement on the gel sample. The first horizontal part on the curve is the experimental baseline when both microelectrodes are placed in distilled water outside the gel. Since both microelectrodes were identical, the base-line is very close to zero potential. As one of the microelectrodes (probe) is inserted into the gel, one can see the potential drop. The potential reaches $-176\,\text{mV}$, and remains constant. When the probe taken out of the gel, the potential immediately returns to the baseline. Such measurements were performed repeatedly on each gel sample. Thus, the average values of potential could be evaluated. In order to confirm the equilibrium nature of the measured potential, the experiments for some samples were repeated after several weeks of storage in distilled water. Values of potential remained consistent. For example, the potential of PMMA gels with cross-link density 1/100, 1/300, 1/500 was respectively $-120, -159, -176\,\text{mV}$, initially. After 12 weeks of storage they became $-115, -160, -180$ respectively, a difference that is within the noise level of the measurements.

Thus, we may assume that we have measured the equilibrium potential of poly(methacrylic acid/potassium methacrylate) gels. The negative sign of the potential is in accordance with the gels' polyanionic nature (Guelch et al., 2000). Below we discuss the nature of this potential in detail. However, it is clear that this potential stems from the non-uniform ionic distribution in the swollen gel/distilled

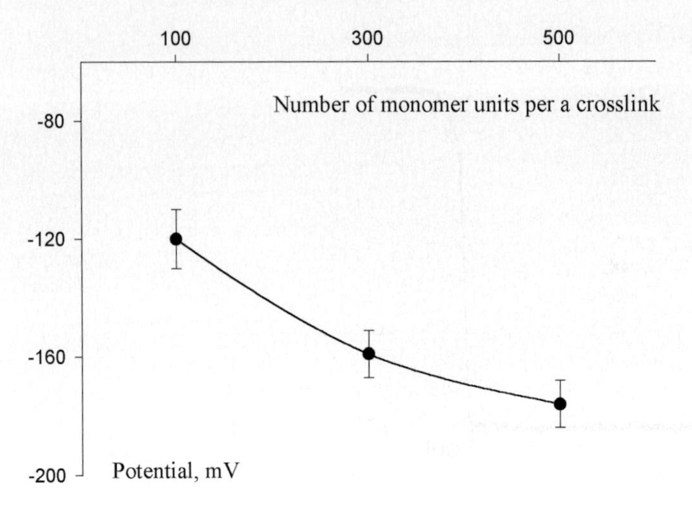

Figure 3. Dependence of Donnan potential in uncharged poly(methacrylic acid) gels on the number of monomer units per a cross-link (n = 22)

water system, and according to cited IUPAC Compendium it should be recognized as a Donnan potential.

Figure 3 represents the dependence of the potential on the cross-link density for the uncharged gels of poly(methacrylic acid) with no salty residues. The value of potential becomes more negative when the number of monomer units per cross-link increases. At first sight it seems odd that weak gels should have larger potential than dense gels. However, this should be analyzed from the viewpoint of Donnan equilibrium in such gels.

Let us consider the binary system: PMAA gel in distilled water. Since there are no other electrolytes, only two ionic equilibria should be taken into account:

(1) $H_2O = H^+ + OH^-$

(2) $M\text{–}COOH = H^+ + M\text{–}COO^-$

where M is a monomer unit in the gel network.

While equilibrium (1) exists both in water and gel phases, equilibrium (2) is restricted only to the gel phase, since carboxylate groups are attached to the network and can not freely move throughout the system. As processes (1) and (2) both are the source of H^+ ions, the concentration of the latter in the gel phase is initially higher than in water, which results in diffusion to the water phase. However, the macroscopic electroneutrality of a phase demands that the diffusion of H^+ ions should be accompanied by the diffusion of free negatively charged ions which can only be OH^-.

The diffusion of OH^- ions results in the difference of their concentration between gel and water phases. Thus, the typical Donnan equilibrium is established when the ionic species are non-uniformly distributed in the system. The condition for equilibrium is the equality of the electrochemical potentials of free ions in coexisting phases:

$$(3) \qquad \mu(H^+)^g + F\varphi^g = \mu(H^+)^w + F\varphi^w$$

$$(4) \qquad \mu(OH^-)^g - F\varphi^g = \mu(OH^-)^w - F\varphi^w$$

where μ – is the chemical potential of an ion in gel (^{g}) and water (^{w}) phases, φ – is the electric potential, F – Faraday number.

The sum of Equation (3) and (4) gives the general condition for Donnan equilibrium:

$$(5) \qquad \mu(H^+)^g + \mu(OH^-)^g = \mu(H^+)^w + \mu(OH^-)^w$$

Further treatment of Equation (6) takes into account the dependence of chemical potential on concentration C_i (more rigorously on the activity) of an ion:

$$(6) \qquad \mu_i = \mu_i^0 + RT\ln C_i$$

Introducing Equation (6) into (5) we get:

$$\mu^0(H^+)^g - \mu^0(H^+)^w + RT\ \ln\frac{[H^+]^g}{[H^+]^w}$$
$$= \mu^0(OH^-)^w - \mu^0(OH^-)^g + RT\ \ln\frac{[OH^-]^w}{[OH^-]^g}$$

$[H^+]$ and $[OH^-]$ being the equilibrium concentrations of ions in corresponding respective phases.

According to the conventional assumption, standard chemical potential of an ion in two water solutions is the same:

$$(7) \qquad \mu^0(H^+)^g = \mu^0(H^+)^w$$

which gives the final equation:

$$(8) \qquad \frac{[H^+]^g}{[H^+]^w} = \frac{[OH^-]^w}{[OH^-]^g}$$

Although Equation (8) is used in the literature (Horkay et al., 2000) to describe Donnan equilibrium in polyelectrolyte gels, its worthwhile to point out that it should not be taken for granted. There is neither theoretical nor experimental evidence that standard chemical potential of an ion in a gel is the same as it is in water, and the validity of Equation (7) is therefore under question.

Assumption (7) is also conventionally made to write the expression for electric potential between coexisting phases. From Equation (3) and (6) it follows:

$$(9) \qquad \varphi^g - \varphi^w = \frac{1}{F}(\mu^0(H^+)^w - \mu^0(H^+)^g) + \frac{RT}{F}\ln\frac{[H^+]^w}{[H^+]^g}$$

Then, using Equation (7) one can finally get for Donnan potential between gel and water:

$$(10) \qquad \Delta\varphi^{g/w} = \frac{RT}{F}\ln\frac{[H^+]^w}{[H^+]^g} = \frac{RT}{F}\ln\frac{[OH^-]^g}{[OH^-]^w}$$

Inside the PMAA gel due to dissociation of carboxylic groups the concentration of H^+ ions is higher than in water, and this provides negative values of $\Delta\varphi^{g/w}$ which are observed in the experiments.

Let us for now accept the conventional assumption (7) and use Equation (10) to analyze the experimental results of Figure 3. The equation can be easily rewritten in common values of pH of the system:

$$(11) \qquad \Delta\varphi^{g/w} = \frac{2.303 \cdot RT}{F}(pH^g - pH^w)$$

While $[H^+]$ in distilled water is invariant $[H^+]$ in the gel depends on the dissociation (2) of carboxylate groups. Since poly(methacrylic acid) is very weak, we may write for the dissociation constant:

$$(12) \qquad K_a = \frac{[H^+][COO^-]}{C_g - [COO^-]} \approx \frac{[H^+]^2}{C_g}$$

where C_g is the molar concentration of polymer in the gel and can be easily evaluated using the equilibrium degree of swelling (Figure 1). Finally one can get the dependence of Donnan potential of the gel on the equilibrium degree of swelling and pK_a of polyacid:

$$(13) \qquad \Delta\varphi^{g/w} = \frac{2.303 \cdot RT}{F}\left(\frac{1}{2}pK_a + \frac{1}{2}\log\frac{\alpha \cdot M}{1000 \cdot d} - pH^w\right)$$

where d is the density of the gel, which is practically equal to the density of water, and M is molar mass of monomer unit.

According to expression (13) negative values of the gel's Donnan potential should decrease with the degree of swelling. This is opposite to the experimental trend presented in Figure 3, taking into account that the degree of swelling is proportional to the cross-link density. At least two possible reasons might be given. First, this might be due to the general invalidity of assumption (7) for gels. Second, it may indicate that pK_a of polymethacrylic acid is not a constant, but is an implicit function of the degree of swelling. Since we have raised questions concerning the classical Donnan consideration let us proceed to analyze the latter possibility. Therefore we can use expression (13) to evaluate the apparent pK_a of PMAA gels with different crosslink density based on the experimentally measured values of Donnan potential. The results are presented in Table 1.

The natural reference level for pK_a of poly(methacrylic acid) is that for hydrated monomer – 2-methylpropionic acid which is 4.85. It is close to the extrapolated

Table 1. Apparent dissociation constant in poly(methacrylic acid) gels with different network density

Average number of monomer units per a cross-link	Equilibrium degree of swelling	Donnan potential, mV	pK_a
100	10.9	−120	9.97
300	51.7	−159	8.22
500	85.0	−176	7.19

results for linear poly(methacrylic acid) in solution at zero degree of ionization (Leyte and Mandel, 1964). However apparent pK_a values for the PMAA gel are much higher than this reference level. Also one can see that the increase in network density results in the increase of apparent pK_a which means that the dissociation of carboxylic groups is lower in the gel than in the monomeric acid and becomes more restrained with the increase in the network density. It is difficult to envision a definite reason for such high values of pK_a. It may be another indication of general invalidity of assumption (7) for gels. The structure of water in the gel phase may be different from bulk water and that would affect the dissociation of carboxylic groups.

Let us now consider the influence of network charge density on the Donnan potential. Figure 4 shows the dependence of the Donnan potential on the fraction of potassium methacrylate monomer units in the network, i.e., degree of ionization.

The negative values of Donnan potential decrease with the increase of network charge density. This is true for gels with any cross-link density. Intuitively one would anticipate the opposite. However, the experimental results are qualitatively consistent with theoretical consideration of the Donnan potential.

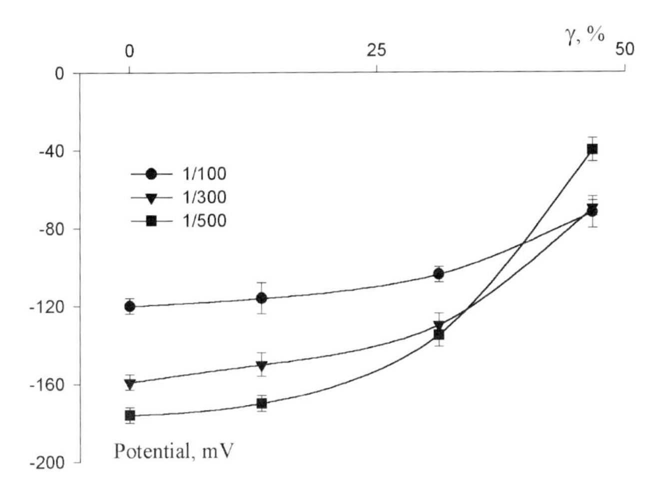

Figure 4. Donnan potential of poly(methacrylic acid) gels with different degree of neutralization

Let us consider PMAA gel with some carboxylic groups neutralized by potassium hydroxide. When some methacrylic acid monomer units in the gel network are replaced by potassium methacrylate units the system corresponds to the solution of a weak acid and its salt. Then we should add the following equilibrium to Equation (1),(2):

$$\text{(14)} \qquad M\text{–}COOK = K^+ + M\text{–}COO^-$$

The general consideration of Donnan potential according Equation (5)–(11) does not change, since conditions of electrochemical equilibrium (3), (4) remain valid. However, the values of $[H^+]^g$ in equation (10) and pH^g in equation (11) will be strongly affected by the dissociation (14) of salty residues. Carboxylate anions that appear in the gel due to the dissociation of potassium methacrylate will strongly suppress the dissociation of acidic residues and decrease the concentration of H^+ ions in the gel. This will result in a decrease in negative values of Donnan potential of the gel according to equation (10). This is qualitatively consistent with experimental results in Figure 4.

It is worthwhile to analyze the experimental values of Donnan potential for the partly charged PMAA gels in a manner similar to that made above for uncharged gels, and estimate the apparent pK_a for methacrylic acid monomer units in the presence of potassium methacrylate monomer units.

The rigorous approach to the dissociation equilibrium of a weak acid in a solution of its salt is well known (Daniels and Alberty, 1961) it includes material balance, condition of macroscopic electroneutrality, and equilibrium constants for equations (1), (2). The result can be taken from reference (Daniels and Alberty, 1961).

$$[H^+] + C_s = \frac{K_a(C_a + C_s)}{K_a + [H^+]} + \frac{K_w}{[H^+]}$$

where C_a and C_s are the concentrations of weak acid and its salt respectively, and K_w is the dissociation constant of water.

One can easily link the concentrations C_a, C_s with the equilibrium degree of swelling of the gel and the fraction of potassium methacrylate residues γ. This gives:

$$\text{(15)} \qquad [H^+] + \frac{1000d \cdot \gamma}{\alpha \cdot M_g} = \frac{K_a \cdot 1000d}{\alpha \cdot M_g(K_a + [H^+])} + \frac{K_w}{[H^+]}$$

where M_g is the average molecular weight of gel monomer units and is related to molecular weights of acidic (M_a and salty residues (M_s):

$$M_g = \gamma M_s + (1 - \gamma) M_a$$

The concentration of H^+ ions in the gel phase can be evaluated according to equation (10) using the experimental values of Donnan potential calculated from the data of Figure 4. Then Equation (15) can be used for the determination of the value of apparent dissociation constant of acidic residues in the partly ionized gel.

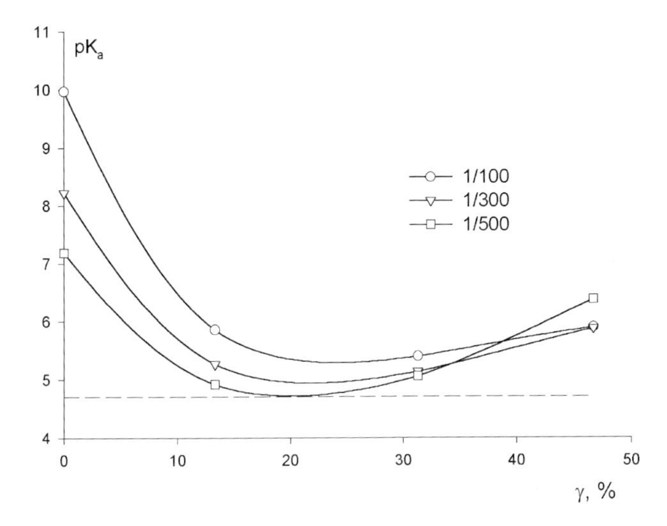

Figure 5. Dependence of apparent pK_a of poly(methacrylic acid) gel on the degree of neutralization

Figure 5 represents the dependence of the apparent value of pK_a on the degree of gel neutralization with different network densities. One can see that all gels show the same behavior: at low degrees of neutralization, pK_a values substantially decrease from relatively high values in the non-neutralized PMMA gel (Table 1) down to the value of pK_a of monomeric carboxylic acid, which is marked by dashed line in the figure. At higher degrees of neutralization, pK_a values slightly increase. The shape of the curve is consistent with the conventional point of view on linear polyacid dissociation at different degrees of neutralization (Morawetz, 1965). The initial descent of the curve, which means the increase of dissociation constant of carboxylic groups, may be associated with the change in gel-water structure due to the increasing number of counterions in the network. At low degrees of ionization the charge density of the chain is not high and carboxylic groups in the gel become rather independent. This makes the apparent pK_a values almost reach that of the monomeric acid. However, when the degree of ionization becomes high, the charge density of the chain increases and suppresses the dissociation of remaining carboxylic groups.

It is not possible to compare the plots of pK_a vs γ for the gel with that for linear PMMA quantitatively because each point at Figure 5 is related to different equilibrium degrees of swelling which corresponds to the different concentrations of polyacid. However the basic trend is the same (Morawetz, 1965).

The influence of network density on pK_a values is more distinct at zero or low degrees of neutralization: more dense gels yield higher values of pK_a in accordance with above-mentioned reasons. However, at high degree of neutralization gels sufficiently swell and the difference between gels with different network density vanishes.

4. CONCLUSIONS

In the present study we have experimentally observed negative electric potentials between swollen PMMA gels and the surrounding water, and have studied the influence of cross-link density and degree of ionization on the value of potential. We used Donnan membrane theory to analyze the experimental results. Despite the fact that no semi-permeable membrane is present in the system, classic Donnan considerations should still remain valid, at least qualitatively. Meanwhile, this approach assumes the condition that the standard chemical potential of water, H^+, and OH^- ions is the same in coexisting phases, which, in general, might be violated in gels due to the additional structuring of water inside the polymer network (see Zheng and Pollack, this volume). We may suppose that this is the reason for several quantitative differences between theory and experiment. However, the unexpected features of electric potential in gels such as the increase in absolute value with decrease in cross-link density and decrease in the degree of ionization have found reasonable theoretical explanation.

REFERENCES

Alberts B, Bray D, Lewis J, Raff M, Roberts K, Watson JD (1994) Molecular Biology of the Cell, 3rd edn, New York: Garland

Daniels F, Alberty RA (1961) Physical Chemistry. New York: Wiley-Interscience

Donnan FG (1924) The theory of membrane equilibria. Chem Rev 1:73–90

(1997) Compendium of Chemical Terminology, 2nd edn, Oxford: Blackwell Scientific Publ

Gao F, Reitz FB, Pollack GH (2003) Potentials in anionic polyelectrolyte hydrogels. J Appl Polym Sci 89:1319–1321

Guelch RW, Holdenried J, Weible A, Wallmersperger T, Kroeplin B (2000) Polyelectrolyte gels in electric fields: A theoretical and experimental approach. Proceedings of the SPIE: Smart Structures and Materials 2000: Electroactive Polymer Actuators and Devices 6/2000 3987:193–202

Guelch RW, Holdenried J, Weible A, Wallmersperger T, Kroeplin B (2001) Electrochemical stimulation and control of electroactive polymer gels. Proceedings of the SPIE: Smart Structures and Materials 2001: Electroactive Polymer Actuators and Devices 7/2001 4329:328–334

Horkay F, Tasaki I, Basser PJ (2000) Osmotic swelling of polyacrylate hydrogels in physiological salt solutions. Biomacromolecules 1:84–90

Leyte JC, Mandel M (1964) Potentiometric behavior of polymethacrylic acid. J Polym Sci Pt A 2:1879–1891

Morawetz H (1965) Macromolecules in solution. Wiley-Interscience: New York.

Osada Y, Gong JP (2002) Electrical behaviors and mechanical responses of polyelectrolyte gels. In: Osada Y, Khokhlov AR (eds), Polymer Gels and Networks. New York: Dekker, pp 177–217

Pollack GH (2001) Cells, gels and the Engines of Life. Seattle: Ebner & Sons

Tasaki I (1999a) Rapid structural changes in nerve fibers and cells associated with their excitation processes. Jpn J Physiol 49:125–138

Tasaki I (1999b) Evidence for phase transition in nerve fibers, cells and synapses. Ferroelectrics 220:305–316

Tasaki I (2002) Spread of discrete structural changes in synthetic polyanionic gel: A model of propagation of a nerve impulse. J Theor Biol 218:497–505

Zheng J-M, Pollack GH (2006) Solute exclusion and potential distribution near hydrophilic surfaces (this volume)

CHAPTER 14

BIOLOGICAL SIGNIFICANCE OF ACTIVE OXYGEN-DEPENDENT PROCESSES IN AQUEOUS SYSTEMS

VLADIMIR L. VOEIKOV*

Department of Bioorganic Chemistry, Faculty of Biology, Lomonosov Moscow State University, 119234, Moscow, Russia

Abstract: Water actively participates in bioenergetics and bioregulation. It is essential for purposeful production of reactive oxygen species (ROS) in cells and extracellular matrix. Due to specific structuring of water it itself may serve the source of free radicals and initiate reactions with their participation. On the other hand water structuring provides for its direct oxidation with oxygen. Processes going on in aqueous systems in which ROS participate are the sources of high grade energy of electronic excitation which is not easily and uselessly dissipated in aqueous milieu of living systems but rather can be accumulated, concentrated, and used as energy of activation for the performance of biochemical reactions. Such processes spontaneously acquire oscillatory character and may serve as pacemakers for biochemical reactions dependent on them. Thus due to its unique structural-dynamic properties water may serve as a transformer of energy from low density to high density form, may accumulate the former and use it for organization and support of vital activity

Keywords: Structured water; Reactive oxygen species; Free radicals; Electronic excitation; Photon emission; Oscillations; Self-organization

1. INTRODUCTION

Albert Szent-Gyorgyi has noted long ago 'The cell is a machine driven by energy. It can thus be approached by studying matter, or by studying energy' (Szent-Gyorgyi, 1968). From the chemical point of view more than 99% of all matter of which cells and intercellular matrix are built is water. Molarity of water is 55 M while molarities

* *E-mail:* vvl@soil.msu.ru

G. Pollack et al. (eds.), Water and the Cell, 285–298.

of even most abundant substances in the internal milieu of an animal, are 2-3 orders of magnitude less. Hence, water should play no less important role in all vital activities than all other biomolecules. Until recently water was neglected as a peer participant in studies of cellular mechanisms, but the situation is changing. More and more evidence appear in favor of the idea that water belonging to living things is an exceptional substance that in unison with low and high molecular weight solids which it embraces determines biological activity at all the levels of organization of living matter.

All the metabolic processes in which living matter participate imply consumption, transformation, or generation of energy. Albert Szent-Gyorgyi was probably the first to claim that 'bioenergetics is but a special aspect of water chemistry' and that '... water arranges an indivisible system with the structure elements (of a cell) making possible electronic excitations which otherwise are highly improbable... in structured water electronic excitation may be surprisingly long-living, and this may be of a paramount importance for the biological energy transfer' (Szent-Gyorgi, 1957).

Energy may be characterized by quantity and by qualities (forms, levels, and orderliness). Levels of energy are subdivided into translational (energy associated with the motion of a molecule in space), rotational and vibrational energy of parts of di- and many-atomic molecules. The highest level of energy relevant to further discussion is energy of electronic excitation (EEE). Levels of energy differ significantly in their density and 'quality' – the higher is the energy level, the more different types of work may be performed by the same quantity of energy and the higher is the efficiency of its utilization. As Mae-Wan Ho noted: 'Life uses the highest grade of energy, the packet or quantum size of which is sufficient to cause specific motion of electrons in the outer orbitals of molecules' (Ho, 1993). However, the idea of the importance of EEE for bioenergetics besides specialized biological functions such as photosynthesis and vision is not still sufficiently absorbed by biological community.

According to the current concept of bioenergetics the overwhelming majority of living organisms gain energy from food burning by oxygen. In a simplified form of this concept specific dehydrogenases abstract 'hot' electrons (plus protons) from "fuel" (sugars and fats) and transfer them to NAD^+ and $NADP^+$. The reduced forms of these carriers donate electrons to the respiratory chain in mitochondria, where their energy is released stepwise and is used for the synthesis of ATP which supports energy requirements of an organism. Oxygen here is the final acceptor (a 'trash box') of electrons that had exhausted most part of their redox potential. As energy portions released in mitochondrial oxidation are equivalent to quanta of middle-far IR-photons ($\leq$0,5 eV, rotational, at most, vibrational energy) this process is analogous to SMOLDERING COMBUSTION. An alternative form of energy gain from oxygen-dependent oxidation is genuine COMBUSTION when direct one-electron oxygen reduction occurs, and quanta of energy equivalent to energy of visible and even UV-photons ($>$1 eV) are generated. One of the classical examples of combustion is direct oxygenation of hydrogen resulting in water production and at which high density energy is released. Combustion, in particular combustion of hydrogen has not been considered until now as relevant for bioenergetics. However, a lot of evidence argues that it should be taken into account as one of the most fundamental processes ensuring vital activity with high grade and well ordered energy. It turns out that water is essential for regular combustion flow in living cells and what is even more surprising, that water can be burnt itself. Here we consider this assertion in more details and try to discuss implication of these newly revealed properties of water for cell physiology.

2. ROS GENERATION IS AN INTRINSIC PROPERTY OF WATER

There is growing understanding that water can not be regarded as some unstructured 'liquid gas'. Many models of water structures are put forward (Bulionkov, 1988; Zenin, 1994; Chaplin, 2000; Maheshwary et al., 2001; see also Chaplin, http). In "real" water structuring is expressed much more than in ideal ultra-pure water because of contribution of multiple interfaces. They include interface between bulk water and walls of a vessel, water/air interface, interfaces with admixtures, etc. Vicinal water with special properties may extend far from the interface which it solvates (Ling, 2003). For example, many layers of structured water extend beyond the protein surface, and induced protein conformational change modifies the extent of non-ideally behaved water (Cameron et al., 1997). Several resilient water molecular layers close to the surface of a solid material immersed in water were detected using atomic force microscope (Jarvis et al., 2000). It was shown by subfemtosecond x-ray absorption spectroscopy that liquid water in a first coordination shell of ice consists of structures with two strong hydrogen bonds of each molecule to its neighbors, resulting in water chains and rings (Wernet et al., 2004). If water contains polymer-like associations, mechanochemical phenomena are expected to take place in it.

Polymers can undergo chemical transformations under the action of mechanical impacts, freezing-thawing and fast temperature variations, action of audible sound and ultrasound, and other low density energy forces that are too weak to induce chemical reactions in monomers or short oligomers. If macromolecules in polymers or their solutions are reluctant to shift along each other due to weak but multiple intermolecular bonds they may accumulate and concentrate mechanical energy to densities that comprise energy quanta enough to excite and break down internal covalent bonds in polymers. That means unpairing of electrons and appearance of a pair of free radicals followed with multiple chemical and physical consequences (Baramboim, 1971).

Basing on the presumption that liquid water contains quazi-polymeric structures the team of Russian physicists headed by G.A. Domrachev started more then 10 years ago to investigate the effects of low density energy physical factors on homolytic water dissociation ($H—O—H \rightarrow HO\bullet + \bullet H$, cf. ionic water dissociation: $H—O—H \rightarrow H^+ + OH^-$). They estimated augmentation of hydrogen peroxide concentration in water because the most probable explanation for its appearance *de novo* is recombination of $HO\bullet$ radicals arising in homolytic water dissociation. It was shown that water freezing-thawing, evaporation-condensation, sonication even with audible sound, filtration through narrow capillaries resulted in an increase of H_2O_2 even in ultra-pure and carefully degassed water. Efficiency of water splitting resulting from evaporation/condensation and freezing/thawing is ~10 times as effective, sonolysis ~70 times and water filtration through narrow capillaries – more than 100 times as effective as its photodissociation with far UV-light (Domrachev et al., 1992, 1993). Yield of H_2O_2 in magnesium sulphate solution (a model of sea water) being in equilibrium with air was much higher than in pure degassed water. What is notable, H_2O_2 concentration continued to grow for some time after resumption of any treatment. About 3% of all energy used for viscous flow of water through capillaries with diameter of 0,2 mkm was used for water splitting.

Japanese authors who were looking for a new way to produce hydrogen by water splitting have shown that powders of NiO, Cu_2O, Fe_3O_4 suspended in distilled water by magnetic stirring, catalytically decompose it into H_2 and O_2. Efficiency of the mechanical-to-chemical energy conversion under these very mild conditions exceeded 4% (Ikeda et al., 1999). Here water splits to the final products because presumably metal oxides instantaneously decompose intermediate peroxides.

In case if a water molecule has dissociated as a mechanically excited polymeric entity:

$$(1) \qquad (H_2O)_n(H_2O\ldots H-|-OH)(H_2O)_m + E \rightarrow (H_2O)_{n+1}(H\downarrow) + (\uparrow OH)(H_2O)_m,$$

the initial products of water splitting are free radicals $H\downarrow$ and $\uparrow OH$ (here we symbolize a given electron as $\uparrow$ or $\downarrow$ to stress their alternative spin states). Indeed, if water is in apparent rest this singlet pair of radicals readily recombine back to water:

$$(2) \qquad H\downarrow + \uparrow OH \rightarrow H_2O$$

However even in such a case this is not just a reverse, equilibrium reaction because water splitting has been achieved under the action of mechanical forces while back recombination of radicals gains an energy quantum of 5,2 eV. In an aqueous system, condensed and organized medium 'electronic excitation may be surprisingly long-living' as A. Szent-Gyorgyi stressed. In fact, long-range energy transfer of electronic and vibrational excitation in water has been demonstrated already in 1930ies-1940ies by J. Perrin, S. Vavilov, Th. Foerster, and others. This phenomenon was confirmed with new techniques recently (Woutersen and Bakker, 1999).

The probability of radicals to move away of each other significantly increases in 'real' water, in which dissolved gases and other molecules and particles are present, especially in cases when multiple layers of water are organized by surfaces which it hydrates and when these layers move along each other with different rates (consider a vortex as an example). This is proved by aforecited data on of the appearance of

H_2O_2 in water filtered through narrow capillaries and H_2 and O_2 in water stirred in the presence of metal oxides. Here the following reactions may proceed:

(3) $HO\uparrow + HO\downarrow \rightarrow H_2O_2$

(4) $H\uparrow + \downarrow H \rightarrow H_2$

(5) $H\bullet + O_2 \rightarrow HO_2\bullet$

(6) $HO_2\uparrow + HO_2\downarrow \rightarrow H_2O_2 + O_2$

(7) $2H_2O_2 \rightarrow 2H_2O + O_2$

The most important of them are the reactions in which oxygen molecules are released ([6] and [7]). It should be reminded that O_2 is unique among other molecules because in its ground state its two electrons are unpaired [$O_2(\uparrow\downarrow)_2\ \uparrow\uparrow$ or $O_2(\uparrow\downarrow)_2\ \downarrow\downarrow$] (besides, an oxygen atom also has two unpaired electrons). Thus, oxygen molecule is a bi-radical (in fact it is a tetra-radical) and it represents a vast store of energy. But it is stable because the laws of quantum physics forbid direct reactions of bi-radicals (they are called also particles in a triplet state) with molecules in which all electrons are paired (singlet state particles). That is why to release its energy reserve oxygen needs to be initially activated.

There are few ways for O_2 activation. It may be excited by an appropriate energy quantum ($\geq 1\,eV$) and turn into a highly reactive singlet oxygen ($O_2(\uparrow\downarrow)_2\ \uparrow\downarrow$, its another symbol, $^1O_2$). A peculiar feature of O_2 is that singlet oxygen may exist only in an electronically excited state from which it may relax only to triplet state. As soon as singlet-triplet transition is "forbidden" by quantum physics laws lifetime of excited singlet oxygen is usually much longer than that of any other molecules in an excited singlet state. Triplet O_2 is also activated by transition metals because in their field its spin state is changed. Finally, triplet oxygen easily reacts with free radicals – atoms and molecular particles possessing an odd number of electrons on their valence orbital. In these reactions oxygen gains or loses an electron, turns into a free radical which can easily take new electrons releasing large portions of energy at each consecutive step of one-electron reduction. Another peculiar feature of free radical reactions in which oxygen participates is that they may easily turn into branching (or run-away) process (Voeikov and Naletov, 1998a), and concentration of free radicals in a reaction mixture grows up exponentially until the rates of their production and annihilation equalize. That is why elevation of H_2O_2 yield in water equilibrated with air in course of its splitting occurs faster, continues for a long time after initial perturbation, and reaches higher levels than in degassed water (Domrachev et al., 1993).

Thereupon it is interesting to speculate that an outcome of water splitting may be significantly influenced by external magnetic fields. There are a lot of reports on the long lasting effects of even a brief treatment of water with magnetic fields, though these effects are not easily reproduced. In principle, magnetic fields may modulate the outcome of free radical reactions. Initial radicals, as mentioned, emerge in a singlet form ($H\uparrow + \downarrow OH$) and they may easily recombine back into water. Under the action of a magnetic field singlet-triplet transition ($H\uparrow + \downarrow OH \rightarrow H\downarrow + \downarrow OH$)

may occur. This prohibits recombination of the radicals favoring the development of the array of reactions 3-7 and others. If water system contains oxygen and some other admixtures development of branching chain reactions in it significantly changes its properties, but as soon as free radical reactions, especially branching chain reactions are highly non-linear, the overall effect should depend drastically upon slight variations of initial conditions.

As it is mentioned above, singlet oxygen belongs to the family of ROS. Recently it was discovered that besides being a source of O_2, water may be directly oxidized with it. This reaction is readily catalyzed *in vitro* by antibodies (immunoglobulins) provided that energy of activation for excitation of molecular oxygen to its singlet state was supplied by dim light illumination of an antibody solution (Wentworth et al., 2000). In other words, antibodies promote water 'burning'. Catalysts do not 'invent' reactions that can not go without them. They organize the reactants in space (and time) so that thermodynamically favorable processes go much faster. Quantum chemical calculations has shown that if two or more water molecules are arranged in space in particular disposition in relation to singlet oxygen and to each other, energy of activation for oxidation of a water molecule with singlet oxygen diminishes to reasonable values and such exotic peroxides as HOOOH, HOOOOH, HOO-HOOO may be produced under mild conditions as intermediates on the way to a more stable H_2O_2 (Xu et al., 2002). Water oxidation goes on very fast in a solution of antibody because its active center provides for the optimal arrangement of water molecules for the process. However, if water is organized in a favorable way by some other means, if singlet oxygen is supplied, for example by the reactions [6] and [7], water oxidation may proceed in aqueous solutions in which water splitting had been initiated. We observed that in the course of branching chain reaction of slow oxidation of amino acids in aqueous solutions initiated with H_2O_2, concentration of H_2O_2 increases to the levels that can be explained only by water oxidation with O_2 (Voeikov et al., 1996). Recently it has been shown that in water containing carbonates and phosphates (Bruskov et al., 2003) or noble gases, such as argon (Voeikov and Khimich, 2002) concentration of H_2O_2 spontaneously increases and its augmentation goes on faster in case of water stirring. Using chemiluminescent methods we also found that such process goes on in aerated mineral waters from natural sources (Voeikov et al., 2003b).

Thus, water – the most abundant substance in any living system, may regularly produce oxygen free radical and another forms of ROS under mild physiological conditions. The fact that a substantial part of organismal water is more or less structured increases the probability of its splitting and oxidation with all the above-listed consequences.

3. COMBUSTION IN A LIVING MATTER

Reactions in which ROS participate has been considered for a long time to be mostly deleterious for cells and tissues, as they may propagate in living matter as chain reactions in which a lot of important bioorganic molecules are corrupted.

They are blamed as a cause of many diseases and as the major cause of aging (Beckman and Ames, 1998). Current studies of oxygen free radicals and other ROS are still to a large extent based on the old idea that they arise as side products of biochemical processes in which oxygen is involved. According to this concept antioxidant enzymes (superoxide dismutase (SOD), catalase, peroxidases) and low molecular weight antioxidants such as ascorbic acid and α-tocopherol should efficiently combat sporadically produced ROS. But when adverse factors induce 'oxidative stress' – excessive ROS generation, antioxidant system can not manage them, and different pathologies arise (Fridovich, 1999).

Recently this seemingly neat theory started to face serious problems. Evidence is accumulating that ROS are purposefully produced in all living organisms. ROS production by cells of immune system in the course of their immune reaction has been known since 1970s (Babior et al., 1973). But only now it became clear, that practically all the cells – animal, plant, cells of unicellular organisms are equipped with the enzymes belonging to NADPH-oxidase family, that directly reduce oxygen. One-electron oxygen reduction is naturally accomplished by many other enzymatic and non-enzymatic mechanisms (Voeikov, 2001).

It turned out that a share of oxygen that undergoes one-electron reduction (actually participates in combustion) is surprisingly high. Any new life begins with egg fertilization, and just after a spermatozoid merges with an ovum oxygen consumption drastically increases. At the cleavage-stage embryos non-mitochondrial respiration accounts for 70% of all oxygen consumption. Only by the stage of blastocysts this share decreases to 30%, however, not at the expense of diminution of direct O_2 reduction, but due to increase in mitochondrial respiration (Trimarchi et al., 2000). Many adult animal organs and tissues use at least 10-15% of all oxygen for generation of ROS (Shoaf et al., 1991), while in intact segments of rat aorta up to 26% of oxygen is directly reduced to superoxide (Souza et al., 2002). White blood cells as well as platelets are actively respiring, and superoxide generation continuously proceeds in whole blood of healthy donors (Voeikov and Novikov, 1997). It is noteworthy that practically all oxygen consumed by neutrophils and eosinophils is one-electronically reduced (Peachman et al., 2001). As mentioned, oxygen is consumed in plasma where immunoglobulins catalyze oxidation of water by it (Wentworth et al., 2000).

What is the purpose of the directional ROS production in living systems? By 1990ies it was already shown that ROS (H_2O_2 and $O_2{\bullet}^-$) regulate carbohydrate and lipid metabolism, poly-ADP-ribosylation, release of calcium from mitochondria, protein kinase and phoshatase activities (Ramasarma, 1990). Now it is known that ROS regulate practically all manifestations of vital processes on molecular, cellular, and tissue levels in all living things (Droge, 2002; Thannickal, and Fanburg, 2000). ROS are shown to have wholesome effects: they promote differentiation of cultured malignant cells into their benign counterparts (Sauer et al., 2001), improve properties of taken out blood (Bocci, 1994), and exercise significant therapeutic effects (Nathan and Cohn, 1981).

Despite the fact that a substantial part of inhaled oxygen is used for ROS generation stationary levels of $O_2^-\bullet$ in cells and tissues do not exceed $10^{-10}-10^{-11}$ M (Niviere and Fontecave, 1995), while that of H_2O_2 in a cell cytoplasm is estimated as $10^{-7}-10^{-9}$ M (Tyler, 1975). ROS are kept at such low levels due to their nearly immediate elimination by the powerful antioxidant system. Thus the probability that ROS may bind specifically to alleged macromolecular 'receptors' except for the enzymes or low molecular weight antioxidants that degrade them is very low. This seems to be puzzling: an organism converts a substantial share of oxygen into ROS and immediately eliminates these particles. How to explain such apparent squandering? And how can these particles, which are so short-lived and practically devoid of *chemical specificity* exercise specific bioregulatory actions?

4. BIOENERGETIC FUNCTIONS OF ELECTRON EXCITED STATES

We suppose that difficulties in comprehension of the real role of ROS in vital activity are related to the attitude to them only as to chemical particles, while they should be considered as participants of continuous flux of oxygen reduction to water: $O_2 + 2H_2 \rightarrow 2H_2O$ (Voeikov, 2001). This reaction consists of several steps:

(8) $4(O_2 + e^- + H^+) \rightarrow 4HO_2\bullet$

(9) $2(HO_2\bullet + HO_2\bullet) \rightarrow 2H_2O_2 + 2O_2$

(10) $H_2O_2 + H_2O_2 \rightarrow 2H_2O + O_2$

----

(11) $4O_2 + 4e^- + 4H^+ \rightarrow 2H_2O + 3O_2$

From such a notation of oxygen reduction (though we could not find similar notation in available literature) several important conclusions follow. First, if oxygen excess over the electrons that reduce it is less than 4-fold, combustion does not go to a final point, and intermediate ROS accumulate, which may initiate chain reactions with bioorganic molecules. Thus, an adequate supply of oxygen is necessary for maintaining low stationary level of ROS and other free radical particles. Second, all these reactions imply recombination of unpaired electrons. This applies also to a reaction [10] where one H_2O_2 molecule may be considered as an electron donor and another as an electron acceptor. Third, all these reactions are sources of energy quanta equivalent to electronic excitation energy. Energy yield in the reaction of dismutation of two superoxide radicals is $\sim$22 kcal/mol, equal to the energy gap between triplet and excited singlet states of oxygen and equivalent to a near IR-photon ($\lambda \sim 1269$ nm). When two singlet oxygen particles transit to triplet state simultaneously, EEE may be 'pooled' and a doubled quantum of energy (equivalent to $\lambda \sim 635$ nm, red light) is released (Cadenas and Sies, 1984). Decomposition of two molecules of H_2O_2 donates an equivalent of 2 eV or $\lambda < 610$nm. When dismutation of $HO_2\bullet$ (reaction [9]) is catalyzed by SOD or decomposition of H_2O_2 [10] is performed by catalase, quanta of high density energy should be generated with some megahertz frequencies due to very high turnover numbers of these two enzymes. This prevents energy from its immediate dissipation into heat and is favorable for energy pooling to even higher quanta.

A key role of EEE and related photon emission in the regulation of vital processes was discovered 80 years ago by A.G. Gurwitsch in the form of the so-called 'mitogenetic radiation' –ultra-weak photon emission in the UV-range of EM-spectrum responsible for triggering cell division (Gurwitsch and Gurwitsch, 1943). This radiation is emitted not only by living cells and tissues, but also by enzymatic (hydrolytic and glycolytic) and chemical reactions including gel-sol transitions in aqueous media. Water splitting and accessibility of active oxygen is a prerequisite condition for the emergence of this radiation (Voeikov, 2003). Ultra-weak photon emission in the range from UV- to near IR of electromagnetic spectrum from living cells and chemical reactions in aqueous media (Slawinski, 1988) affect activity of enzymes (Cilento, 1988), activity and morphology of cells and tissues (Galantsev et al., 1993), regulate locomotion and mutual orientation of cultured cells (Albrect-Buehler, 1995). Back reflected photons emitted during respiratory burst in human blood affect the intensity of this immune reaction by a feed-back mechanism (Voeikov et al., 2003a).

In our opinion regulatory role of ROS is provided by the unique feature of reactions with their participation – generation of electronic excitation energy (EEE) that continuously pumps biophotonic fields of living systems. But if reactions with ROS participation play such a versatile role, they should proceed in all living things including those that are considered to be anaerobic. Indeed, even obligate anaerobic bacteria are equipped with SOD (Hewitt and Morris, 1975) indicating that ROS appear even when molecular O_2 concentration in water is negligibly low. However, the intrinsic property of water to produce oxygen radicals due to its splitting makes their appearance in liquid water practically inevitable.

5. OSCILLATORY NATURE OF ACTIVE OXYGEN DEPENDENT REACTIONS IN AQUEOUS SYSTEMS

Besides serving a role of a source of 'the highest grade of energy, the packet or quantum size of which is sufficient to cause specific motion of electrons in the outer orbitals of molecules' (Ho, 1993), processes in which EEE is generated going on in aqueous systems may automatically acquire oscillatory character and may serve as pacemakers for biochemical reactions dependent on them.

Arousal of ROS in reactions going by in water and generation of EEE provides for the involving of other substances such as nitrogen and carbon dioxide into the process. They may beget amine and carbonyl compounds, and when concentrations of the latter exceed certain thresholds amino-carbonyl (Maillard) reaction develops. In this reaction biologically significant heterocyclic, aromatic, polymeric substances appear (Namiki et al., 1973). Some of them activate oxygen resulting in ROS production and generation of EEE (Voeikov and Naletov, 1998b). We found that profound oscillations of photon emission (Voeikov et al., 2001a) and redox potential (Voeikov et al., 2001b) emerge in Maillard reaction. Oscillations last for many hours and even days and their periods extend from fractions of minutes to tens of minutes. Amplitudes of redox potential variations may reach 0,3 V (from −0,2V to −0,5V).

Intensity of photon emission and amplitude of their oscillations and intensity of oscillations of redox potential as well as periods and phases of oscillations differ in different parts of the reaction system. Mean values of redox potential near the bottom are much more negative and amplitudes of their oscillations are higher than near the air/water interface. On the other hand, the most intense photon emission modulated with profound oscillations comes from the part of the reaction system close to the water/air interface. Here oscillations of photon emission and redox potential are highly correlated. Thus, the bottom part of a reaction system is the source of electrons that reduce oxygen incoming from the air.

High redox potential differences between different parts of the system can not be explained only from uneven distribution of reduced and oxidized forms of organic components because of their low concentrations (few tens of millimolar). It is interesting to speculate that these differences reflect gross changes in reduction and oxidation state of aqueous medium itself.

What is the primary cause of the development of oscillations of ROS production and oscillations of EEE generation? Our experimental data indicates that generation of EEE in reactions with ROS participation is prerequisite for self-organization observed as these processes develop. Initial building up of EEE fosters oxidation and oxygenation of available substrates resulting in an exhaust of dissolved oxygen and accumulation of reducing (easily oxidizable) equivalents. Oxygen continues to diffuse into the system from the air and when its concentration and concentration of reducers reach optimal ratio, a new wave of burning appear followed with the next oxygen depletion until the concentration of diffusing oxygen reaches a threshold value again. Thus, oscillatory behavior naturally emerges in such systems.

It is notable, that oxygen consumption in single neutrophils and other cells that reduce it to ROS using NADPH-oxidase exhibits multimode oscillatory patterns of ROS generation (Kindzelskii and Petty, 2002). Some hormones influence the amplitude of these oscillations, other affect their frequency. In other words, both deepness of respiration of single cells and its rate are related to their functional activity. Respiration rate and deepness (especially in case when oxygen consumption is realized through it one-electron reduction) define in their turn downstream regulatory processes.

Oscillatory behavior is characteristic not only of single cells, but of their populations as well. We observed pronounced oscillations of photon emission from neutrophil suspensions containing hundreds of thousands of cells and even in whole blood, indication of a collective behavior of these big groups of cells related to metabolism of ROS in them (Voeikov et al., 2003a).

Amino-carbonyl reaction proceeding in aqueous systems in which oscillations and waves spontaneously emerge is, in our opinion, the simplest model of arousal

and performance of the respiratory process. Such conditions for the emergence of oscillations of EEE are common to all cells. A steep oxygen gradient between a metabolizing cell and its environment exists. Oxygen is poorly soluble in water, and what is more important, its delivery to a cell may be regulated by interfacial water at a cell-environment boundary. Due to cellular metabolic activity reducing equivalents (e.g., NAD(P)H) accumulate in it. When the ratio of these equivalents to incoming O_2 reaches threshold values energy discharge primarily in the form of EEE occurs. Oxygen is rapidly exhausted, and released energy is directed for metabolic needs. That oxygen in fact taken by single cells in an oscillatory mode has been indeed recently experimentally demonstrated (Porterfield et al., 2000). Oscillations of EEE may play the role of pacemakers for the processes going on on different levels of biological organization. On the other hand oscillatory nature of all these processes provides them the properties of sensible receptors for external electromagnetic and other physical fields.

6. RESPIRATION CYCLE OF WATER: A HYPOTHESIS

It seems trivial that respiration as we know it is a cyclic process. Though it is not so obvious that respiration at a level of a single cell should also be cyclic, experimental evidence supports this conclusion. It can be suggested that cyclic nature of respiration emerges on the one hand from the spatial relationship of oxygen consuming system and its environment and on the other from the orderliness of energy fluxes and high density of energy (EEE) that is generated in the course of oxygen-dependent processes in which ROS participate. Taking into consideration that all the aforementioned phenomena occurred in aqueous systems and that ROS generation is the intrinsic property of water we suggest a hypothesis of the existence of the ‘respiratory cycle of water’. Splitting of water molecules under the action of low density energy (mechano-chemical or mechano-catalytic water decomposition) results in the appearance of oxygen and hydrogen in aqueous systems:

$$(12) \qquad 8H_2O \rightarrow 4O_2 + 16H\bullet$$

Four hydrogen atoms ($H\bullet$) are needed for complete reduction of one oxygen molecule, the rest hydrogen atoms recombine to H_2 molecules: $12\ H\bullet \rightarrow 6H_2 \Uparrow + n(h\nu)$. EEE released may be used, for example, for excitation of oxygen with the appearance of singlet oxygen, for sustaining of an aqueous system in a non-equilibrium, excited state, etc. This sequence of events may be by convention defined as an ‘exhale’ stage because water splitting is accompanied with gas (hydrogen) release.

What may follow afterwards is analogous to an ‘inhale’ stage, as here oxygen is consumed. We remind that for the complete reduction of oxygen molecule a 4-fold excess of oxygen is needed:

$$(13) \qquad 4O_2 + 4H\bullet \rightarrow 2H_2O + 3O_2 + m(h\nu)$$

Energy released in the course of the reactions [12] and [13] is enough to excite oxygen to a singlet state, and under appropriate conditions $^1O_2$ may go on water oxidation:

$$(14) \quad 3O_2^* + 6H_2O \rightarrow 6H_2O_2$$

'Respiration cycle of water' allows to transform low density energy (freezing-thawing, evaporation-condensation, energy of sound, energy of shearing forces of water filtration or its vortexing) into a high density one; at least some part of the latter may accumulate in water in the form of metastable substances such as H_2O_2 and other peroxides as well as in long-living water excitation making it an active physical medium.

As other gases and substances that are present in 'real' water should get involved in the process, respiration cycle should be considered not as a closed loop, but rather as a single convolution of an untwisting helix. Real processes proceeding in water should significantly depend upon the presence of positive and negative catalysts of particular reactions, of substances affecting water structure, upon the nature of interfaces that it solvates, upon the action of external physical factors and fields. Studies of phenomena related to water may help in solving many practical problems of medicine, agriculture, environmental problems, in providing people with healthy drinking water, in optimization of technologies in which water is important.

REFERENCES

Albrect-Buehler G (1995) Changes of cell behavior by near-infrared signals. Cell Motil Cytoskeleton 32:299–304

Babior BM, Kipnes RS, Cumitte JT (1973) Biological defense mechanisms: The production by leucocytes of superoxide, a potential antibactericidal agent. J Clin Invest 52:741–744

Baramboim NK (1971) Mechanochemistry of High Molecular Weight Compounds. Moscow, Chimiya

Beckman KB, Ames BN (1998) The free radical theory of aging matures. Physiol Rev 78:547–581

Bocci V (1994) Autohemotherapy after treatment of blood with ozone. A reappraisal. J Int Med Res 22:131–144

Bruskov VI, Chernikov AV, Gudkov SV, Masalimov ZhK (2003) Activation of reducing properties of anions in sea water under the action of heat. Biofizika 48:1022–1029

Bulionkov NA (1988) Periodic dissipative-module structures of"bound water" – possible constructions defining biopolymere conformations in structures of their hydrates. Krystallografia, Moscow 35:155–159

Cadenas E, Sies H (1984) Low-level chemiluminescence as an indicator of singlet molecular oxygen in biological systems.Methods Enzymol 105:221–231

Cameron IL, Kanal KM, Keener CR, Fullerton GD (1997) A mechanistic view of the non-ideal osmotic and motional behavior ofintracellular water. Cell Biol Int 21:99–113

Chaplin M Available via http://www.martin.chaplin.btinternet.co.uk/index.html

Chaplin MF (2000) A proposal for the structuring of water. Biophys Chem 83:211–221

Cilento G (1988) Photobiochemistry without light. Experientia 44:572–576

Domrachev GA, Rodigin YuL, Selivanovsky DA (1992) Role of sound and liquid water as dynamically unstable polymeric system in mechano-chemically activated processes of oxygen production onEarth. J Phys Chem 66:851–855

Domrachev GA, Roldigin GA, Selivanovsky DA (1993)Mechano-chemically activated water dissociation in a liquid phase.Proc Russ Acad Sci 329:258–265

Droge W (2002) Free radicals in the physiological control of cellfunction. Physiol Rev 82:47–95

Fridovich I (1999) Fundamental aspects of reactive oxygen species,or what's the matter with oxygen? Ann NY Acad Sci 893:13–18

Galantsev VP, Kovalenko SG, Moltchanov AA, Prutskov VI (1993) Lipid peroxidation, low-level chemiluminescence and regulation of secretion in the mammary gland. Experientia 49:870–875

Gurwitsch AG, Gurwitsch LD (1943) Twenty years of mitogenetic radiation: Emergence, development, and perspectives. Uspekhi Sovremennoi Biologii 16:305–334 (English translation: 21st Century-Science and Technology Fall, 1999; 12:41–53)

Hewitt J, Morris J (1975) Superoxide dismutase in some obligately anaerobic bacteria. FEBS Lett 50:315–318

Ho M-W (1993) The Rainbow and the Worm. The Physics of Organisms.World Scientific. Singapore, p 70

Ikeda S, Takata T, Komoda M, Hara M, Kondo JN, Domen K, Tanaka A,Hosono H, Kawazoe H (1999) Mechano-catalysis—a novel method for overall water splitting. Phys Chem Chem Phys 1:4485–4491

Jarvis SP, Uchihashi T, Ishida T, Tokumoto H (2000) Local solvation shell measurement in water using a carbon nanotube probe. J Phys Chem B 104:6091–6094

Kindzelskii AL, Petty HR (2002) Apparent role of traveling metabolic waves in oxidant release by living neutrophils. Proc Natl Acad Sci USA 99:9207–9212

Ling GN (2003) A new theoretical foundation for the polarized-oriented multilayer theory of cell water and forinanimate systems demonstrating long-range dynamic structuring of water. Physiol Chem Phys Med NMR 35:91–130

Maheshwary S, Patel N, Sathyamurthy N, Kulkarni AD, Gadre SR (2001) Structure and stability of water clusters $(H_2O)_n$, n = 8–20: An ab initio investigation. J Phys Chem-A 105:10525–10537

Namiki M, Hayashi T, Kawakishi S (1973) Free radicals developed in the amino-carbonyl reaction of sugars with amino acids. Agric Biol Chem 37:2935–2937

Nathan CF, Cohn ZA (1981) Antitumor effects of hydrogen peroxide in vivo. J Exp Med 154:1539–1558

Niviere V, Fontecave M (1995) Biological sources of reduced oxygen species. Analysis of free radicals in biological systems. In: Favier AE, Cadet J, Kalyanaraman B (eds), Birkhauser: Basel, Boston, Berlin, pp 11–19

Peachman KK, Lyles DS, Bass DA (2001) Mitochondria in eosinophils:Functional role in apoptosis but not respiration. Proc Natl Acad Sci USA 98:1717–22

Porterfield DM, Corkey RF, Sanger RH, Tornheim K, Smith PJS,Corkey BE (2000) Oxygen consumption oscillates in single clonal pancreatic - cells (HIT). Diabetes 49:1511–1516

Ramasarma T (1990) H_2O_2 has a role in cellular regulation. Indian J Biochem Biophys 27:269–274

Sauer H, Wartenberg M, Hescheler J (2001) Reactive oxygen species as intracellular messengers during cell growth and differentiation. Cell Physiol Biochem 11:173–186

Shoaf AR, Shaikh AU, Harbison RD, Hinojosa O (1991) Extraction and analysis of superoxide free radicals $(.O_2\text{-})$ from whole mammalian liver. J Biolumin Chemilumin 6:87–96

Slawinski J (1988) Luminescence research and its relation toultraweak cell radiation. Experientia 44:559–571

Souza HP, Liu X, Samouilov A, Kuppusamy P, Laurindo FR, Zweier JL (2002) Quantitation of superoxide generation and substrate utilization by vascular NAD(P)H oxidase. Am J Physiol Heart Circ Physiol 282:H466–H474

Szent-Gyorgi A (1957) Bioenergetics Academic Press. New York [Back translation from a Russian edition of the book: GIZ Fiz-Mat.Literature, Moscow, 1960, pp 54–56]

Szent-Gyorgyi A (1968) Bioelectronics: A Study in Cellular Regulations, Defense, and Cancer. Academic Press, New York, p 4

Thannickal VJ, Fanburg BL (2000) Reactive oxygen species in cell signaling. Am J Physiol Lung Cell Mol Physiol 279:L1005–L1028

Trimarchi JR, Liu L, Porterfield DM, Smith PJ, Keefe DL (2000) Oxidative phosphorylation-dependent and -independent oxygen consumption by individual preimplantation mouse embryos. Biol Reprod 62:1866–1874

Tyler DD (1975) Polarographic assay and intracellular distribution of superoxide dismutase in rat liver. Biochem J 147:493–504

Voeikov VL (2001) Reactive oxygen species, water, photons, and life. Riv Biol Biology Forum 94:193–214

Voeikov VL (2003) Mitogenetic radiation, biophotons, and non-linear oxidative processes in aqueous media. In: Popp F-A Beloussov L (eds), Integrative Biopysics Biophotonics Kluwer Academic Publishers, Dordrecht/Boston/London, pp 331–360

Voeikov VL, Novikov CN (1997) Peculiarities of luminol- and lucigenin photon emission from nondiluted human blood. SPIE proceedings. In: Benaron DA, Chance B, Ferrari M (eds), Photon Propagation in Tissues III, vol 3194, Italy, San Remo, pp 328–333

Voeikov VL, Naletov VI (1998a) Weak photon emission of non-linear chemical reactions of amino acids and sugars in aqueous solutions. In: Chang J-J, Fisch J, Popp F-A (eds), Biophotons. Kluwer Academic Publishers. Dortrecht, The Netherlands, pp 93–108

Voeikov VL, Naletov VI (1998b) Chemiluminescence development after initiation of maillard reaction in aqueous solutions of glycine and glucose: Nonlinearity of the process and cooperative properties of the reaction system SPIE Proceedings. In: Priezzhev AV, Asakura T, Bries JD (eds), Optical Diagnostics of Biological Fluids III. San Jose, CA, Vol 3252, pp 140–148

Voeikov VL, Khimich MV (2002) Amplification by argon of luminol-dependent chemiluminescence in aqueous NaCl/H_2O_2 solutions. Biofizika 47:5–11

Voeikov VL, Koldunov VV, Kononov DS (2001a) Long-duration oscillations of chemi-luminescence during the amino-carbonyl reaction in aqueous solutions. Russ J Phys Chem 75:1443–1448

Voeikov VL, Koldunov VV, Kononov DS (2001b) New oscillatory process in aqueous solutions of compounds containing carbonyl andamino groups. Kinetics and Catalysis (Moscow) 42:606–609

Voeikov VL, Baskakov IV, Kafkialias K, Naletov VI (1996) Initiation of degenerate-branched chain reaction of glycin deamination with ultraweak UV irradiation or hydrogen peroxide.Russ J Bioorganic Chemistry 22:35–42

Voeikov VL, Asfaramov R, Bouravleva EV, Novikov CN, Vilenskaya ND(2003a) Biophoton research in blood reveals its holistic properties. Indian J Exp Biol 43:473–482

Voeikov VL, Asfaramov R, Koldunov V, Kononov D, Novikov C, Vilenskaya N (2003b) Chemiluminescent analysis reveals spontaneous oxygen-dependent accumulation of high density energy in natural waters. Clin Lab 49:569

Wentworth AD, Jones LH, Wentworth P Jr, Janda KD, Lerner RA (2000) Antibodies have the intrinsic capacity to destroy antigens. Proc Natl Acad Sci USA 97:10930–10935

Wernet Ph, Nordlund D, Bergmann U, Cavalleri M, Odelius M, Ogasawara H, Naslund LA, Hirsch TK, Ojamae L, Glatzel PL, Pettersson GM, Nilsson A (2004) The structure of the first coordination shell inliquid water. Science 304:995–999

Woutersen S, Bakker HJ (1999) Resonant intermolecular transfer of vibrational energy in liquid water. Nature 402:507–509

Xu X, Muller RP, Goddard WA 3rd (2002) The gas phase reaction of singlet dioxygen with water: A water-catalyzed mechanism. Proc Natl Acad Sci USA 99:3376–3381

Zenin SV (1994) Hydrophobic model of water molecules associates. Zhurnal Fizicheskoi Himii 68:634–641

CHAPTER 15

THE COMPREHENSIVE EXPERIMENTAL RESEARCH ON THE AUTOTHIXOTROPY OF WATER

BOHUMIL VYBÍRAL

University of Hradec Králové, Rokitanského 62, 500 09 Hradec Králové, Czech Republic, e-mail: bohumil.vybiral@uhk.cz

Abstract: A description is presented of the research on recently observed phenomenon that we call "the autothixotropy of water". The phenomenon is very weak on macroscopic scale and it appears only if the water is standing still for a certain time. It causes a force of mechanic resistance against an immersed body, arising when it should change its position. Both static and dynamic methods are used: With a static method a moment of force, necessary for a very prominent turn of a stainless steel plate, hung up on a thin filament and immersed in standing water, is measured. With a given angular torsion of the filament, a certain moment of force is reached a (state of stress reaches a critical value) in the water, which is demonstrated by impressive changing of angular position of the plate. When a dynamic methods were used, both oscillations of the plate and a very slow fall of a small ball in standing water were observed. The autothixotropy of water can be explained by a hypothesis of cluster formation by H_2O molecules in standing water. As the phenomenon of autothixotropy does not appear in deionized water, a conclusion can be preliminary drawn, that the phenomenon is determined by a presence of ions in the water

Keywords: Autothixotropy of water; Clusters of water molecules; Standing water; Hysteresis; Deionized water

1. INTRODUCTION

The presented paper follows an article (Vybíral, Voráček, 2003) describing qualitatively recently observed phenomena in clean distilled air-free water, which are an indication of existence of the properties of the water, we call *'autothixotropy of water'*. The static and dynamic behavior of a plate hung up on an elastic filament and immersed in the water was described qualitatively, and the phenomena were hypothetically explained in (Vybíral, Voráček, 2003) through clusters creating chains and networks, made of water molecules.

G. Pollack et al. (eds.), Water and the Cell, 299–314.

This report presents results of experiments with macroscopic bodies, i.e., a flat plate or a little ball, with the help of which some mechanical properties of the clusters and time sequence of their origination were measured. It is evident from the data that 'fresh water' and 'old water' (i.e., 'standing water') behave in different ways. The notion 'autothixotropy' was used, because the described phenomena disappear or decrease considerably by intensive water stirring or boiling. Afterwards, they re-appear spontaneously when the water is left unmoved. Since the phenomenon of autothixotropy is not present in deionized water, we believe it may be determined by the presence of ions in the water.

One can expect that the described water structure can be important in biophysics for description and influence on cell characteristics (see e.g., Pollack, 2001).

It is especially important that two different and mutually independent methods were used for experimental research on the autothixotropy of the water: the static method of torsion described in (Vybíral, Voráček, 2003), as well as dynamic methods (method of torsion oscillations and a method of motion of a ball with laminar flow around it). The basic results of this paper were published on November 11th 2004 at the conference *New Trends in Physics – NTF 2004* in Brno (Vybíral, 2004).

2. STATIC METHOD OF TORSION

2.1 The Principle of the Method

Let us consider a plate made of an indifferent material (for example stainless steel) that is hung up on an elastic filament with torsional rigidity k_τ, and immerse it into the observed clean water (Figure 1). If we twist the upper end of the filament by an angle φ_u, we expect that at the steady state of the water, and in case of an ideal fluid model, the plate will follow the rotation (i.e., an angle φ_d of the bottom end of the filament), so that $\varphi_d = \varphi_u$. According to experiments, carried out already in 1991, this equality – as a consequence of the autothixotropy of water – was not achieved. In a static experiment the increasing consequent changes of angle φ_d, during a very slow change of angle φ_u made step by step, is observed.

With use of the measured angles φ_d and φ_u, we can specify the moment of force M_w, with which the plate influences the water:

$$(1) \qquad M_w = k_\tau (\varphi_u - \varphi_d)$$

If the angle φ_d reaches a *critical value* $(\varphi_d)_{crit.}$, the plate goes into a quick rotation. Due to the moment of force, the deformation of clusters probably reaches such a size that a movement of clusters or their re-modeling is started.

From the dependence of the moment of force (1) on the angle φ_d of plate rotation (if the rotation is reversible), it is possible to determine the torsional rigidity

$$(2) \qquad k_w = \frac{\Delta M_w}{\Delta \varphi_d} = k_\tau \left(\frac{\Delta \varphi_u}{\Delta \varphi_d} - 1 \right)$$

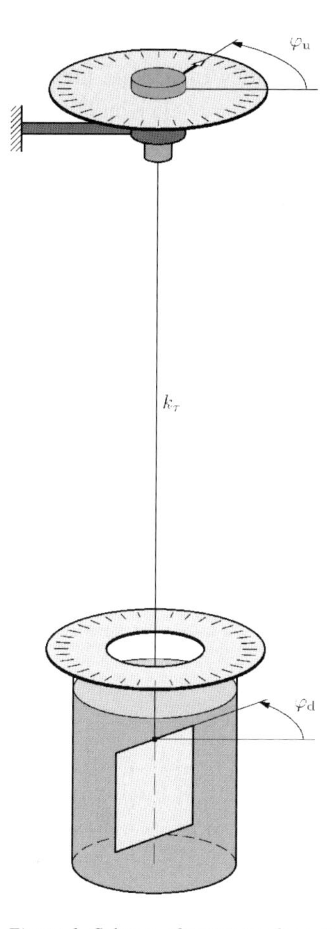

Figure 1. Scheme of apparatus for a static method of measurement

For the plate we used, with this quantity we can characterize the additional elasticity of clusters of the water as a consequence of its autothixotropy (or elasticity of clusters of water molecules).

2.2 Torsional Rigidity of the Filament

For the experiments a first-class phosphor-bronze filament were used with a cross-section of $(0.20 \times 0.025)\,\mathrm{mm}^2$, usually applied in the clock industry. It was proved to be useful in some other experiments, as well (Vybíral, 1987). Nevertheless, it was necessary first to determine exactly its torsional rigidity k_τ. The theory of elasticity, however, does not provide exact results for non-circular sections (see e.g., Dubells, 1956), and moreover, we did not know exactly the shear modulus G for the alloy used. But the linear dependence of torsion angle φ on the moment of force is sufficiently valid also for rectangular sections (see e.g., Dubells, 1956).

The torsional rigidity of the filament used (with length $L = 465$ mm) was determined in an experimental way from torsion oscillations of a plate hung in non-agitated air. On the stainless steel flat plate with mass $m_d = 33.95$ g and dimensions $a = 48.80$ mm, $b = 58.50$ mm, and thickness 1.52 mm, there was lightly and symmetrically glued a straight wire in a horizontal direction. The wire with mass $m_t = 3.92$ g and length $l_t = 153.6$ mm was used for hanging small additional weights. It was possible to hang two small weights symmetrically on the wire at a mutual distance $l = 148.1$ mm, each one with a mass of $m = 10.00$ g. The moment of inertia relative to the rotation axis of the plate with the wire was

$$I_1 = \frac{1}{12}\left[m_d(a^2+b^2)+m_t l_t^2\right] = 2.413 \times 10^{-5}\text{kg.m}^2$$

which was increased with added weights up to

$$I_2 = I_1 + \frac{ml^2}{2}$$

For the angular frequency of free damped oscillations of the plate without weights and with the weights, it is valid that

$$\omega_1^2 = \frac{k_\tau}{I_1} - \delta_a^2, \quad \omega_2^2 = \frac{2k_\tau}{2I_1 + ml^2} - \delta_a^2$$

respectively, supposing that the viscous damping coefficient δ_a of air does not change for small weight dimensions. Then the torsional rigidity of the filament is

$$k_\tau = 4\pi^2 I_1 \left(\frac{2I_1}{ml^2}+1\right)\left(\frac{1}{T_1^2} - \frac{1}{T_2^2}\right) \tag{3}$$

where periods of oscillation were determined by repeated measurements: $T_1 = (99.8 \pm 1.2)$ s, $T_2 = (271.7 \pm 1.3)$ s. From this it follows that the torsional rigidity for length $L = 465$ mm is $k_\tau = (1.01 \pm 0.02) \times 10^{-7}$ N.m/rad. After reduction for length 1 meter we get $k_{\tau 1} = (4.69 \pm 0.07) \times 10^{-8}$ N.m^2/rad.

2.3 Experiment

The equipment that was used for the experiment, has a principle which was demonstrated in Figure 1; in this case the phosphor-bronze filament had a length $L =$ 465 mm and torsional rigidity $k_\tau = (1.01 \pm 0.02) \times 10^{-7}$ N.m/rad. The results shown in this Section (2.3) were carried out for the experiment with a flat stainless steel plate with dimensions: width 38.5 mm, height 60.5 mm, thickness 0.50 mm and mass 8.50 g. Angles φ_u and φ_d were read from the circular scale with an accuracy of 0.5 °.

Water for the experiment was distilled and boiled for 3 minutes before the proper experiment. Its temperature for the experiment was measured and kept within the

interval of 24 °C and 25 °C. Water with volume of approx. 350 ml was in a glass beaker with an inner diameter of 80 mm and a height of 110 mm. The beaker was closed with a paper lid when not in use.

The first experiments that with the help of this static method showed autothixotropy of water were carried out by the author of this paper from September 1991 to March 1992. The author together with Pavel Voráček, who explained this phenomenon with the existence of gradually arising clusters of water molecules, wrote in April 1992 a report (Vybíral, Voráček, 2003) that was published as late as July 2003. From October 2003 to June 2004 the author of this paper performed many other measurements that confirmed and advanced the results received twelve years ago in a more detailed way. The most interesting results of these measurements are presented in Section 2.3/a, b, c, and d.

a) Our attention is mainly concentrated on measurements of the critical angle $(\varphi_u)_{crit}$, i.e., the angle φ_u at which – if reached – the plate began (for a few tens of seconds) to rotate in direction of the rotation, during which time the angle φ_d has been changed considerably (according to the total angle φ_u change it can represent in the case of the plate used ten or a hundred degrees). Here follow some results:

- The plate immersed with 65% of its surface in water that had been standing for 7 days (measurements from January 16th–17th, 2004) – $(\varphi_u)_{crit.}$: $405°, 400°, 395°, 395°, 390°, 400°$; averaged: $398° \pm 3°$.
- Water boiled for a short time, configuration of system kept unchanged (immersion 65%). After cooling $(\varphi_u)_{crit.} \approx 30°$, after two days $(\varphi_u)_{crit.} \approx 115°$.
- Water boiled, the plate entirely immersed (the upper edge 10 mm below water level), critical angle measured on the second and third day – $(\varphi_u)_{crit.}$: $360°, 360°, 358°, 357°, 350°, 348°$; averaged: $356° \pm 3°$.
- The plate immersed only 50% – $(\varphi_u)_{crit.}$: $360°, 355°, 340°, 325°, 335°$; averaged: $343° \pm 8°$.
- The influence of plate immersion on the critical angle $(\varphi_u)_{crit}$ was small (see Table 1).
- Especially it was observed that the period of the water-standing had an influence on the size of the critical angle. With immersion of 85% of the surface of the plate and with a standing period of 17 days, it reached $(\varphi_u)_{crit.} = 1800°$. In consequence of the rupture which followed, the plate rotated through the

Table 1. The influence of plate immersion on the critical angle

Plate immersion	$(\varphi_u)_{crit.}$
100%	350°
93%	350°
68%	335°
68%	325°
23%	300°
2.0%	85°

angle $\Delta\varphi_d = 1430°$. After stabilization of its position, a little change of angle φ_d ('creep') was observed: in 5 minutes, 4° and in subsequent 70 hours, another 32°. During the total immersion such a great critical angle was never reached.

b) The results for a certain configuration of a measurement system hat a good reproducibility when measurements were repeated. For example in the days from March 22nd to April 5th 2004, six measurements of angle φ_d were performed, with the plate totally immersed in water (the upper edge was 3 mm below water level) for equally set of angles φ_u. Sample of means of the measured angle φ_d (including the sample standard deviation) are in Table 2, together with the value of the quantity k_w taken according to the relation (2) and including its standard deviation. For $(\varphi_u)_{crit.}$, probably $k_w \to 0$.

When the critical angle was adjusted to $(\varphi_u)_{crit.} = (239 \pm 2)°$, the plate got into rotation during a few tens of seconds and reached a new equilibrium position $(\varphi_d)_0 = (198 \pm 2)°$.

The water with arisen clusters of molecules behaves like a mechanically elastic system. The plate hung on a filament with torsional rigidity k_τ exercises deformation work through a moment of force (1)

$$W_w = \frac{1}{2} k_\tau (\varphi_u - \varphi_d)^2$$

and exhibits an increase of potential elastic energy of clusters of molecules until the moment when φ_u reaches its critical value. The maximum value of the deformation work presented in the Table 2, was $W_w = (4.0 \pm 0.2) \times 10^{-7}$ J.

c) Measurements for a cyclic change of angle φ_u were carried out, and the results of three measurements are shown in Figure 2 and 3. Figure 2 comprises in its graphs results from April 30th 2004 for a plate entirely immersed in the water that was extensively stirred 17 hours before (the upper edge was immersed 3 mm below level of the water surface). During the change of angle φ_u from the starting equilibrium

Table 2. The results of six measurements with the plate totally immersed in water

φ_u	φ_d	k_w/N.m.rad^{-1}
0°	0° ± 0, 5°	
30°	9.7° ± 0.9°	$(2.1 \pm 0.5) \times 10^{-7}$
60°	19.6° ± 0.7°	$(2.0 \pm 0.4) \times 10^{-7}$
90°	27.4° ± 0.7°	$(2.8 \pm 0.5) \times 10^{-7}$
120°	34.4° ± 0.6°	$(3.3 \pm 0.6) \times 10^{-7}$
150°	41.8° ± 1.3°	$(3.1 \pm 0.8) \times 10^{-7}$
180°	52.3° ± 1.3°	$(1.9 \pm 0.5) \times 10^{-7}$
210°	63° ± 2°	$(1.8 \pm 0.7) \times 10^{-7}$
230°	77° ± 2°	$(0.43 \pm 0.29) \times 10^{-7}$

$(\varphi_u)_{crit.}$: 240°, 240°, 245°, 240°, 235°, 235°; averaged: (239 ± 2)°
$(\varphi_d)_0$: 192°, 198°, 203°, 197°, 199°, 197°; averaged: (198 ± 2)°

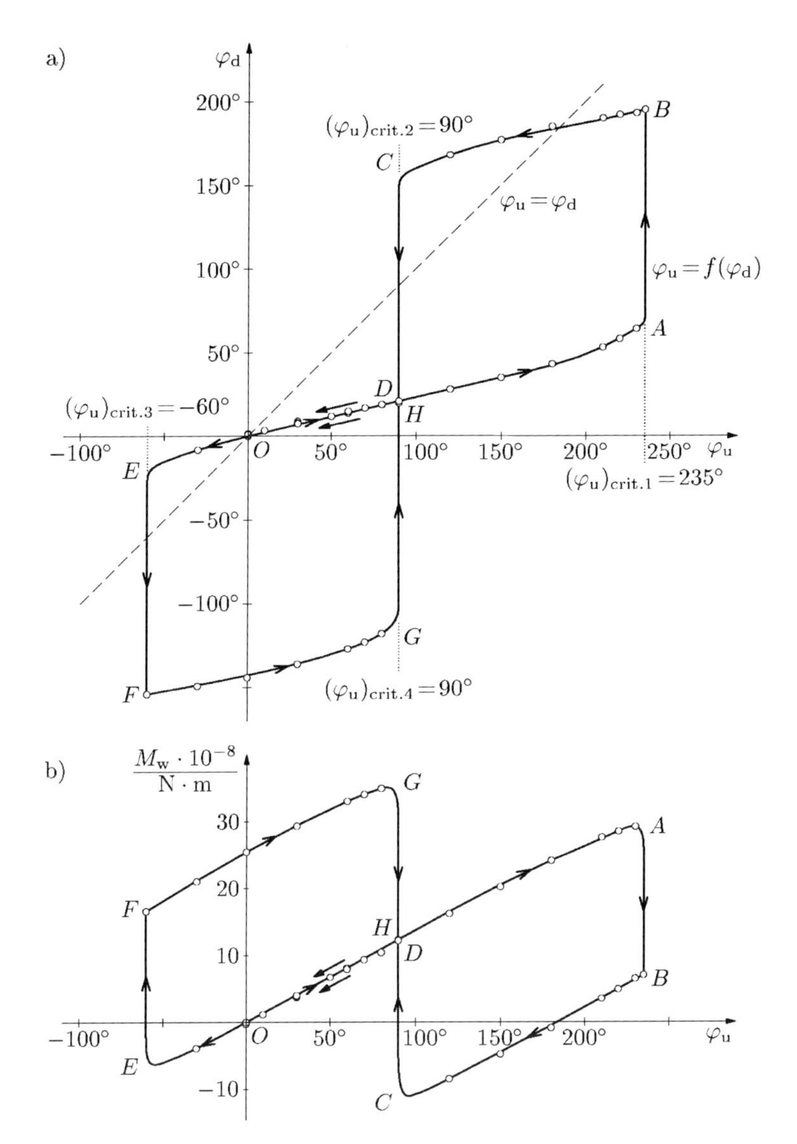

Figure 2. The results of the experiment with the totally immersed plate from April 30th 2004: a) loop of measured changes of an angle $\varphi_d = f(\varphi_u)$ b) loop of changes of a moment of force M_w calculated from the relation (1)

position $\varphi_u = \varphi_d = 0°$, the change of angle φ_d did not follow the ideal straight line $\varphi_u = \varphi_d$, but the curve *O-A*. At point *A* the critical value $(\varphi_u)_{crit.1}$ was reached and them the plate turned into a new equilibrium position: point *B*. With decreasing angle φ_u, the angle φ_d had been being changed according to the curve *B-C* until it reached the second critical value $(\varphi_u)_{crit.2}$ and afterwards the plate turned to another equilibrium position: point *D*. When the angle φ_u was decreasing again, the position of the plate went through the beginning point O to the third critical position: point E and the third critical value $(\varphi_u)_{crit.3}$. Another equilibrium position corresponded to point F and the fourth critical position corresponded to point D. In the fourth critical position (point G) it is valid that $(\varphi_u)_{crit.4} \cong (\varphi_u)_{crit.2}$. The plate having rotated, the fourth equilibrium position followed: point $H \cong D$ again. From there, with decreasing angle φ_u, the position of the plate followed the previous section H-O and for $\varphi_u = 0°$ it reached again the original equilibrium position $\varphi_d \cong 0°$. In Figure 2b the dependence of the moment of force M_w is shown, according to the expression (1), with use of the plate influencing the water in particular periods of the experiment.

In the graphs in Figure 3, the results are of two other experiments with water standing for one week: the loop **a** is for the experiment from May 18th 2004, with a plate totally immersed, and the loop **b** for the experiment from May 17th 2004, for a plate only half immersed; the effect is probably more pronounced than for the plate totally immersed. The loops from the Figure 3 are simpler than those from the Figure 2 and the values $(\varphi_u)_{crit}$ are lower. The reasons can be explained on a microscopic level – the plate probably deformed clusters of water molecules of various dimensions and rigidity.

The demonstrated experiments suggest that the mechanical properties of clusters of water molecules have a certain *hysteresis*. But the hysteresis is limited; e.g., in our experiment it does not appear in situations when a critical angle was not reached. For example, if the configuration of the plate in the section O-A of the graph in the Figure 2a), in front of the point A, and if we begin to decrease an angle φ_u, the change of an angle φ_d will follow the same curve O-A again. In these situations, the cluster seems to behave like an ideal elastic body.

d) During the experiment some adjacent measurements were also made with the view of eliminating other influences on the observed phenomenon of autothixotropy. During the experiment the acidity of the observed sample of water was researched with a potentiometric measurement of pH factor. It did not change considerably in a long term period; for the temperature range 24 °C to 25 °C it moved in the range 7.1 to 6.9. The electric conductivity of entirely fresh water was 5.6 μS/cm and in 5 weeks it was increased to 30.5 μS/cm in 25 °C. The dependence of the observed water properties on this change was not determined. With the use of the Du-Noüyho apparatus it was also possible to observe whether in the course of the experiment some change of surface tension of water appeared. With accuracy of 1%, no measurable change was found.

e) In the second phase of this experiment at the end of the year 2004 distilled deionizated water was used. The experiment showed that in deionizated water there was no phenomenon of autothixotropy, as described in the parts a, b, and c. The same equipment (Figure 1) was used for this experiment and the plate was immersed both to one half of the side-hight and entirely as well. The period of water standing before the measurement was almost 10 days. It was found from the measurements that angle φ_d of plate rotation that was altered in interval $\varphi_d \in (0°, 360°, 0°)$, was equal to angle φ_u of torsion of the upper end of the filament with accuracy of

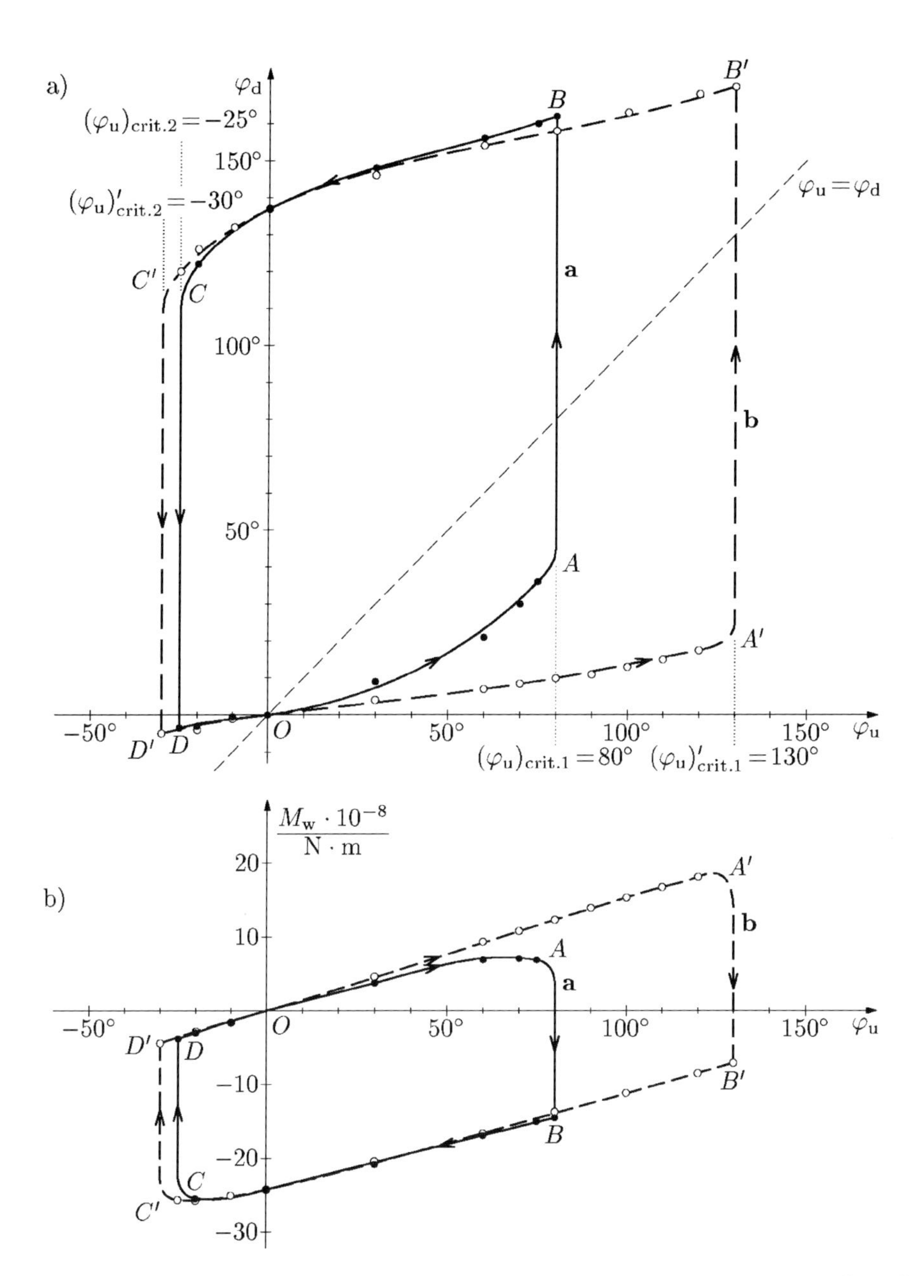

Figure 3. The results of the experiment from May 18th 2004 with the totally immersed plate (loop **a**) and of the experiment from May 17th 2004 with half immersed plate (loop **b**). The same quantities as those in Figure 2 are put into relations here

1.5° evaluated from the repeated measurements. The existence of critical angles $(\varphi_u)_{crit.}$ and a phenomenon of hysteresis were not found. From the experiment carried out, we arrived at the important conclusion that the autothixotropy of water, characterized by non-zero critical angle and hysteresis, is caused by a presence of ions in the water. The described measurements were made in December 2004.

3. THE METHOD OF TORSION OSCILLATIONS

3.1 Theory of Measuring Method

Let us consider a plate with an axis of symmetry (about which it has a moment of inertia I) along the filament with torsional rigidity k_τ from which it hangs. We immerse the plate into the water and describe its torsion oscillations in two situations:

a) In 'fresh' water (i.e. with negligible autothixotropy), with a hypothesis of viscous damping described by coefficient δ, the equation of motion is

$$\ddot{\varphi} + 2\delta\dot{\varphi} + \frac{k_\tau}{I}\varphi = 0$$

and the angular frequency of free damped oscillations is given by

$$\omega_1^2 = \frac{k_\tau}{I} - \delta^2$$

b) In 'standing' water (with autothixotropy) we suppose that it is necessary to add the elastic properties of putative clusters of water molecules to the elastic quantities that we describe for the considered plate with equivalent torsional rigidity k_w. Then,

$$\ddot{\varphi} + 2\delta\dot{\varphi} + \frac{k_\tau + k_w}{I}\varphi = 0$$

and the angular frequency of oscillations of the plate is given by

$$\omega_2^2 = \frac{k_\tau + k_w}{I} - \delta^2 = \omega_1^2 + \frac{k_w}{I}$$

If we measure the periods of oscillation T_1, and T_2 and determine the moment of inertia I, e.g. from the plate dimensions and its mass, we can calculate:

$$k_w = 4\pi^2 I \left(\frac{1}{T_2^2} - \frac{1}{T_1^2}\right) \tag{3'}$$

3.2 Experiment

For the measurement, an aluminium plate with a thickness of 2.95 mm, width (47.59 ± 0.03) mm, height (50.59 ± 0.02) mm and mass 18.70 g, was used. Its moment of inertia was calculated from its dimensions and mass: $I = (7.518 \pm 0.001) \times 10^{-6}$ kg.m^2. The plate was hung up along its longitudinal axis of symmetry on a phosphor-bronze filament with a cross-section of $(0.025 \times 0.2)\ \mathrm{mm}^2$ and a length of 569 mm, and immersed in distilled, boiled water in such a way that the upper edge of the plate was 14 mm above the level of water surface. The water with volume of approximately 400 ml was in a glass beaker with inner diameter 80 mm and height 110 mm while the experiment was carried out in room temperature 23 °C. The period of the damped torsion oscillations was measured three times, first in fresh water. The period of oscillation was evaluated to $T_1 = (101.7 \pm 1.2)$ s. Then the system was left at rest for 7 days, so that a high level of autothixotropy of water could be reached. Afterwards, the plate was carefully rotated from this equilibrium position through 45 ° at which it stayed. In this position, the plate was given a torsion pulse by which damped torsion oscillations were initiated. The period of oscillation was measured ten times and was evaluated for $T_2 = (5.34 \pm 0.06)$ s.

The equivalent torsional rigidity of this system with autothixotropy is, according to the relation (3'), determined to $k_w = (1.04 \pm 0.03) \times 10^{-5}$ N.m/rad.

So, as to judge, the level of autothixotropy of the system, a critical angle $(\varphi_u)_{crit.}$ was measured. The plate was able to rotate from equilibrium position through angle $(\varphi_u)_{crit.} = 340° = 5.93$ rad without coming back. The filament had a torsional rigidity $k_\tau = (8.25 \pm 0.12) \times 10^{-8}$ N.m/rad and thus a corresponding maximal moment of force: $M_{w\,max} = k_\tau \Delta\varphi$, was equal to 4.89×10^{-7} N.m. When the moment was exceeded, the plate came back to the equilibrium position. This measurement of torsion oscillations was made in September 1991.

4. DYNAMIC METHOD OF BALL MOVEMENT

4.1 Theory of Measuring Method

Let us consider a ball with radius r and mass m that we let fall freely in a cylindrical vessel with radius R filled with water. Let the ball have density ρ, only a bit greater than the density of water ρ_w. Its velocity will be small and the flow around it laminar. Besides gravity and buoyant force:

$$\boldsymbol{G} + \boldsymbol{F} = m\boldsymbol{g}\left(1 - \frac{\rho_w}{\rho}\right)$$

there is a hydrodynamic resistance affecting the ball, consisting of force $\boldsymbol{F}_s$ according to Stokes's law adjusted for movement in a limited environment (in a cylindrical vessel with radius R) with dynamic viscosity η and of unknown resistive force $\boldsymbol{F}_a$ – the force of the inner static friction of water caused by its autothixotropy:

$$\boldsymbol{F}_s + \boldsymbol{F}_a = -\left[6\pi\eta r v\left(1 + 2.4\frac{r}{R}\right) + F_a\right]\frac{\boldsymbol{v}}{v}$$

The equation of motion of the ball is as follows

$$\frac{dv}{dt} = A - kv,$$

where

$$A = g\left(1 - \frac{\rho_w}{\rho}\right) - \frac{F_a}{m} \tag{3}$$
$$k = \frac{6\pi\eta r}{m}\left(1 + 2.4\frac{r}{R}\right)$$

The first integral for a movement from equilibrium is

$$v = \frac{A}{k}\left(1 - e^{-kt}\right) = v_b\left(1 - e^{-kt}\right)$$

where v_b is the boundary velocity. The second integral for vertical movement to distance l from the beginning position is

$$l = \frac{A}{k^2}\left(kt + 1 - e^{-kt}\right) \tag{4}$$

From relation (4), with use of the measurement of time t_1, relevant for the path-length l of the ball in the fresh water, when the force F_{a1} is very small, and of time t_2 relevant for the standing water, when force F_{a2} has grown up, it is possible to define a difference of resistive forces caused by autothixotropy in states 2 and 1:

$$\Delta F_a = F_{a2} - F_{a1} = m\left(A_1 - A_2\right)$$
$$= mk^2 l\left(\frac{1}{kt_1 + e^{-kt_1} - 1} - \frac{1}{kt_2 + e^{-kt_2} - 1}\right) \tag{5}$$

4.2 Experiment

The basis of the equipment was a volumetric laboratory one-liter cylinder with inner radius $R = 28.5$ mm and a plastic ball with a non-absorbent surface with radius $R = 28.5$ mm and mass $m = 7.94$ g (average density of the ball was $\rho = 1.019 \times 10^3$ kg.m^{-3}). The ball was initially immersed with the upper edge 30 mm below the water surface and the measured trajectory had length $l = 351$ mm. For the measurement of the time of movement of the ball a universal computer-controlled scaler with optical equipment with phototransistor[1] was used.

The water temperature at the time of the experiment was kept at (22.0 ± 0.2) °C at which the dynamic viscosity and density respectively, are $\eta = 9.57 \times 10^{-4}$ kg.m^{-1}.s^{-1} and $\rho_w = 0.9975 \times 10^3$ kg.m^{-3}. Then the constants (4) are $k = 0.0569$ s^{-1}, and $A_0 = 0.207$ m.s^{-2} (for $F_a = 0$). From these, the boundary velocity of the ball is $v_b = 3.64$ m.s^{-1}. As the path length of the ball l is traversed in time $t \approx 10$ s, the average velocity is $v_m = l/t \approx 3.5 \times 10^{-2}$ m.s^{-1}, i.e. only 1% of the boundary velocity v_b.

[1] The design and development of the equipment and the period measurements were made by Peter Kleiner who was a student of physics teaching at the University of Hradec Králové at that time – see his diploma thesis of May 1994.

Table 3. The results of dynamic measurements

	Sets of measurement	Time of movement t_1, t_2/s	Change of resistive force ΔF_a/N
1	March 2nd 1994	10.01 ± 0.01	
	March 9th 1994	10.33 ± 0.02	$(3.7 \pm 0.3) \times 10^{-6}$
2	March 20th 1994	10.08 ± 0.01	
	April 7th 1994	10.48 ± 0.02	$(4.5 \pm 0.3) \times 10^{-6}$
3	April 11th 1994	10.01 ± 0.01	
	April 24th 1994	10.40 ± 0.02	$(4.5 \pm 0.3) \times 10^{-6}$
4	April 24th 1994	10.27 ± 0.02	
	April 24th 1994 after stirring up	9.85 ± 0.02	$- (5.0 \pm 0.4) \times 10^{-6}$

The experiment by itself started with a preparation of experimental water, i.e., distilled water with volume about 1 litre, was used that had been boiled for about 10 minutes before the experiment. After cooling down to the operating temperature of 22 °C and stabilization, the experiment with 'fresh' water was carried out: ten measurements of the time t_1 of the fall of the ball on the defined length l. After 9 to 18 days the second measurement was taken with the standing water, where, in consequence of autothixotropy, a longer period of ball movement was expected. After repeated measurements (in time intervals after 5 minutes breaks) a defect of molecule clusters probably appeared, and the ball movement-time was rather decreasing, which is likely to be a consequence of the fact that the water was mixed up by the falling ball. For example in the measurement from March 9th 1994 these values of time t_2 were: 10.328 s, 10.271 s, 10.293 s, 10.109 s, 10.129 s, 10.015 s, 9.953 s, 9.890 s, 9.938 s, 9.911 s. For the reason of non-falsification of the whole result of the experiment, only the first time t_2 was included in the evaluation of the experiment, however, it was necessary to consider a greater standard deviation of this time (i.e., 0.02 s).

Three sets of measurements with different times of water standing were performed. The results of measurements and their evaluation by using relation (5) are stated in Table 3. Fresh distilled water was used for each set of measurements. The fourth experiment followed the third one when the measurement in the third set were terminated, the time of ball movement was measured again, then the water was intensively mechanically stirred; after 10 minutes a new measurement was taken for which the time of movement was smaller and a change of resistive force ΔF_a caused by autothixotropy was negative, which was expected.

5. CONCLUSIONS

The results of measurements presented in this paper confirm with sufficient accuracy the phenomena qualitatively described in article (Vybíral, Voráček, 2003), which version comes from the year 1992 (although it was published in 2003). Moreover, the presented work confirms in an experimental way other effects of autothixotropy – both the phenomenon of hysteresis and a phenomenon of the inner friction in water (based on the analysis of the movement of the ball in water). As the probes used (plate, ball) had macroscopic dimensions, it is concluded that the arisen clusters of water molecules must have had such dimensions too. As the phenomenon of autothixotropy is not present in deionizated water, it may be determined by a presence of ions in water.

We have today two diametrically different results sustained by serious observations: According to the first one, the clusters in the water have a duration less than one hundred femtoseconds, while, according to the second one, the clusters are growing to the webs on the time scale of days. We believe that the purity of the water can be a decisive factor, since the webs never arose in the water which was deionized; we must admit that the distilled water used was not perfectly pure and could be significantly contaminated by salt ions, even if only in very minute degree.

Motto a posteriori:
If two different observations seem to be mutually incompatible within the frame of an accepted theory, the most probable explanation is not that one of the observations must be wrong, but the theory is wrong or – at least – incomplete, and the observations just discovered that it were not self-consistent.

On the basis of measured quantities it is possible to formulate these hypotheses about clusters of water molecules:

1. The clusters of water molecules may by of macroscopic dimensions of centimeter range.
2. The clusters of water molecules originate ('grow up') slowly in a range of days and it is possible to destroy them by boiling or intensive stirring.
3. With passing time, the density of network of molecules in the cluster becomes greater on the surface of the water than inside the water volume.
4. Within a given time-interval, the clusters of water molecules do not originate with the same size and density (the proof is e.g., different size of critical angle in static torsion experiments).
5. The clusters of water molecules have a certain level of mechanical properties that are analogous to the properties of solid substances, such as elasticity/rigidity and strength, but these properties are much gentler than in a case of solid substances (the appropriate quantities are of a relative size 10^{-6} and smaller).
6. Mechanical properties of clusters of water molecules show a certain hysteresis (see, e.g., Figure 2 and 3).
7. The water rather deviates from an ideal Newtonian viscous fluid because autothixotropy also is presented in form of a certain internal static friction in water, although it is very weak (see the experiment with the movement of a ball).
8. From comparison of experiments with natural distilled water and deionizated distilled water it is possible to deduce that kernels of macroscopic clusters of water molecules are the ions contained in water.

Note: **Two Observations Supporting the Model Explaining the Autothixotropy by Webs of Water Molecules**

In private communication Pavel Voráček (Lund Observatory, Sweden) reported to me about two observations related to the properties of water:

1. Nodal patterns in water

We put a weak latex suspension into a cylindrical vessel. The suspension was lightly opaque and the water deaerated by boiling. After some hours, we observed that regular patterns began to appear, where the concentration of latex substance grew higher, creating visible lines and surfaces. The patterns were present not only on the surface, but through the whole volume of the water. When we had some regular objects in the water, the patterns were highly complex, most often straight lines, which were divided into other lines, so that the whole pattern was highly symmetrical.

Looking for the explanation of the phenomenon, a similarity can be found in acoustics as Chladny's patterns on oscillating plates. In our situation, the nodal points, lines, and surfaces created, depend on the form of the vessel and the submerged object. This is in accordance with the theory of creation of the webs of water molecules, which are oscillating with different magnitudes of amplitude in different places in the water.

Note: When a needle is carefully stuck into the water in the vicinity of the pattern and then moved very slowly in any direction orthogonal to the needle, the pattern follows the needle, within a region of millimeters; the observation is quite consistent with our theory of explanation of water-autothixotropy.

2. Toxicity of water

The containers of drinking water on rescue boats have an instruction to shake the container thoroughly before use. The reason for this is that people have gotten seriously ill – and in some cases even died – after drinking water that has not been shaken.

The generally accepted explanation is that water kept in containers for long periods of time becomes deaerated. We made a simple experiment: We deaerated one liter of water by boiling, and after it cooled down, we drank it without any consequences. This means that the right explanation rather would be the autothixotropy of the water when the long parts of the water molecule web are blocking the inner walls of the gastro-intestinal tract, causing the illness.

It is traditionally known that the water found in puddles or crevices is toxic even if there is no biological life in it, and such water is often called 'dead water'.

REFERENCES

Dubells Taschenbuch für den Maschinenbau, zweiter berichtigter Neudruck (1956) Springer-Verlag, Berlin/Göttingen/Heidelberg

Pollack G (2001) Cells, gels and the engines of life. Exner and Sons Publisher, Seattle WA, USA

Vybíral B (1987) Experimental verification of gravitational interaction of bodies immersed in fluids. Astrophys Space Sci 138:87–98

Vybíral B, Voráček P (2003) "Autothixotropy" of water – an unknown physical phenomenon. Available via http: //arxiv.org/abs/physics/0307046

Vybíral B (2004) Experimental research of the autothixotropy of water. Proceedings of the conference new trends in physics – NTF 2004 Brno. University of Technology, Czech Rep, Brno, pp 131–135

CHAPTER 16

NON-BULK-LIKE WATER ON CELLULAR INTERFACES

IVAN L. CAMERON[1,*] AND GARY D. FULLERTON[2]

[1] *Department of Cellular and Structural Biology, University of Texas Health Science Center at San Antonio, San Antonio, TX, USA*

[2] *Department of Radiology, University of Texas Health Science Center at San Antonio, San Antonio, TX, USA*

Abstract: Given that a major fraction of cellular water is non-bulk-like in its physical properties the question arises: What is the molecular basis of this non-bulk water? As proteins are, by mass, the major solute in the cell it is natural to suspect proteins as the main factor responsible for the non-bulk properties of water in cells. This report reviews possible theories and facts on the origins of this non-bulk-like cellular water. After review of theories in the literature it is concluded that native globular proteins in their free and polymerized state can account for the major fraction of cell water as being non-bulk-like in its physical properties

Keywords: Cell water; Interfaces; Protein conformation; Osmosis; Hydration; Intracellular water; Hydrophobic; Hydrophilic

1. INTRODUCTION

One of the theories to explain non-bulk-like water in cells concerns the existence of multilayers of water molecules on the interfacial surface of proteins. This water is structured differently than bulk water. Such non-bulk-like water molecule structuring is thought to slow the motion of the water molecules and change its physical properties (i.e. density, specific heat, thermal expansion, sound conductance, heat conductivity, viscosity, energy of activation, solvation for ions, ionic

* Corresponding author. Department of Cellular and Structural Biology, University of Texas Health Science Center at San Antonio, 7703 Floyd Curl Drive, Mail Code 7762, San Antonio, Texas 78229-3900, USA. Tel.: 210-567-3817; Fax: 210-567-3803. *E-mail address*: cameron@uthscsa.edu

G. Pollack et al. (eds.), Water and the Cell, 315–323.

conductivity, dielectric relaxation, proton NMR relaxation time (for examples see Table 1 from Clegg and Drost-Hansen, 1991; Drost-Hansen et al., 1991; Pollack, 2001). This non-bulk water also differs in its colligative properties, i.e., osmotic behavior, vapor pressure, boiling point and freezing points. All these parameters have biological significance.

1.1 Introduction to the Theory of Polarized Multilayered Water Adsorption

Ling's polarized multilayer (PML) theory of water (2003) emphasizes the importance of properly spaced positive and negative charge groups on surfaces to acquire a maximum extent of multilayering of water molecules. Positive (P) and negative (N) charge sites spaced at the distance of a water molecule, ~3.1Å diameter, gives the greatest potential for polarized multilayers of water (PML) to form at the surface (Figure 1). Water in the PML condition cannot be frozen and has an extremely high boiling point.

Surface charge spacing and other types of charge distribution, i.e,) P-O, P-O, (O=neutral) or N-O, N-O may also polarize multilayers of water if the charge group is property-spaced. However this type of PML of water has weaker effects and shorter residence time of the water molecule. A properly spaced pattern of N and P changes on the surface or between two NP-NP surfaces give longer water residence times.

In cells most proteins have a backbone of positively charged NH groups (P sites) and negatively charged CO-(N sites) however in globular proteins most NH and CO groups are thought to be neutralized and shielded in the form of intramolecular H bonds. To serve as a proper NP surface the protein backbone must be opened and the N and P sites exposed for PML of oriented water to form.

A major concern with this PML theory is that almost all cellular proteins are globular however a recent report indicates that some intracellular proteins exist in an open non-globular form (Uversky, 2002). The question then is do such unfolded

Table 1. Comparison of some properties of bulk, and, non-bulk or vicinal water (from Clegg and Drost-Hansen, 1991)

Property	Bulk	Vicinal
Density (g/cm^3)	1.00	0.97
Specific heat (cal/kg)	1.00	1.25 +/- 0.05
Thermal expansion coefficient ($°C^{-1}$)	$250 \cdot 10^{-6}$(25 °C)	$300–700 \cdot 10^{-6}$
(adiab.) Compressibility coefficient (Atm^{-1})	$45 \cdot 10^{-6}$	$60–100 \cdot 10^{-6}$
Excess sound absorption ($cm^{-1} \cdot S^2$)	$7 \cdot 10^{-17}$	$\sim 35 \cdot 10^{-17}$
Heat conductivity ($(cal/sec)/cm^2/°C/cm$)	0.0014	~0.01–0.05
Viscosity (cP)	.089	2–10
Energy of activation ionic conduction (kcal/mol)	~4	5–8
Dielectric relaxation frequency (Hz)	$19 \cdot 10^9$	$2 \cdot 10^9$

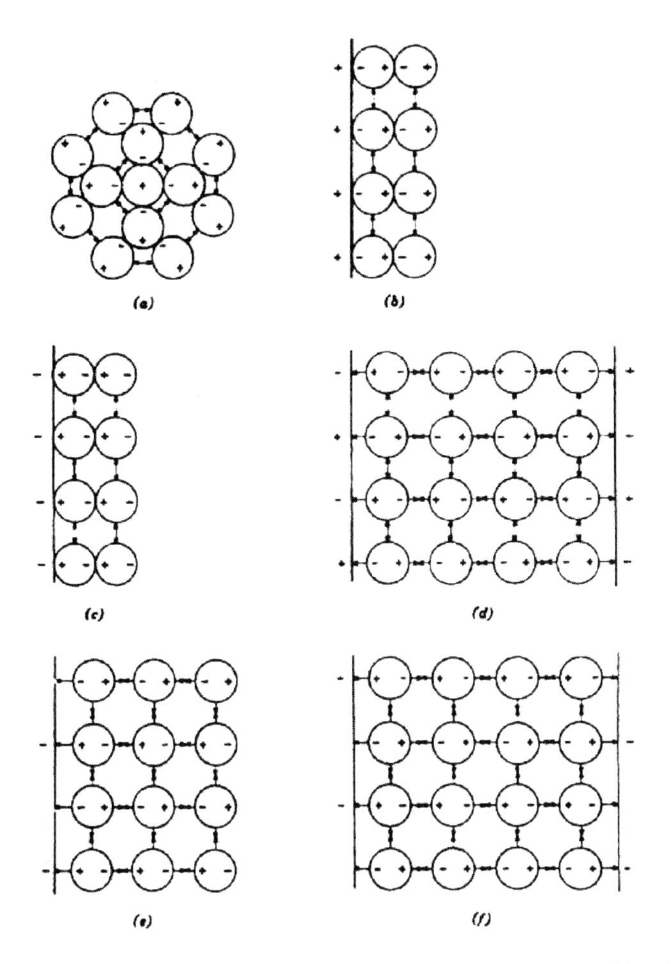

Figure 1. Diagrammatic illustration of the way that individual ions (a) and checkerboards of evenly distributed positively charged P sites alone. (b) or negatively charged N sites alone. (c) polarize and orient water molecules in immediate contact and farther away. Emphasis was, however, on uniformly distanced bipolar surfaces containing alternating positive (P) and negative (N) sites called an NP surface. When two juxtaposed NP surfaces face one another, the system is called an NP-NP system. (d) If one type of charged sites is replaced with vacant sites, the system would be referred to as PO or NO surface. (e) Juxtaposed NO or PO surfaces constitutes respectively an PO-PO system or NO-NO system. (f) Not shown here is the NP-NP-NP system comprising parallel arrays of linear chains carrying properly distanced alternating N and P sites. Note how directions of paired small arrows indicate attraction or repulsion (modified after Ling, 1972; reprinted by permission of John Wiley & Sons Inc.)

intracellular proteins provides properly spaced N and P sites to support PML of water? Although the PML of water is strongly supported by data from inanimate systems and from filamentous proteins, like gelatin with open and exposed N and P sites, the existence of properly space NP site proteins within a living cell remains an open question.

Another unanswered question is does the surface of any globular proteins in the cell have an NP or NO or PO spacing pattern that would allow PML of water? Also might allosteric/conformational change in the protein structure or change site modifications allow for PML of water to occur *in vivo*? Thus it may be possible that there are appropriate areas on cellular surfaces with potential for PML of water to form. Certainly denatured and open arrangements of N and P sites on proteins can provide appropriate sites for PML of water to form as observed with gelatin. It seems possible that PML formation may occurs in the glycocalix or in slime.

On the other hand Drost-Hansen et al., (1991) posits that proximity of water close to any solid surface, regardless of its specific chemical nature, produces a multiple layer of 'vicinal water molecules' that has effects on many of the water's physical properties (Table 1). Use of atomic force microscopy (AFM), gives direct evidence for up to six water shell hydration layers over a hydrophilic surface (Jarvis et al., 2000). The application of this AFM to biological interfaces is planned. Ling (2003) disagrees with Drost-Hansen citing evidence that proper spacing of NP sites between AgCl plates does not allow water to freeze even at −176 °C but water between plates of CaF_2 or fluorite where the NP spacing is less than optimal does freeze without difficulty.

1.2 Other Models to Explain Non-Bulk-Like Water on Proteins and in Cells

What has been discussed so far is the theory of polarized multilayers of water on charged hydrophilic surfaces where water molecules can form hydrogen bonds due to water's dipolar-like charge distribution. On the other hand hydrophobic surfaces preclude hydrogen bonding of water with the hydrophobic surface, but induces bonding between adjacent water molecules over the hydrophobic surface. The angle of the two hydrogen atoms with the oxygen atom of the water molecule is 104 degrees which makes it possible for an easy arrangement of five water molecules in a pentagonal ring of water molecules planar to the hydrophobic surface (Urry, 1993). Having more or fewer water molecules in a planar ring of water molecules would require more or less change in angle than the 104 degrees needed to form a planar five membered water molecule pentagonal ring structure. Multiple water pentagons can therefore form a network or clathrate of pentagonal structures over hydrophobic surfaces (Figure 2). Whether these pentagonal rings can extend away from the hydrophobic surface to form three dimensional clathrate structures is an open question. What is known is that the placement of a charge group like: phosphate, carboxylate, or sulfate at the hydrophobic surface will disrupt the water pentagonal structure in the area of the charge site (Urry, 1993, Figure 2).

Martin Chaplin (2004) has recently presented a new theory on the structuring of water in the cell that switches from low-density clusters of water to high density non-clustered water that is modulated by key proteins which in turn are controlled by the energy status and ionic content of the cell. His theory is summarized here: Chaplin proposes that a rotating globular protein, like G-actin, has a mixed environmental

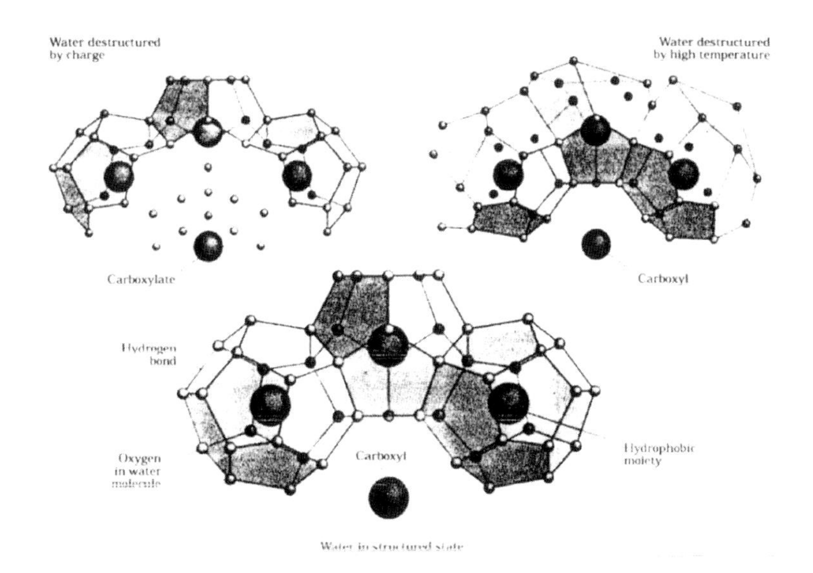

Figure 2. Pentagonal water structure is the result of hydrogen bonds between water molecules adjacent to hydrophobic molecular groups. This pentagonal structure is stable at low temperatures (center) but becomes progressively more unstable at higher temperatures (right). Charged, hydrophilic molecules can destroy this pentagonal structure (left) by forcing water molecules to line up around them. Pentagonal water surrounding an amino acid chain of hydrophobic amino acids can prevent the chain fro shortening because energy is required to break the hydrogen bonds so that the water can move out of the way (Reproduced from Urry, 1995a, reproduced from Elastic biomolecular machines, Urry DW, © 1995 by Scientific American, Inc. All rights reserved)

surface with both a clathrate-like structure of water molecules over hydrophobic surface areas and a H-bonded water molecules at charged, i.e., carboxylate, sites. Both types of protein surfaces exist and provide a first layer of structured water with properties that differ from bulk water. Increased diffusive motion of the protein will cause changes in the clathrate structured water (disruption) outside the first layer of structured water. When ATP is added to G-actin, the major protein in the majority of eukaryotic cells, it undergoes a conformation change which causes conversion of an α-helix to a β-turn to form F actin polymers. This G to F transition causes an increase in the amount of intracellular low density water clustering as the F-actin is less mobile. Some proteins may cross link (aggregate/polymerize) and trap water which will have decreased entropy. Chaplin goes on to propose that F actin allows greater structuring of water molecules with lower density. He also states that enclosures of water in a meshwork of polymers may have capillary action forming "stretched" confined water which is more highly structured and less dense than bulk water (also see Wiggins 1995). When a static protein is freed to rotate it loose some of its outer shell of low density water molecules as well as its associated ions. In summary of Chaplin's model static/polymerized protein compared to rotating/free protein has more low density water, lower carboxylate O charge, greater $-CO_2$-K^+ ion pairs, lower ionic strength, more H_2PO_4, and more water clathrate clustering structure. Thus water clathrate networks with low density facilitates K^+ ion binding to glutamate and aspartic acid groups. This theory is compatible with Pollack's sol to gel transition as the gel state would convert to lower density water clustering around K^+-carboxylate ion pairs. Raised levels of Na^+ and/or Ca^{2+} ions in a cell, as occurs during cell signaling, will destroy low-density clathrate water structure and replace some of the bound K^+ ions. Chaplin concludes: rotating proteins have zones of higher density water which changes to low density clathrate water clusters as rotation decreases, thus low density clathrate water favors solution of K^+ vs. Na^+, conversely static proteins with more clathrate water prefer K^+ ion pairs over freely soluble K^+ ions (Chaplin, 2004).

Here are some pros and cons on Chaplin's clathrate cluster water theory. Actin is present in significant amounts in most eukaryotic cells. Bundles of F or filament actin, as observed in electron micrographs, have a clear zone of up to 1000 nm from the bundle surface which implies high water-structuring capability (Pollack, 2001). If these clear zones either are or are not an artifacts of tissue processing for electron microscopy is not known. Also actin can cross link with alpha-actinin to form a gel which when exposed to critical level of ATP contracts to about 10% of its original volume (in Pollack, 2001). Conversely polymerization or aggregation of globular proteins, like G actin and hemoglobin, or oligomer formation of monomer proteins causes loss of water accessible surface area (Miller et al., 1987a) with loss of osmotically unresponsive water as the globular proteins dock to one another (Fullerton, 2006c; Bogner et al., 2005). In the case of sickle cell hemoglobin massive polymerization results in osmotic pressure change and cell volume decrease presumably due to loss of osmotically unresponsive water from the hydration shell of unpolymerized hemoglobin (Prouty et al., 1985, also see Fullerton et al., 1987, Bogner et al., 2005 and Fullerton et al., 2006c). The extent, structure half-life, and physical properties of the clathrate cluster water theory of Chaplin deserves critical evaluation (see Cluster-Quackery, 2004).

1.3 An Alternate Model to Explain the Extent of Non-Bulk-Like Water in Cells

Given that the majority of intracellular water has non-bulk like motional, colligative and other physical properties (see table 1 for a list of physical properties plus Ling, 1972, 1984, 2001, 2003, Pollack, 2001, Cameron et al., 1997, 2006) then the question is: what is/are the mechanism(s) responsible for all of this non-bulk water in cells? The first proposition, as discussed above was that multilayers of water molecules are oriented differently than water molecules under bulk water conditions and that these layers of oriented water molecules form over the surface of cellular proteins when, according to Ling, 2003, the proteins exist in an unfolded state which exposes the positively-charged NH groups and the negatively-charged CO groups. In contrast there is much evidence that most cellular proteins exist in a globular state where their NH and CO groups are thought to be neutralized and shielded from forming hydrogen bonds with water molecules.

Studies on the extent of osmotically unresponsive water (OUW) on a globular protein like bovine serum albumin (BSA), done both before and during unfolding to the open or linearized form, to expose the proteins backbone and to provide the N and P site conditions needed supports Ling's PML theory, have been accomplished (Zimmerman et al., 1995, Kanal et al., 1994, Fullerton et al., 2006c). In these protein unfolding studies the environment of the globular BSA was modified by a series of changes in pH and in salt (NaCl) concentration. The extent of OUW varied from 1.4 g water/g dry BSA when the protein was in its most tightly packed conformation up to values of 9 to 12 g water/g dry BSA when the protein was unfolded. Molecular calculations of solvent (water) accessible surface area, of native globular BSA, based on the method of Miller et al. (1987b), indicates enough area for a monolayer of water of at least $1.4\,gH_2O/g$ dry protein (Fullerton et al., 2006b). Now with a fully unfolded globular BSA molecule protein the calculated surface area is increased enough to account for $\sim 2.8\,g\,H_2O/g$ dry BSA (Zimmerman et al., 1995). Given the maximum OUW values of Kanal et al., 1994, Zimmerman et al., 1995 and Fullerton et al., 2006c (in the range of 9 to 12 g water/g dry mass), there is enough OUW on the unfolded/linearized BSA for about 3-4 layers of water. These findings give evidence in support of Ling's PML theory but the evidence for a sufficient amount of unfolded or linearized proteins in the living cell still remains an open question. An alternate explanation for this large extent of OUW in unfolded proteins is that unfolded proteins can cross-link, often with disulfide linkage bounds, to form a gel-like network. Thus a crosslinked three dimensional network of protein fibers can have cavities of water molecules not in fast enough exchange with the surrounding bulk water to be included in the bulk water compartment. Such cavity water has escaped from the surrounding bulk water compartment into the network of the cross-linked fibrous protein gel. The cavity water may be "stretched" and non-bulk in its physical properties or may be bulk like but sufficiently separated from the outside bulk water. This would make cavity water appear to be OUW. Note that the globular protein, in its native physiological conformation still has 4g water/g dry protein mass as OUW (Kanal et al., 1994). This 4g of water per 1g dry cell mass would account for all of the water in a cell with 80% water. Thus there maybe no need to propose the existence of unfolded proteins to account for the known extent of OUW observed in those cell types where it has been measured (Fullerton et al., 2006c).

It is therefore concluded that globular proteins in their native structural state can alone account for most if not all of the non-bulk water in living cells. Here the reader is referred to the reports of Fullerton et al., 2006b,c. These reports demonstrate that a native globular protein has an osmotically unresponsive water fraction of 4g water/g dry protein. This amount of water has escaped from the surrounding bulk water to become non-bulk water or protein hydration water and this amount of non-bulk water can account for the majority of the water in most cells as being non-bulk water. However in these same studies (Fullerton et al., 2006a,b,c) it was also demonstrated that induction of globular protein conformation changes and aggregation by modification of their environment (i.e., pH, salt, urea Denaturation,

Table 2. Mass water/mass BSA pumped in and out of a cell (dialysis cassette) under varying conditions used to manipulate protein conformation and aggregation (from Cameron et al., 2006)

Method	Maximum % Change ($\Delta M/M_c$ x 100%)	Mechanism
Temp. Annealing (25 °C to 70 °C)	−7% (Water out)	Unfold and cross-link Decrease N Increase ASA
Salt – NaCl (2000 mmolal to 1 mmolal)	−7% (Water out)	Unfold Increase N Increase ASA
Urea – 150 mmolal NaCl (8 molal to 0 molal)	−14% (Water out)	Fold Decrease AN Decrease ASA
pH – 200–300 mosmol NaCl (pH = 5.4 to pH = 9.0)	+14% (Water in)	Disaggregation above IEP and then unfold Increase N Increase ASA Increase AN

N = number of particles
ASA = accessible surface area
AN = apparent number of particles (segmental motion)
IEP = isoelectric point of BSA

temperature) can change the protein's water accessible surface area for interfacial interactions with water molecules and thus increase or decrease the extent of water which is osmotically unresponsive or non-bulk like in its physical properties. Thus, protein conformation, aggregation and diffusional mobility play a role in regulating the extent of non-bulk water cell (Table 2).

The common assumption of cell physiologists is that essentially all intracellular water is bulk-like in its physical properties. This assumption is not true and is no longer warranted. Clearly this fact complicates most current textbook dogma but at the same time opens new avenues to our understanding of cell function.

REFERENCES

Bogner P, Misetta A, Berent Z, Schwarz A, Kotex G, Repa I (2005) Osmotic and diffusive properties of intracellular water in camel erythrocytes: The effect of hemoglobin crowdedness. Cell Biol Int (2005)29:731–736

Cameron IL, Kanal KM, Fullerton GD (2006) Role of protein conformation and aggregation in pumping water in and out of a cell. Cell Biol Int 30:78–85

Cameron IL, Kanal KM, Keener CR, Fullerton GD (1997) A mechanistic view of non-ideal osmotic and motional behavior of intracellular water. Cell Biol Int 21:99–113

Chaplin M (2004) Water structure and behavior. Available via www.isbu.ac.uk/water/cell

Clegg JS, Drost-Hansen W (1991) On the biochemical and cell physiology of water. In: Hochachka, Mommsen (eds), Biochemistry and Molecular Biology of Fishes, vol 1, Elsevier, NY

Cluster Quackery Available via www.chem/.com/CQ/cluqk.html

Drost-Hansen W, Singleton J, Lin _ (1991) Our aqueous heritage: evidence for vicinal water in cells. In: Bitner EE (ed), Fundamentals of Medical Cell Biology, vol 3A. JAJ Press Inc
Fullerton GD, Finnie MF, Hunter KE, Ord VA, Cameron IL (1987) The influence of macromolecular polymerization on spin-lattice relaxation of aqueous solutions. Magn Reson Imaging 5:353–370
Fullerton GD (2006a) Evidence that collagen, tendon and cellular proteins have monolayer water coverage in the native state. Cell Biol Int 30:56–65
Fullerton GD, Nes E, Amurao M, Rahal A, Krasnosselskaia L, Cameron IL (2006b) An NMR method to characterize multiple water compartments on mammalian collagen. Cell Biol Int 30:66–73
Fullerton GD, Kanal KM, Cameron IL (2006c) Osmotically unresponsive water fraction on proteins. Cell Biol Int 30:86–92
Fullerton GD, Ord VA, Cameron IL (1986) An evaluation of the hydration of lysozyme by an NMR titration method. Biochem Biophys Acta 869:230–246
Kanal KM, Fullerton GD, Cameron IL (1994) A study of the molecular source of nonideal osmotic pressure of bovine serum albumin solutions as a function of pH. Biophys J 66:153–160
Jarvis SP, Uchihoshi T, Ishida T, Tokumoto H (2000) Local solvation shell measurements in water using a carbon nanotuble probe. J Phys Chem 104:6091–6094
Ling GN. (1972) In: Horne A (ed), Water and Aqueous Solutions: Structure, Thermodynamics and Transport Processes. Wiley-Interscience, NY, pp 663–699
Ling GN (1984) In Search of the Physical Basis of Life, Plenum, NY
Ling GN (2001) Life at the Cell and Below-Cell Level Pacific Press, NY
Ling GN (2003) A new theoretical foundation for the polarized-oriented multilayered theory of cell water and for inanimate systems demonstrating long-range dynamic structuring of water molecules. Physiol Chem Phys Med NMR 35:91–130
Miller S, Janin J, Lesk AM, Chothia C (1987a) The accessible surface area and stability of oligomeric proteins. Nature 328:834–836
Miller S, Janin J, Leek AM, Chothia C (1987b) Interior and surface of monomeric proteins. J Mol Biol 196:641–656
Pollack GH (2001) Cells, Gels and the Engines of Life, Ebner and Sons, Seattle, WA
Prouty MS, Schechter AN, Parsegian VA (1985) Chemical potential measurements of deoxyhemoglobin S polymerization. Determination of the phase diagram of an assembled protein. J Mol Biol 184:517–528
Urry DW (1993) Molecular machines: how motion and other functions of living organisms can result from reversible chemical changes. Angewandte Chemie Intl Ed Engl 32:819–841
Urry DW (1995) Elastic biomolecular machines: Synthetic chains of amino acids, patterned after those in connective tissue, can transform heat and chemical energy into motion Sci Am 64–9
Uversky VN (2002) What does it mean to be natively unfolded? Eur J Biochem 269:2–12
Wiggins PM (1995) Micro-osmosis in gels, cells and enzymes. Cell Biochem Funct 13:165–172
Zimmerman RJ, Kanal KM, Sanders J, Cameron IL, Fullerton GD (1995) Osmotic pressure method to measure salt induced folding/unfolding of bovine serum albumin. J Biochem Biophys Methods 30:113–131

CHAPTER 17

THE PHYSICAL NATURE OF THE BIOLOGICAL SIGNAL, A PUZZLING PHENOMENON: THE CRITICAL CONTRIBUTION OF JACQUES BENVENISTE

YOLÈNE THOMAS*, LARBI KAHHAK AND JAMAL AISSA
Laboratoire de Biologie Numérique, 32 rue des Carnets, Clamart 92140, France

Abstract: Making a brief history of what is named the 'Memory of Water' is obviously not an easy task. Trying to be as fair and accurate as possible is hampered by two main difficulties: 1) one of the main actors, Jacques Benveniste, recently passed away and 2) cutting edge science creates many controversies, especially with those whose lifetimes have been spent pursuing an unorthodox track. High dilution experiments and memory water theory may be related, and may provide an explanation for the observed phenomena. As Michel Schiff said: 'the case of the memory of water may or not contribute to the knowledge about water structure. Perhaps the tentative interpretation Jacques suggested will finally have to be modified or even abandoned. Time and further research will tell, provided that one gives the phenomena a chance (Schiff, 1995, p 45)'

Keywords: human neutrophil; guinea pig heart; coagulation; water; audio-frequency oscillator; computer-recorded signals

Abbreviations: EMF: electromagnetic field; PMA: phorbol-myristate-acetate; ROM: reactive oxygen metabolites; ACh: acetylcholine; H: histamine; DTI: Direct Thrombin Inhibitor; d-X: digital EMF signal from the molecule

1. INTRODUCTION: THE EARLY HISTORY OF HIGH DILUTIONS EXPERIMENTS / HISTORICAL CONTEXT

Jacques Benveniste gained an international reputation as a specialist on the mechanisms of allergies and inflammation with the 'Platelet Activating Factor' (paf-acether) discovery in 1972 (Benveniste et al., 1972, 1974). Benveniste's

* *present address*: Institut Andre Lwoff IFR89, 7, rue Guy Moquet-BP8, 94 801 Villejuif Cedex, France. email: yolene@noos.fr

G. Pollack et al. (eds.), Water and the Cell, 325–340.

research into allergy has taken him deep into the mechanisms which create such responses. Understanding that the smallest amount of a substance affects the organism - 'A person can enter a room two days after a cat has left it and still suffer an allergic response' – led Benveniste in the mid-eighties, to research how homeopathic dilutions appear to have a real and material effect upon immune system cells called basophils. After 5 years of research he and his collaborators empirically observed that highly dilute (i.e., in the absence of any physical molecule) biological agents triggered relevant biological systems. It is worth recalling that at that time, two papers were submitted and published in peer review journals, the *European Journal of Pharmacology* and the *British Journal of Clinical Pharmacology* (Davenas et al., 1987; Poitevin et al., 1988). Here, the work was treated as conventional research like many other manuscripts from peer-reviewed journals which can be found in the scientific literature on the effect of high dilutions (Schiff, 1995, p 150; Elia et al., 2004).

In 1988, Benveniste's laboratory (I.N.S.E.R.M U 200) and three external laboratories announced that their research showed that highly diluted antibodies could cause the degranulation of basophils and that water has a memory. Briefly, the experimental dilution (anti-IgE) and the control one (anti-IgG) has been prepared in exactly the same manner, with the same number of dilution and agitation sequences. They co-authored an article, which was submitted to *Nature* (Davenas et al., 1988). *Nature*'s referees could not find any fault in Benveniste's research. It was G. Preparata. and E. Del Guidice (quantum physicists working at Milan University) at a conference organized a few months before the *Nature* 'affair' erupted, who brought the theoretical basis for what is known as 'the memory of water'. They have hypothesized that interactions between the electric dipoles of water and the radiation fields of a charged molecule generate a permanent polarization of water which becomes coherent and has the ability to transmit specific information to cell receptors, somewhat like a laser (Del Giudice et al., 1988). Two weeks after publication, the three-man fraud squad (John Maddox, James Randi and Walter Stewart) sent by *Nature* spent 5 days in the laboratory. The investigation concluded that Benveniste had failed to replicate his original study (Maddox et al., 1988). This marked the beginning of the 'Water Memory' saga, which placed him in a realm of 'scientific heresy'. As Michel Schiff remarked: 'INSERM scientists had performed 200 experiments (including some fifty blind experiments) before being challenged by the fraud squad. The failure to reproduce (Maddox et al., 1988) only concerned two negative experiments (Schiff, 1995, p 88, 151). Benveniste replied (Benveniste, 1988) and reacted with anger: ' – not to the fact that an inquiry had been carried out, for I had been willing that this be done – but to the way in which it had been conducted and to the implication that my team's honesty and scientific competence were questioned. The only way definitely to establish conflicting results is to reproduce them. It may be that we are all wrong in good faith. This is not crime but science – ' In rebuttal, we simply refer the reader to the article confirming the initial findings in *Nature*, which appeared in the

Comptes Rendus de l'Académie des Sciences de Paris in 1991 (Benveniste et al., 1991), reporting the results of subsequent blind experiments entirely designed and run by Alfred Spira, and his research I.N.S.E.R.M Unit of independent statistical experts.

To date, since the Nature publication in 1988, several laboratories have attempted to repeat Benveniste's original basophils experiments. Importantly, a blind multi-center trial of four independent research laboratories in France, UK, Italy and Holland, confirmed that high dilutions of histamine modulate basophil activity (Belon et al., 1999, 2004; Brown et al., 2001) Histamine solutions and controls were prepared independently in three different laboratories. This trial was coordinated by an independent laboratory led by M. Roberfroid at Belgium's Catholic University of Louvain, who coded all the solutions and collected the data, but was not involved in the experiments. In addition, an independent statistician analyzed the resulting data. Not much room, therefore, for fraud or wishful thinking. Three of the four labs involved in the trial reported a statistically significant inhibition of the basophil degranulation reaction by high dilutions of histamine compared with the controls. The fourth lab gave a result that was almost significant, so the total result over all four labs was positive for histamine high dilution solutions. 'We are,' the authors say in their paper, 'unable to explain our findings and are reporting them to encourage others to investigate this phenomenon.' Benveniste may well have been right all along.

In the meantime, between the repetitions, Benveniste and his team, of which we were members, found the time to do their part: research aimed at understanding the physical nature of the biological signal. In particular, we asked ourselves questions concerning the nature of the biological activity in high dilutions. We suspected some sort of ordering involving electromagnetism. Indeed, in collaboration with an external team of physicists (Lab. Magnetisme C.N.R.S.-Meudon Bellevue, France), we showed in twenty four blind experiments that the activity of highly dilute agonists was abolished either by heating (70° C, 30 min) or exposure to a magnetic field (50 Hz, $15 \times 10-3$ T, 15 min) which had no comparable effect on the genuine molecules (Hadji et al., 1991). We could thus speculate that transmission of this ordering principle was electromagnetic (EM) in nature. Furthermore, it is not insignificant that a growing number of observations suggest the susceptibility of biological systems or water to electric and low-frequency electromagnetic fields (Tsonga, 1989; Frey, 1993; Blanchard et al., 1994; Novikov et al., 1997; Vallée et al., 2005). *Together, these considerations informed exploratory research which led us to speculate that biological signalling might involve low frequency waves potentially transmissible to cells or water by purely electromagnetic means.*

For the sake of simplicity, we shall present here only three salient biological models. The detailed descriptions of the different models have also been reported in publications, technical reports and patents, most of which are available on the digibio website (www.digibio.com).

2. MATERIALS AND METHODS

2.1 Reagents

Ultra-pure water (W), phenol red-free Hank's balanced salt solution (HBSS) were obtained from Biochrom; cytochrome c (horse heart, type III), 4-phorbol-12-b-myristate-13-acetate (PMA), acetylcholine (ACh), histamine (H), bovine thrombin and bovine fibrinogen were obtained from Sigma Chemicals. PMA was dissolved in DMSO at 10 mM and stored at −20° C. Vehicle (DMSO from the same batch) was also aliquoted and stored at −20° C. Immediately before use, the stock solutions was diluted to appropriate working concentrations in W. Vehicle consisted of DMSO at the same concentration as that present in the respective PMA solutions.

ACh and H was dissolved in water at 1 μM and stored at −20° C. Bovine thrombin (1 U/ml) and bovine fibrinogen (24 mg/ml) were dissolved in W and NaCl 0.9% respectively, then aliquoted and stored at −20° C. All plastic materials were sterile and purchased from Becton-Dickinson.

2.2 Preparation of Human Neutrophils

Human blood from consenting healthy donors was anticoagulated with citric acid-dextrose. Blood was sedimented for 30-45 min in 0.3% final gelatin. The supernatant was layered on Ficoll-Hypaque and centrifuged. The cell pellet was resuspended in 1 ml of washing buffer (HBSS supplemented with 0.25% (v/v) BSA, 1ng/ml LPS and 20mM HEPES). Erythrocytes were lysed by adding 3 vol. distilled water to the cell suspension, followed 40s. later by 1 vol. of NaCl 3.5% (w/v). Cells were then washed twice, resuspended in washing buffer and counted. All preparations contained at least 98% neutrophils as determined by microscopic observation after staining with May Grünwald-Giemsa (Leyravaud S et al., 1989). Before transmission or addition of molecular agonists, neutrophils were suspended at 1×10^6/ml in washing buffer and Ca^{2+} (1.3 mM), Mg^{2+} (1mM) and cytochrome c (80 uM) were added to the cell suspension which was then aliquoted (1 ml) into Eppendorf tubes (Thomas et al., 2000). Reactive oxygen metabolites (ROM) production was measured as the reduction of cytochrome c using a spectrophotometer at 550 nm.

2.3 Heart Preparation (Figure 1)

Isolated hearts were perfused according to the classical Langendorff method (Benveniste et al., 1983; Kim et al., 1983). Acetylcholine (ACh), histamine (H) or water (W) was injected via a catheter just above the aorta. Variation in coronary flow (CF) was measured every min for 30 min. During the same time, other mechanical parameters (min. and max. tension, heart rate) were recorded using a dedicated software (Emka Technonologies, Paris, France). Percent (%) increase in CF was calculated as follows: [1 -(CF maximal value / CF time 0 value)] $\times$ 100.

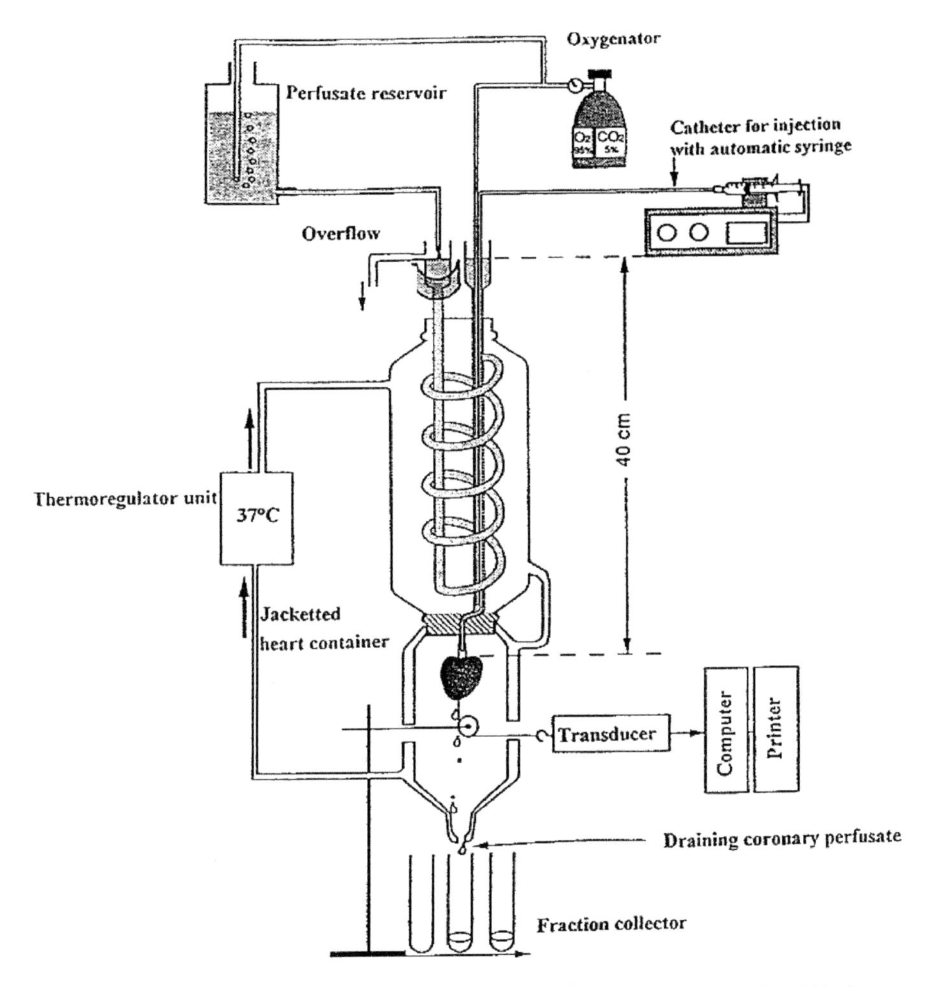

Figure 1. Langendorff heart perfusion system. Isolated hearts (male Hartley guinea-pigs, 300 g) were perfused using Krebs-Henseleit buffer (pH 7.4) gassed with O2/CO2, 95/5%, at a pressure of 40 cm H2O at 37°C. Samples are injected (2 ml) via a catheter just above the aorta

2.4 In Vitro Coagulation

During blood coagulation there is a complex series of molecular interactions. Two of the molecules are thrombin and fibrinogen. These two can interact alone in water without any of the other players normally found in the formation of a clot (Greenberg et al., 1985). Thrombin is a serine proteinase that converts fibrinogen to fibrin. At room temperature and within a short time, a clear clot will form. Addition of a Direct Thrombin Inhibitor (DTI), such as melagatran (Gustaffson et al., 2003) can delayed or even blocked entirely the thrombin–fibrinogen reaction. Coagulation

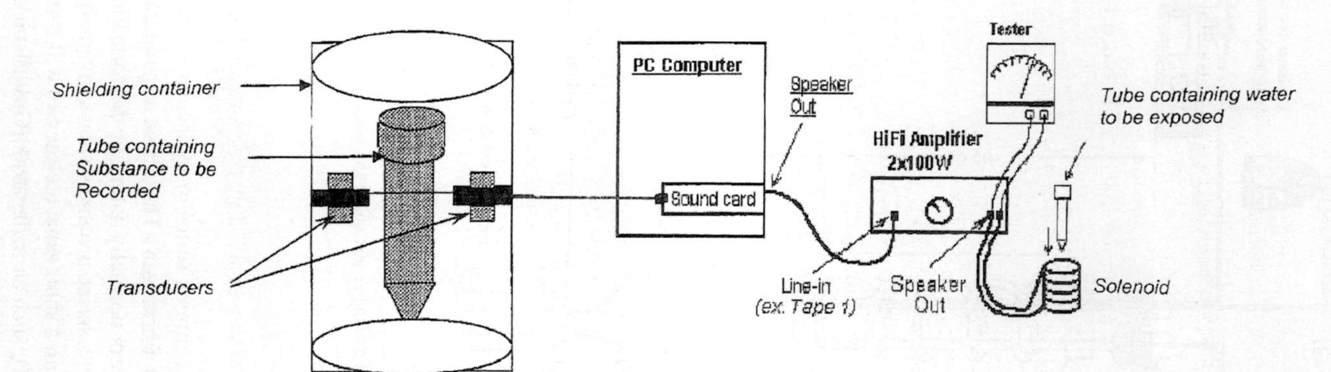

Figure 2. Schematic drawing of the computer-recorded signals: capture, storage and replay. Shielded cylindrical chamber: composed of three superposed layers: copper, soft iron, permalloy, made from sheets 1 mm thick. The chamber has an internal diameter of 65 mm, and a height of 100 mm. A shielded lid closes the chamber. Transducers: coil of copper wire, impedance 300 Ohms, internal diameter 6 mm, external diameter 16 mm, length 6 mm, usually used for telephone receivers. Multimedia computer (Windows OS) equipped with a sound card (5KHz to 44 KHz in linear steps), (Sound Blaster AWE 64, CREATIVE LABS). HiFi amplifier 2x100 watts with an 'in' socket, an "out" socket to the speakers, a power switch and a potentiometer. Pass band from 10 Hz to 20 kHz, gain 1 to 10, input sensitivity +/− V. Solenoid coil: conventionally wound copper wire coil with the following characteristics: internal diameter 50 mm, length 80 mm, R = 3.6 ohms, 3 layers of 112 tums of copper vire, field on the axis to the centre 44 10^{-4} T/A, and on the edge 25 10^{-4} T/A. All links consist of shielded cable. All the apparatus is earthed

is assessed by spectrophotometry at OD_{620}. Percent (%) inhibition coagulation was calculated as follows: $[1-(OD_{620}\,DTI/OD_{620}\,W)] \times 100$.

2.5 Transmission Apparatus: Audio-Frequency Oscillator

The device used for transmission comprised a standard audio amplifier (Kemo kit 105, West Germany) with magnetic coils connected respectively to the input and output (impedance 8 ohms). Tubes whose contents were to be transmitted were placed on the input coil and cells or water on the output coil. When the amplifier was not connected to the output coil, its output, as viewed with an oscilloscope, appeared to be noise with some 50 Hz contaminations from the French power grid. However, when the amplifier was connected to the output coil, it behaved as an audio-frequency oscillator and signal analysis revealed the emission of a stable square wave with a frequency of about 3 kHz and voltage of approximately 7 V. In the presence of a weak, mV range signal not only the amplitude but also the frequency of the wave were modulated (WO patent-94-17406). During the transmission procedure, the various parameters such as power, voltage, capacitance and impedance remained constant, the nature of the source tube being the only variable.

2.6 Computer-Recorded Signals: Capture, Storage and Replay

The characteristics of the designed apparatus are described in Figure 2 and in the US patent-03-6541978. Briefly, the process is to first capture the electromagnetic signal from a biologically active solution and store this digitized signal on a computer's hard drive: Thus, tubes containing ACh, H, DTI at 1 μM or W were used as source. After recording (6 sec, 16 bits in mono mode, 44 kHz) the signal is then "played back" for 10 mins from the computer sound card through a solenoid coil containing a tube of water (tension of 4 Volts). The digital signals were standard Microsoft sound files (*.wav). The order of the conditions and their repetitions was always randomized and blinded. For ease in the discussion, the terminology d-X refers to the digital EMF signal from the molecules.

3. RESULTS

3.1 Mimicking the Effects of Molecules Using a Transmission Apparatus: Audio-Frequency Oscillator

Between 1991 and 1996, using a standard audio amplifier that, when connected to another coil, behaves as an audio-frequency oscillator, we performed a number of experiments showing that we could transfer specific molecular signals to water or directly to cells. For instance, we investigated whether molecular signals associated with PMA could be transmitted by physical means to human neutrophils to modulate reactive oxygen metabolite (ROM) production. Briefly, neutrophils were placed in

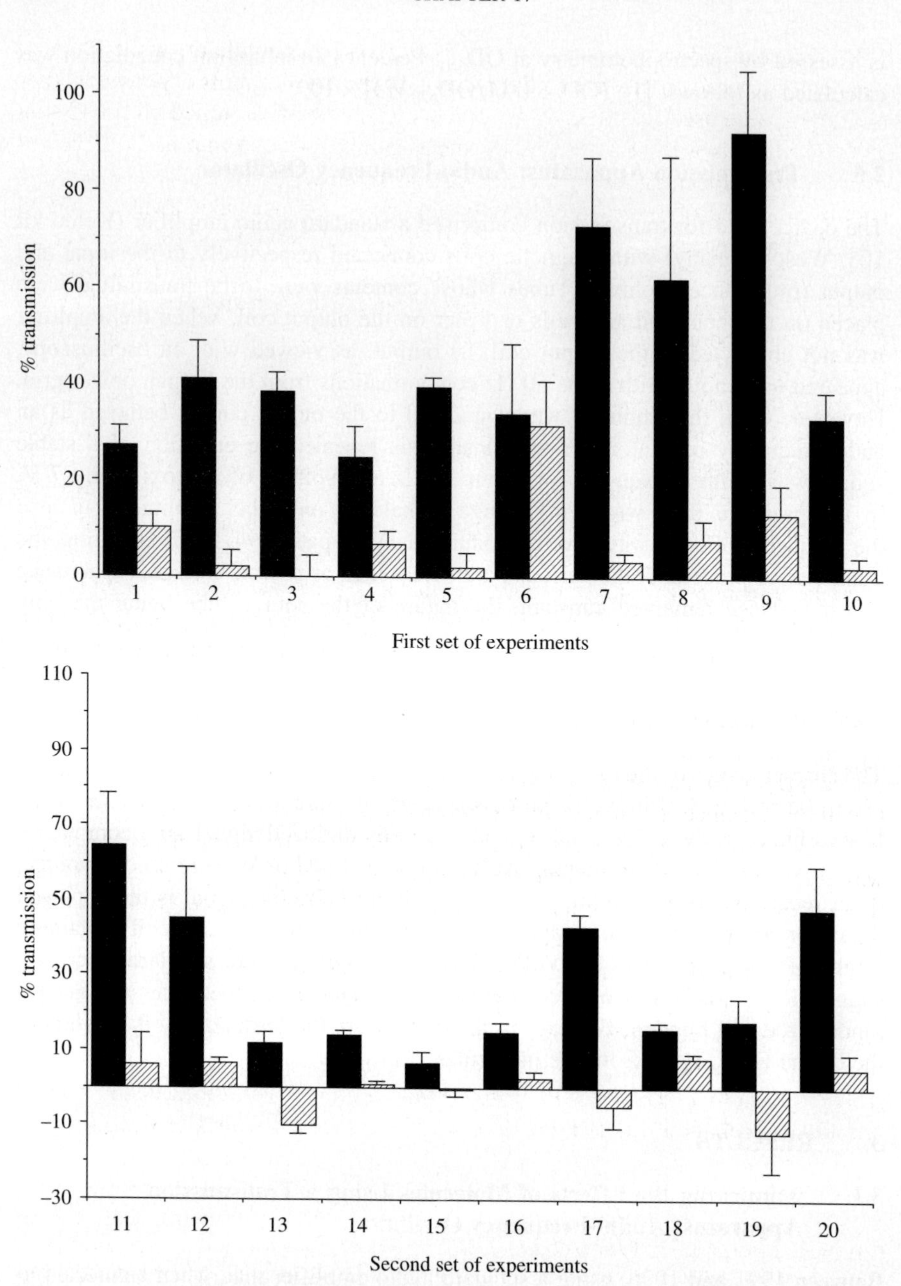

Figure 3. Effect of transmitted phorbol-myristate-acetate on neutrophil ROM production. For each transmission sequence to neutrophils, the input coil coupled to the amplifier was operated at room temperature, while the output coil was placed in a 37°C humidified incubator. The tube containing PMA, (1 uM) or vehicle was placed on the input coil, and tubes (duplicate) containing neutrophils on the output coil. The oscillator was then turned on for the 15 min transmission period. In each experiment, 4 simultaneous

a 37° C humidified incubator on one coil attached to the oscillator, while PMA or vehicle was placed on another coil at room temperature. For most experiments four oscillators were used simultaneously. The oscillator was then turned on for 15 min after which cells were usually further incubated for up to 45 min at 37° C before OD_{550} measurement. Additional check consisted of unexposed cells. The positive control consisted of neutrophils directly stimulated by molecular PMA (1 pM to 10 uM). The procedure and the results of twenty consecutive blind experiments are shown in Figure 3. One of the two series of experiments was performed in a different laboratory, with randomization and coding of source tubes being performed by the head of the laboratory (Dr. F. Russo Marie, INSERM U332). Exposing cells to transmitted PMA (T-PMA) resulted in an OD increase of 37 ± 4% (mean ± S.E.M, 40 transmissions) compared to unexposed cells. By contrast, exposing cells to transmitted vehicle (T-vehicle) resulted in a 4.1 ± 1.8% change. In the absence of cells, transmission of PMA or vehicle alone was without effect on cytochrome c reduction. The effect of transmitted PMA was roughly equivalent to that of 0.1 nM molecular PMA. Additional experiments indicate that ROM were not induced when 4 α-phorbol 12,13-didecanoate (PDD), an inactive PMA analogue, was transmitted in the same manner as PMA. The observation that T-PMA but not T-PDD stimulated ROM production suggested the involvement of Protein kinase C (PKC), the main target of PMA. Indeed, the impact of transmitted PMA was substantially reduced in cells pretreated with two PKC inhibitors, GF109203X or H-7 (Thomas et al., 2000).

We next attempted to block the transmission effect: one parameter of the basic design was modified in half of the transmissions. Either: 1) the oscillator was turned off or 2) the PMA solution or the cells were shielded with Mu-metal (an alloy designed to inhibit magnetic fields down to low frequencies). Data of 12 independent experiments indicate that PMA transmission effect (42 ± 8%) was essentially suppressed when the amplifier was turned off (−1.8 ± 1.4%) and when either the PMA solution or the neutrophils were shielded with Mu-metal (−4.3 ± 2.7%,).

The statistical significance of the experiments was analyzed using the Student's t-test. Percent transmission (as defined in the legend of Figure 3) was computed for each set of cells (cells exposed to T-PMA, T-vehicle, T-PDD or T-PMA oscillator off). Differences between cells exposed to T-PMA and other experimental groups (cells exposed to T-vehicle, T-PDD or T-PMA oscillator off) were calculated at 60 min (total incubation time). T-PMA cells were associated with a 33.6 ± 3.4% OD

◄

Figure 3. transmissions were performed, using 4 source tubes (2 PMA and 2 vehicles). These 4 source tubes were prepared, randomized and blinded by coding at the beginning of each experiment. After transmission, the oscillators were switched off and all cells were left in the incubator for the additional 45 min post-transmission incubation period, before OD measurement. Additional check consisted of unexposed cells. Viability of all samples was assessed by trypan blue exclusion both before and after incubation. For each individual experiment, percent (%) transmission was calculated as: 100× (OD_{550} exposed cells - OD_{550} unexposed cells) / OD_{550} unexposed cells. Each error bar corresponds to the standard error estimated from 4 OD values of exposed cell-tubes. (black bar) T-PMA cells; (hatched bar) T-vehicle cells

increase, in contrast to $2.3 \pm 1.3\%$ ($n = 58$ transmissions, $p < 10^{-3}$) for T-vehicle, T-PDD and T-PMA oscillator off (Thomas et al., 2000).

Although, the precise physical mechanism(s) involved remain(s) unknown, together, these results suggest that PMA molecules emit signals that can be transferred to neutrophils by artificial physical means in a manner that seems specific to the source molecules. Along this line are other studies showing transmission of thyroxine signal via electronic circuit using water as target for the transmitted signal (Endler et al., 1995). Part of this work was published (Thomas et al., 2000). Appended to this article were two affidavits, one from a French laboratory testifying that they supervised and blinded the experiments we did in this laboratory; the other from an US laboratory (W. Hsueh, Department of Pathology, Northwestern University, Chicago) testifying that they did some preliminary experiments similar to ours, without any physical participation on our part, and detect the same effect as we described.

3.2 Mimicking the Effects of Molecules Using a Computer-Recorded Signal

Because of the material properties of the oscillator and the limitations of the equipment used, it is most likely that the PMA signal is carried by frequencies in the low kilohertz range. Theses considerations led to the establishment in 1995 of a new procedure for the recording and retransmission of the molecular signals (Figure 2). Briefly, the process is to first capture the EM signal from a biologically active solution and store this digitized signal on a computer's hard drive. The EM signal is then "played back" through a sound card to a solenoid containing a tube of water.

One of the biological systems, which can be used to detect digital files endowed with biological activity, is the measurement of coronary flow (CF) in isolated perfused guinea-pig hearts (Fig. 1). In particular, we investigated the effect of digital EMF signals of acetylcholine (d-ACh) and histamine (d-H). Digital EMF signal of water (d-W) and ACh or H, similarly were applied as negative and positive controls respectively. The procedure and the results of consecutive blind experiments performed between November 21, 1997 and April 14, 1998 are shown in Table 1.

d-ACh, ACh, d-H and H increase CF compared to d-W. d-W induced effects that were indistinguishable from spontaneous flow variations. The two comparisons d-ACh vs d-W and d-H vs d-W are both significant ($p < 0.05$, Student's t test for unpaired variates, Sigma plot 40, Jandel Scientific Corte, Madena, CA). Interestingly, atropine, an ACh inhibitor, inhibited both the effects of the ACh and d-ACh but not those of H and d-H. Mepyramine, an H1 receptor blocker, inhibited both H and d-H but not ACh and d-ACh.

In 1996, a team from Northwestern University at Chicago recorded a group of biological signals, either from bioactive solutions (ACh, Ovalbumin (OVA), ...) or control (water), on a computer with a sound card, (using a recording instrument

Table 1. Effects of digital acetylcholine and histamine on the coronary flow in isolated guinea-pig hearts (Consecutive blind experiments performed: November 21, 1997-April 14, 1998)

Exp.	d-W	d-ACh	ACh 1uM	d-H	H 1uM
A. Buffer	4.6 ± 2.1 [28]	19.5 ± 7.4 [21]	26.6 ± 8.3 [16]	14.3 ± 2.5 [14]	21.1 ± 8.4 [5]
B. Buffer + atropine	4.2 ± 1.3 [12]	7.3 ± 2.8 [10]	8.8 ± 3.3 [3]	14 ± 2.1 [3]	23.6 ± 4.3 [4]
C. Buffer + mepyramine	5.9 ± 2.0 [9]	19.1 ± 3.9 [3]	29.5 ± 4.2 [5]	5.8 ± 1.8 [5]	8.2 ± 2.9 [6]

Acetylcholine (ACh), histamine (H) and water (W) were recorded as in Fig. 2. Files were digitally amplified and the signal of digital EMF ACh (d-ACh), H (d-H) or W (d-W) was replayed as described in Materials and Methods. Atropine is used to inhibit the action of ACh, and mepyramine, to inhibit the action of H.

A. Water, appropriately exposed to d-ACh or d-H, was then infused to isolated hearts. d-W, ACh or H at 1 uM were infused as negative and positive controls respectively.

B. Water, appropriately exposed to d-ACh or d-H, was then infused in the presence of atropine (2 mg/ml), to isolated hearts. d-W, ACh or H at 1 uM were infused as negative and positive controls respectively.

C. Water, appropriately exposed to d-ACh or d-H, was then infused in the presence of mepyramine (5 mg/ml), to isolated hearts. d-W, ACh or H at 1 uM were infused as negative and positive controls respectively.

Results are expressed as percent (%) increase in CF as defined in Materials and Methods. Data are presented as mean $\pm$ SD, nb of experiments.

provided by us), and transmitted them to us, blinded, via Internet. Several months of "fine tuning" the methodology by both teams (including determining the optimal time interval and amplification of recording settings, the optimal settings for playing back the signal, the way of handling the samples, sending the file via e-mail one file a time, rather than sending all files together, using the same stock solutions, etc.) had to be done in order to eliminate the variables which might interfere with the recording and transmission of electromagnetic molecular signals. Although the possibility exists that we were not completely successful in removing these interfering variables, we could detect the transmitted biological activities with high accuracy (% increase in CF). For instance: d-OVA: 24.0 ± 1.4, $n = 30$ compared to d-water 4.4 ± 0.3, $n = 58$ ($p = 4.5\ e^{-17}$, Student's t test for paired variates). OVA $0.1 \mu M : 28.9 \pm 3.7$, $n = 19$ is not statistically different compared to d-OVA.

In 1999, the Team developed an other biological system: inhibition of fibrinogen coagulation by a Direct Thrombin Inhibitor (DTI). The hypothesis tested was whether the reaction rate for coagulation between thrombin and fibrinogen could be modulated by d-DTI. d-W and DTI (1uM) were used as negative and positive controls respectively. As illustrated in a representative experiment (Figure 4), addition of DTI and d-DTI result in a slower reaction rate as compared to W or d-W. The results of twenty-two consecutive blind experiments performed between April 16 and June 26, 2004 are shown in Table 2. In the majority of the experiments d-DTI prolongs the clotting compared to d-W although to a lesser extent than 1 uM

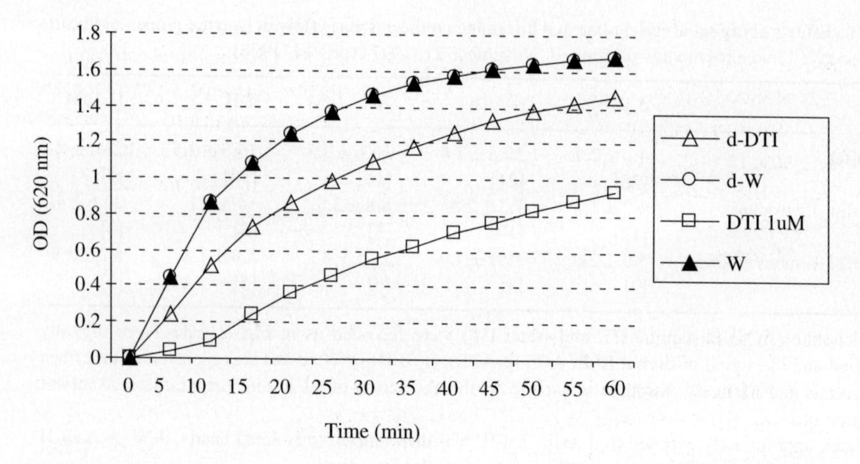

Figure 4. Effects of a Direct thrombin inhibitor on thrombin induced fibrinogen coagulation. Direct thrombin inhibitor (DTI) and water (W) were recorded. Files were digitally amplified and the signal of digital EMF DTI (d-DTI) or W (d-W) was replayed for 10 min, as described in Materials and Methods. Water, appropriately exposed to d-DTI is added to fibrinogen along with thrombin (Thr). W, d-W and DTI (1 uM) were used as negative and positive controls respectively. After different time periods, coagulation is assessed by spectrophotometry and expressed as OD_{620}. One representative experiment is shown

Table 2. Effects of direct thrombin inhibitor on thrombin induced fibrinogen coagulation (Consecutive blind experiments performed: April 16–June 26, 2004)

		Mean ± SD [n]
d-W	0, 0, 1, 0, 0, 0, 0, 0, 0, 0, 0, 0, 0, 0, 0, 0, 0, 0, 0 0, 0, 1	0.09 ± 0.29 [22]
d-DTI	38, 25, 39, 31, 36, 61, 44, 40, 35, 35, 65, 30, 26, 26, 31, 24, 28, 26, 81, 16, 35, 20	36.00 ± 15.36 [22]
DTI 1uM	65, 70, 68, 72, 75, 75, 69, 71	70.62 ± 3.42 [8]

Water, appropriately exposed to the digital EMF signal of DTI (d-DTI) is added to fibrinogen along with thrombin. Water (W), digital EMF water (d-W) and DTI (1 uM) were used as negative and positive controls respectively. Coagulation is assessed by spectrophotometry at OD_{620}. Results are presented at 30 min and expressed as percent (%) inhibition coagulation as defined in Materials and Methods. Data are mean ± SD, nb of experiments.

DTI. The comparison d-DTI vs d-W is highly significant ($p = 3.7\ e^{-10}$, Student's t test for unpaired variates).

These results suggest that at least some biologically active molecules emit signals in the form of electromagnetic radiation of less than 44 kHz that can be recorded and digitized. The digitized signal can be replayed to water, target cells or organs in a manner that seems specific to the source molecules.

However, our attempts to replicate these data in four other laboratories yielded mixed results. We then realized the difficulty in 'exporting' a method, which is very far from conventional biology. This may reflect key variables like, for instance, the purity of water, its conductance, the purity of the chemicals, electromagnetic environmental conditions. Also, individual variations of the operator's performance could explain some erratic results. In order to eliminate these uncontrolled parameters, the same reagents are always used and two shielded robots were built in order to eliminate the distorting effects of human intervention. An external laboratory where a team of scientists is currently attempting to replicate the experiments is using one of those.

4. DISCUSSION: THE CURRENT STATE OF KNOWLEDGE

Among the various theoretical problems associated with such a signal, three appear particularly pertinent. The first relates to background noise. Given the level of electromagnetic noise present in the environment, it is necessary to postulate ways in which the signal-to-noise ratio or the detection of specific signals, or both, are enhanced. In fact, an appropriate level of noise enhances a specific periodic signal rather than overwhelming it, a phenomenon known as stochastic resonance (Wiesenfeld et al., 1995; Astumian et al., 1995; Pickard, 1995). The relevance of this concept to the phenomena reported here remains to be determined. Second, the limitations of the equipment used here, suggest that the signal is carried by frequencies in the low kilohertz range, many orders of magnitude below those generally associated with molecular spectra (US patent-03-6541978). The 'beat frequency' phenomenon may explain this discrepancy, since a detector, for instance a receptor, will 'see' the sum of the components of a given complex wave (Banwell, 1983). Third, how to explain the ability of water to carry and memorize biological signals? Will Quantum ElectroDynamics (QED) provide these answers (Del Giudice et al., 1988; Preparata et al., 1995)?. QED-based long-range electromagnetic communication between molecules may represent the founding theory able to unravel the nature of the molecular signal and the role of perimolecular water in its transmission. The best is to let Preparata explain it himself (excerpts from the proceedings of the meeting (14/12/1999) at the Institute of Pharmacology, University of Rome 'La Sapienza', *The role of QED in medicine* : 'The space-time order in biochemistry cannot be the product of the chemical interactions whose range is too short (a few Angstroms) to allow the molecules to detect each other from afar and, moreover, when they are inside a crowd of other molecules, not involved in the specific biochemical sequence. QED solves this problem completely, since, within a coherent medium, molecules may interact through their common coupling to the electromagnetic field and the intensity of the force depends inversely upon the difference of their oscillation frequencies, so that molecules whose oscillation frequencies are significantly different ignore each other, whereas resonant molecules attract themselves strongly. We get thus a selective recognition code based on the electromagnetic resonance, which could provide the dynamic basis to the biochemical codes. Electromagnetic fields have a long range and then are able to produce a recognition at a distance, also in a crowd of non-resonating molecules'.

Alternative hypotheses have been proposed for explaining water memory. For instance, one hypothesis predicts changes in the water structure by forming more or less permanent clusters (Fesenko et al., 1995). Louis Rey using a technique that measures thermoluminescence points to the unusual properties of water under certain treatments suggesting that water does have a memory of molecules that have been diluted away (Rey, 2003). Clearly, more theoretical and experimental work is needed to unveil the physical basis of the transfer (and storage?) of specific biological information either between interacting molecules or via an electronic device.

5. CONCLUDING REMARKS

This story, exemplifies the fact that most if not all researchers, nowadays and in the past, were misguided to apply existing reasoning and methods to a completely new domain of research.

The debate on the memory of water started in 1988 and in 2005, i.e., 17 years late, the majority of the scientific community rejects it, even though an increasing number of scientists report they have confirmed the basic results made by Jacques Beneveniste 'group.

As Isaac Behar, who has worked closely with Jacques Benveniste, pointed out: 'a parallel can be drawn between the polemics on memory of water, presuming that the action of molecules are mediated by an electromagnetic phenomenon, and the polemics on the transmission of nerve influx. This debate started in 1921 with the first experiments performed by Otto Loewi. The polemic was still active in 1949 i.e., 28 years after the first test assuming that transfer of nerve influx through synapses are mediated by specific molecules, the neurotransmitters (Bacq, 1974).'

Since the very beginning we have placed a great deal of emphasis on carrying out our work under the highest standards of methodology and great effort has been made to isolate it from environmental artifacts. More difficulties most probably lie ahead. Now that Jacques Benveniste is no longer with us, the future of the 'digital biology' is in the hands of those who have been convinced of the reality of the basic phenomena. Most likely they will succeed if they combine full biological and physical competences to understand the nature of the biological signals (Ninham, 2005).

ACKNOWLEDGEMENTS

The authors express their sincere appreciation to the members of the laboratory staff, past, present and future, whose valuable contributions have been essential to the success of this scientific adventure. A special mention is given to Françoise Lamarre who for 30 years has served as the executive secretary. Now she is continuing her part through the 'Association Jacques Benveniste pour la Recherche' (http://jacques.benveniste.org). We are deeply grateful to supporters and financial investors who have enabled the "Laboratoire de Biologie Numerique" to carry on the work thus far. We are also indebted to Dr. Wei Hsueh (Northwestern University, Department of Pathology, Chicago, USA) for her valuable scientific contributions and collaborations.

REFERENCES

Astumian RD, Weaver JC, Adair RK (1995) Rectification and signal averaging of weak electric fields by biological cells. Proc Natl Acad Sci USA 92:3740–3743

Bacq Z (1974) Les transmissions chimiques de l'influx nerveux. Villars G (ed), Paris, France

Banwell CN (1983) Fundamentals of Molecular Spectroscopy. McGraw-Hill Publ UK, pp 26–28

Belon P, Cumps J, Ennis M, Mannaioni PF, Sainte-Laudy J, Roberfroid M, Wiegant FA (1999) Inhibition of human basophil degranulation by successive histamine dilutions: results of a European multi-centre trial. Inflamm Res 48 Suppl 1:S17–S18

Belon P, Cumps J, Ennis M, Mannaioni PF, Roberfroid M,Sainte-Laudy J, Wiegant FA (2004) Histamine dilutions modulate basophil activation. Inflamm Res 53 (5):181–188

Benveniste J, Henson PM, Cochrane CG (1972) Leukocyte-dependent histamine release from rabbit platelets. The role of IgE, basophils, and a platelet-activating factor. J Exp Med 136:1356–1377

Benveniste J (1974) Platelet-activating factor, a new mediator of anaphylaxis and immune complex deposition from rabbit and human basophils. Nature 249:581–582

Benveniste J (1988) Dr Jacques Benveniste replies. Nature 334:291

Benveniste J, Davenas E, Ducot B, Cornillet B, Poitevin B, Spira A (1991) L'agitation de solutions hautement diluées n'induit pas d'activité biologique spécifique. Comptes-Rendus de l'Académie des Sciences de Paris 312:461–466

Benveniste J, Bowllet C, Brink C, Labat C. 1983. The actions of PAF–acether on guinea.pig isolated heart preparations. Br J Pharmacol. 80(1):81–83

Blanchard JP, Blackman CF (1994) Clarification and application of an ion parametric resonance model for magnetic field interactions with biological systems. Bioelectromagnetics 15(3):217–238

Brown V, Ennis M (2001) Flow-cytometric analysis of basophil activation: inhibition by histamine at conventional and homeopathic concentrations. Inflamm Res 50(Suppl 2):S47–S48

Davenas E, Poitevin B, Benveniste J (1987) Effect of mouse peritoneal macrophages of orally administered very high dilutions of silica. Eur J Pharmacol 31, 135(3):313–319

Davenas E, Beauvais F, Amara J, Oberbaum M, Robinzon B, Miadonna A, Tedeschi A, Pomeranz B, Fortner P, Belon P, Sainte-Laudy J, Poitevin B, Benveniste J (1988) Human basophil degranulation-triggered by very dilute antiserum against IgE. Nature 333:816–818

Del Giudice E, Preparata G, Vitiello G (1988) Water as a free electric dipole laser. Phys Rev Lett 61:1085–1088

Elia V, Niccoli M (2004) New Physico-chemical properties of extremely diluted aqueous solutions. J Therm Anal Calorimetry 75:815–836

Endler PC, Pongratz W, Smith CW, Schulte J (1995) Non-molecular information transfer from thyroxine to frogs. Vet Hum Toxicol 37:259–263

Fesenko EE, Gluvstein AY (1995) Changes in the state of water,induced by radiofrequency electromagnetic fields. FEBS Lett 367:53–55

Frey AH (1993) Electromagnetic field interactions with biologicalsystems. FASEB J 7:272–281

Greenberg CS, Miraglia CC, Rickles FR, Shuman MA (1985) Cleavage of blood coagulation factor XIII and fibrinogen by thrombin during in vitro clotting. J Clin Invest 75(5):1463–1470

Gustaffson D, Elg M (2003) The pharmacodynamics and pharmacokinetics of the oral direct thrombin inhibitor ximelagatran and its active metabolite melagatran: A mini-review.Thromb Res 109(Suppl 1):S9–S15

Hadji L, Arnoux B, Benveniste J (1991) Effect of dilute histamineon coronary flow of guinea-pig isolated heart. Inhibition by amagnetic field. FASEB J 1. 5:A–I583

Kim DH, Akera T, Kennedy RH (1983) Ischemia-induced enhancement of digitalis sensitivity in isolated guinea-pig heart. J Pharmacol Exp Ther 226(2):335–342

Leyravaud S, Benveniste J (1989) Regulation of cellular retention of PAF-acether by extracellular pH and cell concentration. Biochim Biophys Acta 1005:192–196

Maddox J, Randi J, Stewart WW (1988) High-dilution'experiments a delusion. Nature 334:287–290

Ninham BW, Boström M (2005) Building bridges between the physical and biological sciences. Cell Mol Biol (The scholars who talk to the wind, Thomas Y & Mentre P (eds)) in press

Novikov VV, Karnaukhov AV (1997) Mechanism of action of weak electromagnetic field on ionic currents in aqueous solutions of amino acids. Bioelectromagnetics 18:25–27

Pickard WF (1995) Trivial influences: A doubly stochastic Poisson process model permits the detection of arbitrarily small electromagnetic signal. Bioelectromagnetics 16(1):2–8 and 9–19

Poitevin B, Davenas E, Benveniste J (1988) In vitro immunological degranulation of human basophils is modulated by lung histamine and Apis mellifica. Br J Clin Pharmacol 25:439–444

Preparata G (1995) QED Coherence in Matter. Singapore: WorldScientific

Rey L (2003) Thermoluminescence of ultra high dilutions of lithium chloride and sodium chloride. Physica A 323:67–74

Schiff M (1995) The Memory of Water. Thorsons (ed), UK

Thomas Y, Schiff M, Belkadi L, Jurgens P, Kahhak L, Benveniste J (2000) Activation of human neutrophils by electronically transmitted phorbol-myristate acetate. Medical Hypotheses 54:33–39

Tsonga TY (1989) Deciphering the language of cells. Trends Biochem Sci 14:89–92

Vallee Ph, Lafait J, Mentré P, Monod MO, Thomas Y (2005) Effects of pulsed low frequency electromagnetic fields on water using photoluminescence spectroscopy: Role of bubble/water interface? J Chem Phys 122:114513–114521

Wiesenfeld K, Moss F (1995) Stochastic resonance and the benefits of noise: From ice ages to crayfish and SQUIDS. Nature 373:33–36

CHAPTER 18

FREEZING, FLOW AND PROTON NMR PROPERTIES OF WATER COMPARTMENTS IN THE TEMPOROMANDIBULAR DISC

CHRISTINE L. HASKIN[1,*], GARY D. FULLERTON[2] AND IVAN L. CAMERON[3]

[1,*]*University of Nevada Las Vegas and School of Dental Medicine*
[2]*University of Texas Health Science Center at San Antonio, Department of Radiology*
[3]*University of Texas Health Science Center at San Antonio, Graduate School of Biomedical Sciences, Cellular and Structural Biology*

Abstract: The temporomandibular joint (TMJ) disc is a loaded tissue that is subjected to pressure during virtually every functional movement. To understand the biomechanical properties of the TMJ disc requires a detailed understanding of how water is bound to and organized around the macromolecular components of the disc. Specifically, how much of the water in the disc is unbound to the macromolecular components and free to flow with the same characteristics of bulk water?

The combined data from three different methods (flow rate, proton NMR dehydration and freezing point characteristics) lead to the conclusion that all or almost all of the water in the intact TMJ disc is bound water and does not have properties consistent with free or bulk water. Two major non-bulk-like fractions of water were identified and their amounts in g water/g dry mass were determined. The inner water compartment has 1.13–1.30 g water/g dry mass while the outer water compartment has 0.90–0.99 g water/g dry mass. That all three methods yielded similar water compartment values indicate these two water compartments have distinct physical properties

Keywords: Hydration, temporomandibular disc, proton NMR

* Corresponding author. 1001 Shadow Lane, School of Dental Medicine, University of Nevada Las Vegas, Las Vegas, NV 89106-4124, USA. Tel.: 1-702-774-2676; Fax: 1-702-774-2651; *E-mail address*: christine.haskin@ccmail.nevada.edu.

G. Pollack et al. (eds.), Water and the Cell, 341–351.

1. INTRODUCTION

The temporomandibular joint (TMJ) disc is between 95 and 98% extracellular connective tissue composed primarily of collagen with a dry weight fraction from 75% to 90% (Milam et al., 1991; Minarelli and Liberti, 1997; Detamore and Athanasiou, 2003), proteoglycans with a dry weight fraction from 10 to 15% (Nakano and Scott, 1989; Kobayashi, 1992), and glycosaminoglycans with a dry weight fraction from 0.5% to 10% of the disc (Detamore et al., 2005). The macromolecular components account for approximately 15–35% of the wet weight of the disc, and the total water by mass of discs varies from 71%–77% (Haskin, 1995; Nakano and Scott, 1996; Sindelar et al., 2000; Tanaka and van Eijden, 2003) and varies significantly between different regions of the disc (Detamore et al., 2005). Along the mediolateral axis the medial region had the highest water content (75.3%) with the central and lateral regions having significantly lower water content (71.3%). Along the antero-posterior axis, the anterior band had 74.5%, the intermediate zone had 73.7% and the posterior band 70.1%. (Detamore et al., 2005). Thus, the most abundant component of the disc is water, with water accounting for about 2.2 g H_2O/g dry mass.

Since water is essentially incompressible under boundary conditions, it has been assumed that the compressive stiffness of the disc is primarily due to the ability of the highly anionic sulphated proteoglycans to trap water within the matrix. However, since water can be demonstrated to flow from the disc under compressive loading (Haskin, 1995; Haskin et al., 2005), it is reasonable to argue that the incompressibility of water is not an adequate explanation of compressive stiffness of the disc. Indeed, the very presence of deformation, indentation, elasticity and plasticity under biomechanical loading would support a view of the disc as being a microporous material (Beek et al., 2003) and that the flow of water between regions of the disc is necessarily a consequence of indentation loading, compressive loading (Beek et al., 2000; del Pozo et al., 2002), and dissipation of strain energy under tensile loading (Tanaka et al., 2003b). All responses to various types of biomechanical load as reviewed by Tanaka and van Eijden (Tanaka and van Eijden, 2003) are evidence that boundary conditions for water are not present in the disc and thus, the incompressibility of water under boundary conditions simply cannot be an explanation of biomechanical properties of the disc.

Thus, a detailed understanding of how water is bound to and organized around the macromolecular components of the disc is needed. It is not sufficient to simply measure the total water content and the area specific distribution of water. Specifically, how much of the water in the disc is unbound to the macromolecular components and free to flow with the same characteristics as bulk water? How much of the water of hydration is perturbed from ordinary bulk water by interaction with macromolecular components but able to contribute to non-boundary properties of the disc? How much of the water content of the disc is so tightly bound to macromolecular components that it is essentially immobile even under sustained compressive loading? In this study the physical properties of the water within the temporomandibular disc have been measured using: (1) pulsed proton nuclear magnetic resonance during sequential dehydration; (2) NMR proton spectra of freezing point depression of the water of hydration over temperature ranges of +20 °C to −98 °C; and (3) determination of sequential loss of water under centrifugal loading of 4.0–5.0 MPa.

2. MATERIALS AND METHODS

TMJ articular discs were harvested from juvenile (6–8 month old) pigs that were killed under general anaesthesia as part of a gastroenterology study. Animal care and procedures were performed in facilities approved by the American Association for the Accreditation of Laboratory Animal Care and according to the institutional guidelines for the use of laboratory animals. Disc samples used in this study appeared normal without macroscopic evidence of degeneration, disease or joint damage. Articular discs of the TMJ in the baboon *Papio cynocephalus* were removed from animals that were exsanguinated under general anaesthesia (ketamine sedation followed by intravenous sodium pentothal). Two adult males were euthanized for failing health documented in the protocols for NIH Grants HL28973, HV53030. Disc samples appeared normal with no evidence of degeneration or damage.

2.1 Resistance to Fluid Flow under Centrifugal Loads

The resistance to loss of water under centrifugal loads was measured during centrifugation. Discs were obtained from 6 to 8 month old pigs, dissected from surrounding soft tissues and sectioned into 1 mm^3 pieces. The initial weight was determined for each sample. The resistance to fluid flow was measured for the disc as a whole (randomized 1 mm^3 pieces) and for specific areas of the disc. Approximately five to six 1 mm^3 from each of the selected areas of the disc were loaded into microcentrifuge tubes containing a filter membrane placed over a bed of filter paper to contact blot the water separated from the disc during the braking period of the centrifuge. Samples were centrifuged for 5–10 min intervals for a total of 120 min on a 5 cm rotor at a total force of 13 000 g to provided stress of 4.0–5.0 MPa, assuming the individual sample weights represented the maximum tare subjected to the deepest part of the sample. Intermediate weights of the disc tissue were recorded at each centrifugation interval. Following centrifugation, all samples were dried to weight equilibrium at 100 °C in a vacuum oven and the final dry weight was measured to allow the water content of each sample to be expressed as grams of water per gram dry weight.

2.2 Determination of Bound Water Compartments by NMR Dehydration

Pulsed proton nuclear magnetic resonance (NMR) titration was performed on randomly orientated 1 mm^3 diced samples of the tendon while slowly dehydrating the samples. Assuming fast exchange between water compartments, dehydration will sequentially remove unbound or free water and then more tightly bound water. The measured T_1 relaxation times at multiple steps during the dehydration provided a weighted average of the relaxation rates of each water fraction and defines a series of lines corresponding to the hydration compartments found in the tissues. After weighting the tissue at each step in the dehydration procedures, proton relaxation time (T_1) were determined by NMR, as previously described (Fullerton et al., 1986; Fullerton and Cameron, 1988), thus allowing the size of each water compartment to be computed once the final dry mass was known. The total dry mass of the tissue samples was determined after drying the specimens in a vacuum oven at 100 °C for three days. Because water that is associated with fat is not in fast exchange, three sequential ether extractions were done to gravimetrically determine the fat content of each of the tissue samples. Fat content was then subtracted and a final dry mass computed.

2.3 Determination of Water of Hydration Compartments by Proton NMR and Freezing Characteristics

Based on the observations that water that is frozen does not produce a measurable NMR proton spectral signal, that those water molecules interacting with polar or charged sites are energetically and conformationally unavailable for ice crystal formation at a given temperature, and that all water molecular that are not participating in ice crystal formation at a given temperature contribute to the NMR proton spectral signal. Thus, pulsed proton NMR has been used to quantify hydration compartments in biological tissues (Kiyosawa, 1988). Randomly orientated, 1 mm^3 pieces of baboon disc samples were placed in 10 mm diameter NMR tubes. A 300 MHz GE-NMR spectrometer was used to measure the integrated amplitude of the proton NMR signal as a function of temperature over the range of +20 °C to −98 °C. Thirty minute intervals were allowed for equilibration at each temperature. The initial wet weight of the specimen and the final wet weight of the specimen after NMR measurement showed no significant loss of water. To express data as grams of water per gram dry mass, the specimen were dehydrated for 3 days in a vacuum oven at 100 °C. The integrated spectral area (signal amplitude) was then converted to g water/g dry mass as a function of temperature. Analysis of the plot of g water/g dry mass vs. temperature allows determine of the amount of water present in different hydration compartments. Because the water from the least tightly bound water compartment enters the ice crystal lattice first, followed by water from more tightly bound water compartments, breakpoints or changes in the slope of the plot delineate the size of the different water of hydration compartments.

3. RESULTS

3.1 Resistance to Fluid Flow and the Water-Holding Capacity under Compressive Load

In order to more directly compare the NMR analysis of hydration compartments, the centrifugation experiments were done on discs sectioned into 1 mm^3 pieces. The resistance to fluid flow was measured for the disc as a whole (randomized 1 mm^3

pieces) and for specific areas of the disc. The total water in the sample was 1.77 g water/g dry weight, and 0.88 g water/g dry weight was forced from the disc after 120 min (in 5–10 minute intervals) at a stress of 4.0 MPa.

The cumulative grams of water forced from the disc was expressed as grams water per gram dry mass and then plotted against time under centrifugal load, such that the slopes of the curves defined the rate of water loss (i.e. flow rate through and out of the tissue). Thus, the slope provided a numerical estimate of the fluid flow. As illustrated in Figure 1, curve fit analysis demonstrated two major water compartments. The data points from the outer water compartment (the least tightly bound water and therefore the first water to be removed from the tissue) fit logarithmic curves with an r^2 value of 0.985. The data points from the innermost water compartments (the most tightly bound water) fit simple linear lines with an r^2 of 0.980. Curve fit analysis (Figure 3) demonstrated that the disc had two water compartments-an inner, tightly bound water compartment with a lower flow rate, and an outer, more loosely bound water compartment with a higher flow rate.

3.2 Analysis of Water of Hydration using NMR Dehydration

Data were taken using the Praxis II pulsed NMR analyser and a saturation recovery (90–T–90) pulse sequence. The difference in PID pulse height for long delay time ($t >> 5T_1$) and for short delay time was measured for a series of 30 sequential t-values. The T_1 value, or spin-lattice relaxation time, a measurement of the mean relaxation time of all protons remaining in the tissue, was calculated as the inverse

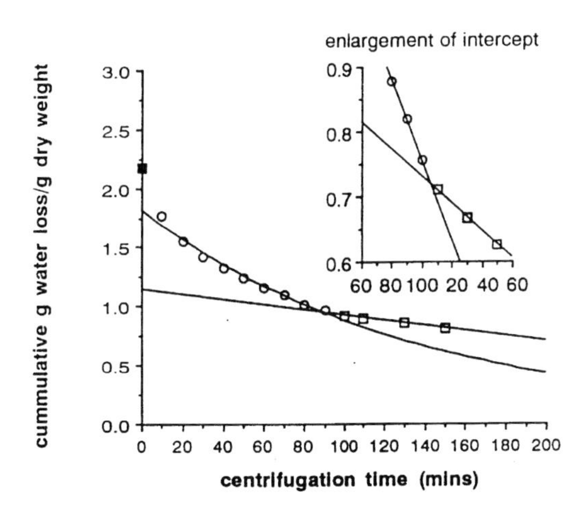

Figure 1. Resistance to fluid flow of the TMJ disc. The cumulative grams of water loss per gram dry weight are plotted against time under compressive load and describe a logarithmic line such that the slope is the log of water loss. The slope of the line derived for each location is therefore a numerical expression of the resistance to fluid flow under compressive load

of the slope of the best fit linear regression for all data points. By weighing the tissue after each NMR measurement during dehydration, the size of each water compartment in the porcine TMJ disc could be computed once the final dry mass was known, as is done in Figure 2.

The NMR analysis of water of hydration compartments in porcine temporomandibular disc indicated that the discs contained no detectable bulk water as explained below. The inverse of the proton spin-lattice relaxation rate ($1/T_1$) was plotted against the ratio of mass solute/mass water. When all of the water from the least tightly bound water compartment was removed due to dehydration, the slope of the plot changes, such that the breakpoints delineate the size of the different water of hydration compartments. After solving for the (x,y) intercepts of the lines, the total amount of water in each compartment was calculated once the dry mass of the specimen was known. In addition, the data plot provides information about the structure of the water compartments. Extrapolation of the first line of the plot in Figure 1 to the y-intercept (Ms/M water = 0) gave a y-intercept of 1.57, equivalent to a relaxation rate of $T_1 = 0.637$ ms for the outer most (least bound) water compartment. If bulk water were present, the y-intercept would have been about 0.37 (for a $T_1 \approx 2.700$ ms). Thus, all of the water in the pig temporomandibular disc differed in its T_1 from that expected if it were bulk water. The absence of free water in the disc was also confirmed by NMR measurement of freezing point depression below.

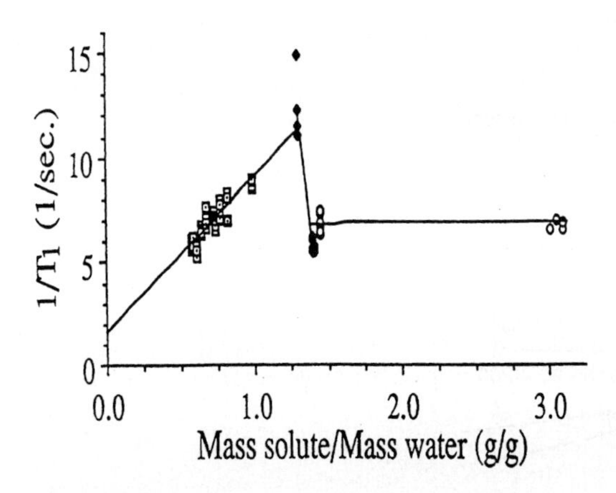

Figure 2. NMR analysis of water of hydration compartments in the pig TMJ disc. Results of dehydration studies of the proton spin-lattice relaxation rate (1/Ta) are plotted against the ratio of mass solute/mass water. Extrapolation of the plot to the y-intercept (Ms/M water $= 0 : 1/T_1 = 1.5670$) gives a relaxation time of 0.638 ms for the outer most (least bound) water compartment. If bulk water were present the intercept would be approximately 0.37. This shows that all of the water in the pig TMJ disc is structured water. The breakpoints in the slope of the plot delineate the different water of hydration compartments and by solving for the (x,y) intercept of each line the total amount of water in each compartment can be calculated based on the dry mass of the specimen

3.3 Freezing Characteristics of Water in the TMJ Disc

Proton NMR spectra were measured for adult baboon temporomandibular disc over the temperature range of 4 °C to −98 °C and plotted as seen in Figure 3 (A). The integrated signal spectra for mature baboon discs were converted to grams water/gram dry mass and plotted against temperature, as shown in Figure 3 (B). Three water compartments can be differentiated. The first break in the slope of the plot occurred at −12 °C when there was a 60% loss in signal amplitude. A second change in slope occurred at −72 °C, dividing the most tightly bound water into two water compartments. There was a detectable signal at the lowest temperature assayed, indicating a small compartment of very tightly bound water. The amount of water in the first two water freezing compartments is summarized in Table 1.

Although centrifugation studies of each area of the disc clearly demonstrated two bound water of hydration compartment as evaluated by intercepts of curve fit analysis, there is also an indication that there is water of hydration component that may have properties equivalent to free or bulk water. In each curve fit analysis, by area and for the disc as a whole, the initial data point lies above the best curve fit line. Calculation of this water compartment (i.e. subtraction of the initial data point from the x-axis intercept value) results in an average value of 0.47 g H_2O/g dry weight, with a range of 0.3 g H_2O/g dry weight in the posterior band portion of the disc; 0.4–0.475 g H_2O/g dry weight in the anterior band; and, 0.52–0.69 g H_2O/g dry weight in the intermediate zone.

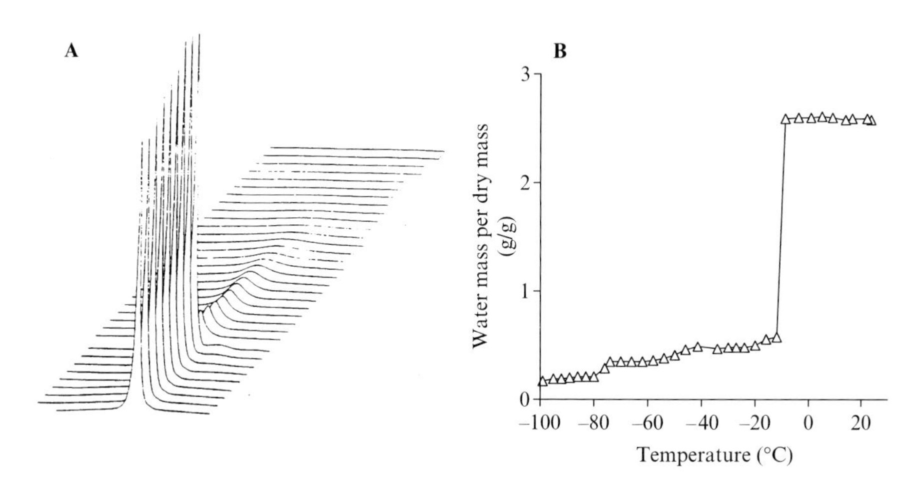

Figure 3. (A) (B) NMR proton spectra as a function of temperature for adult baboons' articular discs. Proton spectra were recorded using a 300 MHz spectrometer, and were measured over the temperature range of +20° to −98 °C. Above 0 °C, the water peak consists of a single Lorentzian component with a little shoulder on the upfield side of the spectrum. The proton signal shows a 60–78% reduction upon freezing at −12 °C. The amount of water in each hydration compartment is summarized in Table 1

Table 1. Comparison of size of water compartments in g water/g dry mass in the temporomandibular joint disc as determined by three experimental methods

		Total water g water/g dry mass	Inner water g water/g dry mass	Outer water g water/g dry mass
Pig, 6–8 months	Flow analysis	2.14*	1.16*	0.99*
Pig, 6–9 months	NMR analysis	2.20	1.30	0.90†
Baboon, 13–16 years	Freezing analysis	2.24	1.13‡	0.90§

* Average of anterior, intermediate and posterior regions.
† Does not have relaxation time characteristic of bulk water.
‡ Evidence of at least two inner water compartments. There appears to be a separate inner water compartment of 0.20 g water/g dry mass that did not freeze at −75 °C to −98 °C.
§ No evidence for bulk water.

4. DISCUSSION

A separate chapter of this monograph dealt with the composition and organization of the temporomandibular joint (TMJ) disc (Haskin et al., 2005). Examination of various load bearing regions for fluid flow, element content and distribution of sulphated glucosaminoglycans led to the investigation of the water of hydration of the disc. The data in this study leads to the following conclusions. Compressive loading of the TMJ disc revealed evidence of a bulk water fraction, a faster flowing, outer water compartment and a slower flowing inner water compartment. The free bulk water compartment was calculated by subtracting the initial value (the first centrifugation measurement) from the axis intercept value. If the method of calculation for this compartment is to use the initial water content minus the log fit intercept at the ordinate, the method as used by Ling and Walton (Ling and Walton, 1975), than there is a bit smaller percent water value (Table 2).

Thus, by this one method there is a free water component that does reduce the size of the outer (non-bulk) water compartments measured by the two NMR methods,

Table 2. Free Water in TMJ Disc Based on Fluid Flow at Given g Force (Centrifugation × Time)

Specimen ID	g/g initial reading	Intercept by log fit to ordinate	Change in g/g	First value method	Change in g/g
A	2.17	1.8	.37	1.75	.42
F	2.15	1.6	.55	1.59	.56
B	2.20	1.85	.35	1.81	.39
G	2.40	2.10	.30	2.08	.32
C	2.23	1.77	.46	1.74	.45
E	1.96	1.42	.54	1.42	.54
H	2.11	1.75	.36	1.71	.40
D	1.90	1.59	.36	1.49	.41

as listed in Table 1, but does not affect the estimated size of the inner (non-bulk) water compartment. Ling and Walton, using a similar centrifugation and weighing technique found about the same free water fraction in the extracellular space in seven different tissues in the frog (Ling and Walton, 1975).

The presence of a bulk water fraction in a tissue sample that has been dissected and diced prior to subjecting the sample to a functional compressive load should be viewed as possibly incidental to procedural protocol. Clearly, cutting the disc into 1 mm^3 pieces and then subjecting the tissue to compressive load would not be representative of the boundary condition of the poroelastic material existing in intact tissue. *In vivo*, when the disc receives excessive impulse compression, structural changes occur in the disc including surface fissuring, shearing on a visible level and shearing and separation of collagen fibers at a microscopic level (Tanaka et al., 2003b). Clearly, the integrity of the collagen fibers of the disc are one of the primary determinates of resistance to compressive loading, shear and high strain (Tanaka et al., 2003a). Damage of these collagen fibers ultimately leads to tissue failure characteristic of pathology. Thus, the loss of water of hydration under compressive loading of dissected and minced specimens may represent the type of response one might expect of pathological tissue already undergoing fatigue failure due to the disruption of the integrity of collagen fibers.

To further characterize the physical characteristics of water compartments within the disc, two additional methods were applied. In the NMR techniques used to measure water of hydration compartments, no functional stress was applied to diced material, thus allowing conditions that would more clearly reflect that found *in vivo*. Measurement of proton NMR relaxation time during sequential dehydration of the whole disc of pig TMJ revealed two water compartments. Likewise, sequential measurement of the amplitude of the proton NMR signal at temperatures ranging from $+20\,^{\circ}C$ to $-100\,^{\circ}C$ demonstrated significant freezing point depression characteristic of tightly bound water unavailable for sequestration into ice crystal formation. This second method also revealed multiple water freezing point fractions. As no proton NMR signal change occurred until $-12\,^{\circ}C$, the data was interpreted as there being no evidence of water with the freezing characteristics of bulk water, i.e. freezing point at $0\,^{\circ}C$.

What seems most interesting about the existence of the two non-bulk-like water compartments measured by the three different methods: flow rate, proton NMR dehydration and freezing point depression (Table 1) is the agreement in the amount of water in each compartment. Thus, one can conclude that neither the outer nor the inner water compartment of water has water with bulk water characteristics. The centrifugation studies indicate that it is possible that there might be a small fraction of water with bulk water characteristics present in the TMJ disc but such a fraction was not detected by the measurement techniques use NMR proton spectra and may be a result of tissue damage necessarily a result of the experimental protocol.

Multiple non-bulk water compartments have been reported for collagen and lysozyme (Fullerton, 2006; Fullerton et al., 2005). Both the collagen and lysozyme specimens reveal three water compartments as measured by proton NMR relaxation time during dehydration. The size of each of the three compartments is: (1) about 0.05–0.07 (2) 0.18–0.27 and (3) 1.4–1.6 g water/g DM. Fullerton and co workers (Fullerton, 2006; Fullerton et al., 2005) conclude that all of the water is interfacial monolayer water. If this also holds true for the TMJ disc, then all of the water in the inner water compartment (1.13–1.30 g water/g DM) may be interfacial monolayer water, with the water in the outer water compartment being a second shell of water of hydration on the surface of the macromolecules.

5. SUMMARY AND CONCLUSIONS

Three different methods (flow rate, proton NMR dehydration, and freezing point characteristics) lead to the conclusion that all or essentially all of the water in the intact TMJ disc is bound water and does not have properties consistent with free water. Two major fractions of non-bulk water were identified and their amounts in g water/g DM were determined. The inner water compartment had 1.13–1.30 g water/g DM while the outer water compartment has 0.90–0.99 g water/g DM. That all three methods yielded similar water compartment values indicate these two water compartments have distinct physical characteristics.

REFERENCES

Beek M, Koolstra JH, van Eijden TMGJ (2003) Human temporomandibular joint disc cartilage as a poroelastic material. Clin Biomech 18:69–76

Beek M, Koolstra JH, van Ruijven LJ, van Eijden TMGJ (2000) Three-dimensional finite element analysis of the human temporomandibular joint disc. J Biomech 33:307–316

Del Pozo R, Tanaka E, Tanaka M, Okazaki M, Tanne K (2002) The regional difference of viscoelastic property of bovine temporomandibular joint disc in compressive stress-relaxation. Med Eng Physics 24:165–171

Detamore MS, Athanasiou KA (2003b) Structure and function of the temporomandibular joint disc: implication for tissue engineering. J Oral Maxillofac Surg 61:494–506

Detamore MS, Orfanos JG, Almarza AJ, French MM, Wong ME, Athanasiou KA (2005) Quantitative analysis and comparative regional investigation of the extracellular matrix of the procine temporomandibular joint disc. Matrix Biol 24:45–57

Fullerton GD (2006) Evidence that collagen, tendon and cellular proteins have monolayer water coverage in the native state, in press

Fullerton GD, Cameron IL Relaxation of biological tissues. In: Wehrili FW, Shaw D, Kneeland JB (eds) Biomedical Magnetic Resonance Imaging—Principles, Methodology, and Application. VCH Publisher Inc., New York, pp 115–155

Fullerton GD, Nes E, Amurao M, Rahal A, Krasnosselskaia L, Cameron IL (2005) An NMR method to characterize multiple water compartments in mammalian collagen. Water in tendon: orientational analysis of the free induction decay. Magn Reson Med 54:280–8

Fullerton GD, Ord VA, Cameron IL (1986) An evaluation of the hydration of lysozyme by an NMR titration method. Biochem Biophys Acta 869:230–246

Haskin CL (1995) Adaptation in the temporomandibular joint: cellular, structural and molecular response to mechanical forces. Dissertation UTHSCSA Graduate School of Biomedical Sciences

Haskin CL, Fullerton GD, Cameron IL (2005) Molecular basis of articular disc biomechanics: fluid flow and water content in the temporomandibular disc as related to distribution of sulfur (in publication)

Kiyosawa K (1988) Precise expression of freezing-point depression in aqueous solutions. In: Lauger P, Packer L, Vasilescu V (eds) Water and Onions in Biological Systems. Birkhauser-Verlag, Boston, pp 425–432

Kobayashi J (1992) Studies on matrix components relevant to structure and function of the temporomandibular joint. Kokubyo Gakkai Zasshi 59:105–123. Translated (Abstract) only

Ling GN, Walton CL (1975) Simultaneous efflux of K+ and Na+ from frog sartorius muscle freed of extracellular fluids: evidence for rapidly exchanging Na+ from the cells. Physiol Chem Phys 7:501–515

Milam SB, Klebe RJ, Triplett RG, Herbert D (1991) Characterization of the extracellular matrix of the primate temporomandibular joint. J Oral Maxillofac Surg 49:381–391

Minarelli AM, Liberti EA (1997) A microscopic survey of the human temporomandibular joint disc. J Oral Rehabil 24:835–840

Nakano T, Scott PG (1989) Proteoglycans of the articular disc of the bovine temporomandibular joint. I. High molecular weight Chondroitin sulfate proteoglycans. Matrix 9:277–283

Nakano T, Scott PG (1996) Changes in the chemical composition of the bovine temporomandibular joint disc with age. Arch Oral Biol 41:845–853

Sindelar BJ, Evanko SP, Alonzo T, Herring SW, Wight T (2000) Effects of intraoral splint wear on proteoglycans in the temporomandibular joint disc. Arch Biochem Biophys 379:64–70

Tanaka E, Hanaoka K, van Eijden T, Tanaka M, Watanabe M, Nishi M, Kawai N, Murata H, Hamada T, Tanne K (2003a) Dynamic shear properties of the temporomandibular joint disc. J Dent Res 82:228–231

Tanaka E, Kawal N, van Eijden T, Watanabe M, Hanaoka K, Nishi M, Iwabe T, Tanne K (2003b) Impulsive compression influences the viscous behavior of porcine temporomandibular joint disc. Eur J Oral Sci 111:353–358

Tanaka E, van Eijden T (2003) Biomechanical behavior of the temporomandibular joint disc. Crit Rev Oral Biol Med 14:138–150

INDEX